Advances in Intelligent Systems and Computing

Volume 260

T0142209

Series editor

Janusz Kacprzyk, Polish Academy of Sciences, Warsaw, Poland
e-mail: kacprzyk@ibspan.waw.pl

For further volumes:
http://www.springer.com/series/11156

About this series

The series "Advances in Intelligent Systems and Computing" contains publications on theory, applications, and design methods of Intelligent Systems and Intelligent Computing. Virtually all disciplines such as engineering, natural sciences, computer and information science, ICT, economics, business, e-commerce, environment, healthcare, life science are covered. The list of topics spans all the areas of modern intelligent systems and computing.

The publications within "Advances in Intelligent Systems and Computing" are primarily textbooks and proceedings of important conferences, symposia and congresses. They cover significant recent developments in the field, both of a foundational and applicable character. An important characteristic feature of the series is the short publication time and world-wide distribution. This permits a rapid and broad dissemination of research results.

Salvatore Gaglio · Giuseppe Lo Re
Editors

Advances onto the Internet of Things

How Ontologies Make the Internet of Things Meaningful

Editors
Salvatore Gaglio
Giuseppe Lo Re
Dipartimento di Ingegneria Chimica,
Gestionale, Informatica, Meccanica
Università di Palermo
Palermo
Italy

ISSN 2194-5357 ISSN 2194-5365 (electronic)
ISBN 978-3-319-03991-6 ISBN 978-3-319-03992-3 (eBook)
DOI 10.1007/978-3-319-03992-3
Springer Cham Heidelberg New York Dordrecht London

Library of Congress Control Number: 2013957383

Printed on acid-free paper

Springer is part of Springer Science+Business Media (www.springer.com)

Preface

The title of this book is a pun on the use of the preposition "onto" with the aim of recalling "Ontology," the term commonly adopted in the computer science community to indicate the study of the formal specification for organizing information about objects and entities.

The Ontology notion originates from philosophy, where it refers to the metaphysical study of the nature of being and existence. In computer science and more specifically in the field of knowledge engineering, ontologies are used for a quite different purpose, that is, for modeling concepts and relationships on some domain.

Year 2013 celebrates the twentieth anniversary of the World Wide Web. The simple network of hypermedia has transformed the world of communications with enormous implications on the social relationships; however, it soon showed its main drawback that, in the opinion of its creator—Tim Berners-Lee—is the lack of meaning in exchanged data when artificial agents are involved. Contents were designed to be read by humans and not to be meaningfully manipulated by computer programs.

With the introduction of the semantic web the meaningful contents are opportunely structured, in order to allow software agents roaming from page to page to readily carry out sophisticated tasks.

From an infrastructural perspective the traditional World Wide Web has experienced a further extension represented by the Internet of Things (IoT), today feasible thanks to the integration of the pervasive technology of sensor networks. Sensor networks are composed of several devices capable of sensing environmental phenomena, of performing small on-board computations, and of communicating with each other in order to cooperate.

Two different aspects, observable at two separate layers, characterize the Internet of Things. The physical devices connected to the *network* and the data they are able to collect and transmit constitute the raw infrastructure, deployed all over the globe.

The semantics of the collected data, the meaning of the actions they are able to trigger, their exploitation in ever-more challenging applications capable of dramatically changing everyday life, represent the real knowledge that human beings and even computers themselves may acquire.

The most important contribution of IoT regards the possibility of enabling more efficient machine-to-machine cooperation. To such aim, ontologies represent the most suitable tool to enable transfer and comprehension of information among computer applications, even those designed and developed by unrelated people in different places.

It is thus possible to surf the sea of information available today in digital form without the intervention of a human being, thus accomplishing a real web of things.

This new paradigm uncovers new horizons for the development of visionary and challenging applications. Such new services, bridging the virtual and physical worlds, span various domains such as energy efficiency, health care, precision agriculture, and infrastructure monitoring.

This book proposes a collection of contributions illustrating different applications following these directions and that are the outcomes of real experiences developed in the context of research projects.

A relevant portion of the book contains papers illustrating the Internet of Things in the specific domain of Ambient Intelligence (AmI). AmI is a recent research field that considers the user as the focus of an environment equipped with pervasive devices, with the main goal of satisfying his requirements, and of assisting him in daily activities. The complexity of such a domain imposes the adoption of formal methods of knowledge representation; in such context, ontologies represent a useful instrument for overcoming the intrinsic difficulties arising from heterogeneity and dynamicity, thus effectively making the Internet of Things fully exploitable.

Papers presented in the first part of the book (1–15) fall within this group and have been discussed during a project workshop held in Palermo on October 29, 2013.

Furthermore, in order to reach a broader audience, we collected some other interesting contributions devoted to illustrate other compelling application fields, ranging from the tourism market to the public administration, from the thermo-solar plants to the multi-risk assessment.

We would like to thank all the authors for their contributions, which we believe represent interesting and stimulating advances in this cross-disciplinary field.

We also would like to thank all the colleagues for their invaluable support in reviewing the papers, and finally Dr. Alessandro Perricone for his help in the final editing.

November 2013

Salvatore Gaglio
Giuseppe Lo Re

Contents

An Ontology-Based Autonomic System for Ambient Intelligence Scenarios

Alessandra De Paola

Abstract Pervasive computing and Ambient Intelligence (AmI) demonstrate that computer systems which directly interact with users are characterized by increasing size and complexity, so that the human user will still not be able to adequately manage them for a long time to come. As a response to this trend, the Autonomic Computing paradigm aims to design and develop systems able to self-configure and self-manage. The research reported here is part of an AmI project that proposes a multi-tier cognitive architecture for aggregating sensory information at different levels of abstraction. In such an architecture, a central reasoning component is able to understand the environmental state and the user's preferences and consequently to plan the opportune actions to be performed. This chapter describes an ontology able to provide a formal representation of the environment in which the AmI system is placed, as well as a representation of the system itself and of its interaction with the environment. By exploiting this knowledge, the AmI system can develop consciousness of itself and of its cognitive processes, and consequently the capability of autonomously managing its own functioning. In particular, this task is performed by a rule-based planning module, integrated within the multi-level architecture, and capable of managing and configuring the sensory infrastructure. By means of this module, the AmI system can manage its own monitoring activity to obtain a good understanding of the context while minimizing system energy consumption.

1 Introduction

Ambient Intelligence (AmI) [2, 8] is based on the integration of the Internet of Things (IoT) [1] and Artificial Intelligence. This paradigm defines an application scenario where the user is the focus of a pervasive environment augmented with sensors and

A. De Paola (✉)
University of Palermo, Viale delle Scienze, ed 6, 90128 Palermo, Italy
e-mail: alessandra.depaola@unipa.it

S. Gaglio and G. Lo Re (eds.), *Advances onto the Internet of Things*,
Advances in Intelligent Systems and Computing 260, DOI: 10.1007/978-3-319-03992-3_1,

actuators, enabling an intelligent system to monitor the environmental conditions and to perform actions aimed at satisfying user requirements [25].

Although the main goal of AmI is not the development of pervasive sensory and actuator devices, their availability is necessary to develop AmI systems, because such devices enable the development of a pervasive sensory infrastructure, while maintaining a low degree of intrusiveness, and with low costs of production, deployment and maintenance.

The adoption of such sensory infrastructure poses new challenges, involving both the information management process and the analysis of data gathered to detect anomalous behavior [9]. In order to take full advantage of the raw information gathered by pervasive devices, information has to be properly represented to extract meaningful knowledge. For this purpose, the work reported here adopts a multi-level cognitive architecture [5], whose sensory infrastructure is based on Wireless Sensor Networks (WSN) [29, 31]. Sensor nodes gather environmental data and forward them towards a central intelligent engine, where high-level reasoning occurs. Sensory information is then aggregated and processed inside the reasoning engine by a stack of modules, adopting the most appropriate representation according to the level of abstraction at which information is processed. The knowledge achieved in this way is exploited to perform high-level inferences about the perceived context and about user's preferences and needs, in order to plan the most suitable actions to be performed to meet system goals.

Within such a scenario, this work proposes an ontology capable of supporting the design and the development of AmI systems based on the adopted multi-tier architecture. A methodology for developing autonomic behavior to self-manage the monitoring system is also proposed. The ontology described here provides a formal representation both of the specific application domain and of the AmI system itself. In addition to driving the process of knowledge abstraction, from raw sensory data up to higher-level concepts, it enables the AmI system to gain consciousness of itself, of its own interactions with the environment and of its own cognitive activity, thus enabling the autonomic management of its own behavior.

An ad-hoc module exploits the system's ontological representation to self-configure its monitoring activities. In particular, by means of an introspective analysis of its own state and of its cognitive activity, the system is able to define symbolic plans to be translated into commands to be given to the actuators, to self-configure the sensory infrastructure.

The case study selected to verify the potential of the proposed approach is a Building Management System (BMS) for controlling ambient conditions of an office environment, in terms of heating, ventilation, air conditioning (HVAC) and lighting, with the goal of maximizing user comfort while reducing energy consumption of the sensory infrastructure.

The chapter is organized into six sections. Section 2 introduces AmI and Autonomic Computing paradigms, highlighting the ways they relate to each other, and briefly describing some relevant approaches proposed in the literature. Section 3 describes the main features of multi-tier architecture adopted, specifying the role of the proposed ontology and introducing self-management features. Section 4 explains the proposed ontology and examines some architectural details of the system. Section 5 describes the component devoted to the dynamic self-configuration of the sensory infrastructure, exploiting a rule-based reasoning engine. Finally, Sect. 6 sets out the conclusions drawn by the author and outlines possible future developments.

2 Ambient Intelligence and Autonomic Computing

Ambient Intelligence and Autonomic Computing are two emerging paradigms, which have many features in common. Some relevant desirable features for AmI systems, such as self-management capability, adaptivity to highly dynamic scenarios, context-awareness, monitoring and analysis capability, represent the essence of Autonomic Computing. The development of autonomic systems can therefore be considered a technological prerequisite for designing AmI applications.

Many approaches presented in the AmI literature exploit artificial intelligence (AI) techniques to manage the huge set of devices deployed in the environment, to cope with the intrinsic uncertainty and imprecision of environmental models, and to adapt the system to changes in environmental conditions and user behavior.

Among the most widely adopted AI techniques, are neural networks, fuzzy systems and Bayesian networks [14]. The authors of [7] propose a non-supervised learning method for a fuzzy system devoted to the control of environment actuators, such as artificial lighting, windows and HVAC systems. The system is able to monitor environmental quantities such as light and temperature, besides the user interactions with actuators. In their Neural Network House, the authors of [22] propose the use of neural networks, together with rules for occupancy detection, to predict the binary occupational state of monitored sites. Input data for the neural network is provided by sensory readings from binary motion sensors. The authors of [20] use a Bayesian network to identify the sequence of actions to be performed on the actuators in order to carefully imitate a user's past behavior.

Independently of the approaches adopted, an AmI system has to include knowledge of itself and the environment in which it acts. Some works [13, 19, 24] have focused on the forms of representation and communication of this knowledge, and on the semantic enrichment of data processed by the system through the use of ontologies. This self-consciousness enables complex systems to autonomously perform configuration, maintenance, management and optimization, which are all tasks typically assigned to human operators. This development trend is desirable because AmI systems are becoming increasingly complex, distributed and heterogeneous, and consequently designers are no longer able to predict in advance all possible patterns of interaction between components [16]. This philosophy is driven by the initiative launched by IBM in 2001 [18], which called it "Autonomic Computing", in reference to the human autonomic nervous system.

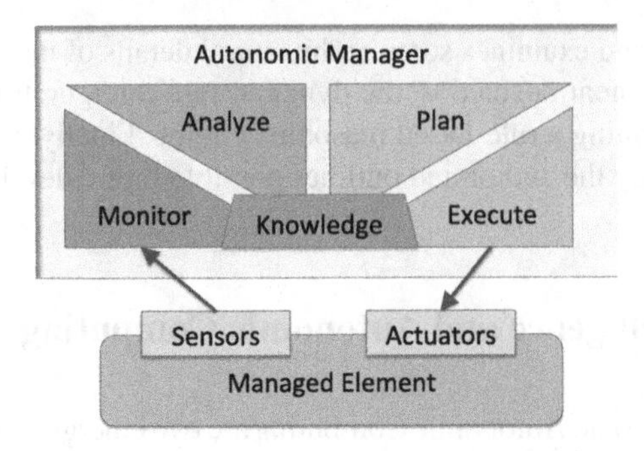

Fig. 1 The monitor-analyze-plan-execute cycle of autonomic computing [32]

The main Autonomic Computing functionalities are listed below:

- *self-configuration*: the system is able to self-configure according to high-level policies defined by human operators, autonomously composing new diverse subsystems;
- *self-optimization*: the system is able to tune its own parameters in order to optimize its own behaviour;
- *self-healing*: the system is able to identify problems and causes of failure, and to solve them , thus returning the system to normal functioning;
- *self-protection*: the system is able to protect itself against malicious attacks and to anticipate possible failures.

In order to enable those functionalities, a system needs a sensory infrastructure to monitor its behavior, analysis modules to detect relevant events from sensory data, and reasoning modules to plan the sequence of actions to be performed using a set of actuators [15, 16, 32]. The resulting monitor-analyze-plan-execute cycle is showed in Fig. 1.

The features and goals of Autonomic Computing grow in importance as Ambient Intelligence becomes more pervasive and dynamic. In such a scenario, in order to make the AmI system invisible to the user, the system needs to be capable of autonomously interacting with the environment, of self-managing only on the basis of high-level policy, and of dynamically learning user preferences.

It is thus evident how autonomic capabilities are a basic requirement for making future AmI systems capable of effectively coping with diverse contexts, without become a burden for human operators and users. In other words, AmI represents a vision of a future world in which the IoT and the Artificial Intelligence cooperate in a scenario which requires complex features, such as dynamism, adaptability, non-intrusiveness and self-management, all facilitated by the Autonomic Computing paradigm, which thus becomes a necessary condition for its realization.

3 A Multi-Level Approach for AmI Applications

This chapter proposes the adoption of the cognitive multi-tier architecture for AmI systems described in [4, 5]. Such an architecture adopts a distributed sensory infrastructure for gathering environmental and context information, and different levels of abstraction for representing and processing knowledge. In such a framework, intelligent modules make it possible to understand user needs, to plan the sequence of actions to be performed, and to self-configure system behavior.

The architecture adopted consists of four subsystems, each of them characterized by a modular and configurable structure, as defined below:

- *Sensing subsystem*: based on Wireless Sensor Networks (WSNs), it perceives relevant ambient information, and sends raw data toward the understanding subsystem;
- *Understanding subsystem*: it processes data through different levels of abstraction to extract high-level knowledge; at the highest level it provides a concise and meaningful description of the current context;
- *Planning subsystem*: it exploits the ambient description provided by the understanding subsystem to plan the most appropriate sequence of actions to be performed to satisfy user needs and optimize the system's behavior;
- *Actuation subsystem*: consisting of all actuators able to modify the environment status, i.e., heating, ventilation, air conditioning (HVAC) and lighting systems; available actuators receive control commands from the planning subsystem.

The understanding subsystem is split into multiple levels, according to a multi-tier structure of interconnected modules for representing knowledge. In particular, there are three types of module, namely "subsymbolic", "conceptual", and "symbolic" modules. Knowledge flows through those tiers, assuming the suitable form required by each module. A more detailed view of the system structure and of the role played by each component, is provided in Sect. 4 together with a description of the ontology proposed here.

In the case study considered here, the *sensing subsystem* is composed of a WSN, whose nodes are deployed in different locations inside the controlled premises, for monitoring relevant physical quantities, namely, temperature, humidity and lighting level. Moreover, a set of sensors on actuators make it possible to monitor user activity and obtain implicit feedback relating to comfort levels. The sensory data gathered are processed by the *understanding subsystem*, which builds a concise representation of current environmental condition and context, such as information about user presence or activity, about current user comfort, and about energy consumption. To take one example, let's consider a set of sensors able to catch the switching on/off of the artificial lighting system. This raw information allows the system (i) to infer the current status of the artificial lighting system and to correlate it to the corresponding lighting level, (ii) to learn the correspondence between the actuator status and its current energy consumption, and (iii) to infer the appreciation level of users about the current lighting conditions.

4 The Proposed Ontology

As highlighted by a considerable body of research in the literature, an AmI system needs to acquire, implicitly or explicitly, the knowledge of the scenario being considered and of its own interaction with it [13, 19, 24]. Moreover, if the autonomic capability of self-management is required, the AmI system needs to have an internal representation of its own structure and of interactions between its own modules [16, 27, 30, 32]. A formal and explicit representation of such knowledge can be obtained by defining an ontology, which represents an unambiguous vocabulary consisting of the definitions of classes in the domain of interest and of relations among them [12, 26].

The ontology proposed here serves several purposes:

- to provide an unique definition of the concepts on which the system is based, to support the design phase;
- to make it possible to define a generic framework, easily configurable for different application of the AmI system;
- to allow the system administrator to express only high-level goals and neglect low-level details;
- to facilitate automatic interaction between system modules able to describe their own services;
- to support the development of the autonomic capabilities of the systems.

Thus, the ontology defines the environment and its properties, the user and his interaction with it, as well as system components, their interconnections and their relationships with the environment. For instance, it is possible to represent how data flows within the system, or what relationship exists between sensory devices and the environmental properties perceived by them.

The representation employed here can be sub-divided into two ontologies: the *General Ontology* and the *Domain Ontology*. The former represents the structure of a generic system and its possible relationships with the environment. It is not tied to a specific application scenario, and does not therefore have a specific configuration of system modules. The *Domain Ontology*, on the other hand, includes knowledge about an instance of the system for a specific scenario, in our case the BMS for environmental comfort in an office.

Both ontologies are described by the OWL DL language, and the domain ontology is anchored to general one through the import mechanism allowed by the OWL [21]. The choice of OWL DL allows the designer to exploit the services of automatic control of consistency, of classification and inference, and of verification of the correctness of the coded knowledge.

A set of logical rules related to the proposed ontology makes it possible to infer new knowledge directly by processing the ontology. These rules are expressed by means of the Semantic Web Rule Language (SWRL) [17], as new OWL axioms, with the classic form `antecedents` → `consequent`. SWRL rules can be used to define a property as a composition of other properties, thus expressing the idea that a set of basic properties imply a composite property. They can also be used to transfer the value of a given property from certain individuals to another related individual.

The use of a rule engine capable of processing ontologies and their logical rules, makes it possible to perform automated reasoning on the domain. A tool for this purpose is the inference engine Jess [10] integrated with the development environment Protégé [11] used to define the ontology. This inference engine, integrated into the symbolic planning module, allows the system to reason both about the environment and the system itself, and then to take the appropriate actions.

4.1 The General Ontology

The *General Ontology* defines classes and properties required to describe the components of the multi-tier architectures, their interconnections, and the data flow inside the system. Moreover, it describes the basic elements constituting the sensory and actuator subsystem.

4.1.1 Environmental Properties and Physical Devices

What the AmI system is able to monitor is defined as an `AmbientProperty`; this class is further specialized into three separate subclasses: `PhysicalProperty`, representing physical observable phenomena such as temperature and humidity, `Status`, representing the status of a particular environmental element, such as a door in the `CLOSED` state, and `Event`, representing observable events such as a device fault or the entry of the user into a monitored room.

Physical devices controllable by the AmI system, and composing the sensory and actuator infrastructure, are modeled by means of the `Device` class. This class is subdivided into three specialized subclasses, namely `Sensor`, `Actuator`, and `Node`. Each device is deployed in a specific room, and this topological relationship is also represented inside the ontology.

The Planning subsystem controls a generic device by sending an opportune command. This dependency is expressed inside the ontology which also models the data flow inside the system by means of a set of properties. In particular, the property `hasInputFrom` makes it possible to code the fact that a `Device` is able to receive input only from an `ActuatorModule`, that is a specific module of the multi-tier architecture. Analogously, the ontology models the fact that a `Sensor` is able to produce input data for a specific type of module, namely for a `SubsymbolicModule`.

Besides these properties, the ontology models other characteristics of a sensor, such as sampling rate, continuity of monitoring, energy consumption, and the node over which the sensor is installed. Similarly, each actuator is also characterized by the set of commands that it is able to receive.

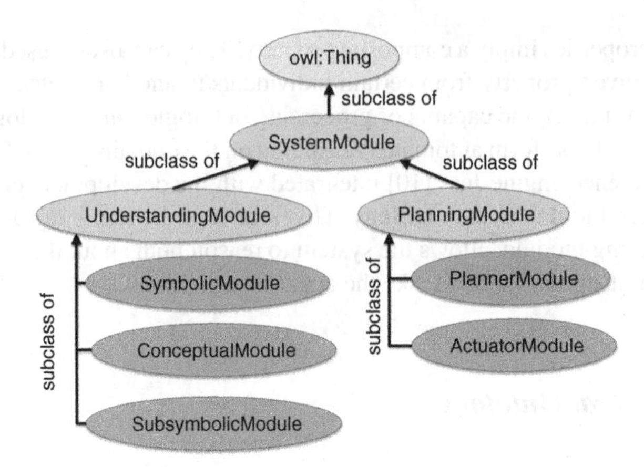

Fig. 2 Taxonomy of the `SystemModule` class in the proposed ontology. Graph nodes represent ontology classes, and *arrows* represent properties, going from a class of the property domain toward a class of the property co-domain

An important relationship is the connection between an ambient property and sensors able to perceive it. This relation is exploited by a SWRL rule to determine which properties are observable or not:

```
AmbientProperty(?x) ∧ Sensor(?y) ∧ senses(?y, ?x)
→ ObservableProperty(?x)
```

A similar but specular relationship exists between actuators and ambient properties. i.e., `actsOn`, making it possible to model which actuators are able to modify a particular environmental condition.

4.1.2 Architecture Modules

Architecture modules are modeled inside the ontology by means of the `SystemModule` class, whose descending taxonomy is shown in Fig. 2. The main distinction is functional, distinguishing between `UnderstandingModule` and `PlanningModule`, as described in Sect. 3.

The interconnections between modules are coded by means of the `hasInputFrom` and `hasOutputTo` properties. The knowledge representation for each module, in terms of coding for input and output, is described by means of the `hasInputData` and `hasOutputData` properties, with instances of `DataType` classes as their dominion.

The `UnderstandingModule` is further specialized in the following classes: `SubsymbolicModule`, `ConceptualModule`, and `SymbolicModule`. The organization into levels of increasing abstraction is constrained by the `hasOutputTo` property, which can take values from the `SubsymbolicModule` class for the `Sensor` and `Node` classes, from the `ConceptualModule` class for

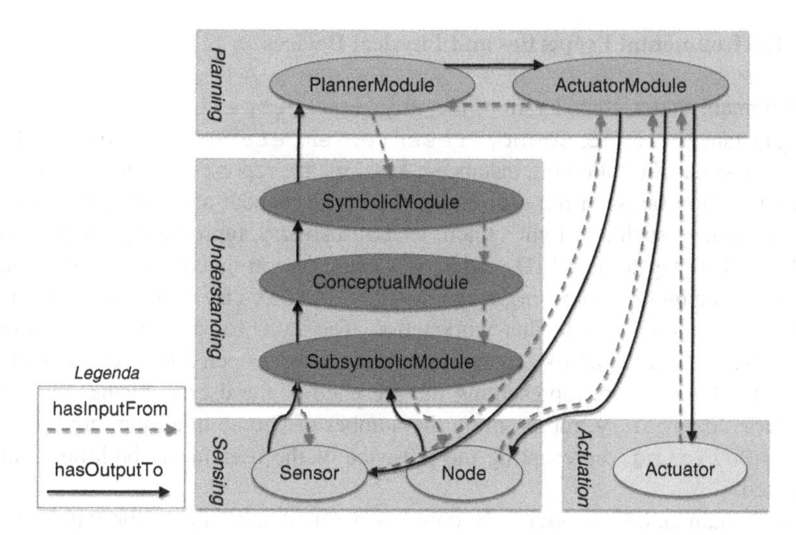

Fig. 3 Three-tier organization of the understanding subsystem, as coded by the proposed ontology

the `SubsymbolicModule` class, and from the `SymbolicModule` class for the `ConceptualModule` class. Data produced by a `SymbolicModule` are directly usable by a `PlannerModule`. This type of module acts at the highest level of abstraction, with the aim of reasoning on symbolic representations of environment and context, and thus producing high level plans designed to achieve system goals. Plans are processed by `ActuatorModules`, which translate them into low-level commands for `Actuators`, to modify environmental conditions, and for `Nodes` and `Sensors`, to modify the behavior of the sensing subsystem. This interconnection pattern is shown in Fig. 3.

4.2 The Domain Ontology

The *domain ontology* imports and extends the *general ontology*, by defining subclasses and individuals to describe a specific instance of the AmI system. The application scenario considered in this chapter is *Sensor9k* [6], a testbed, designed for an office environment, and for reasoning about user comfort and energy saving. Different nodes are equipped with several types of sensors capable of monitoring the environment. The understanding subsystem is capable of estimating the number of users in the monitored office, and evaluating the current lighting level. This information is then combined to evaluate the adequateness of the lighting level. A simple planning module uses such high level evaluation to determine which action needs to be performed in order to achieve an adequate lighting level with minimum energy consumption.

4.2.1 Environmental Properties and Physical Devices

In the domain under consideration, the `PhysicalProperty` class has the following instances: `Light`, `Sound`, `Pressure`, `Temperature`, `Humidity`. The `Event` class has the following instances: `Activity`, representing the occurrence of an interaction between the user and some of the manual actuators deployed in the environment, such as a light switch; `RFIDPassing`, representing the proximity of an RFID tag to an RFID reader (this event is correlated with the presence of the user in a monitored room); `WorkstationActivity`, representing the interaction of the user with his/her workstation. The `Status` class has the following instances: `DoorStatus`, representing the closed / open / locked status of a door; `UserInOffice`, representing the presence of the user in his/her own office; `RoomOccupancy`, representing the number of people in a monitored room; `UserInBuilding`, representing the presence of the user in the building being monitored.

The domain ontology obviously contains a set of individuals for topological classes, and the correct values for properties that make it possible to specify the placement of each device. Moreover, each sensor or actuator is linked with the ambient property that it is able to monitor or modify.

4.2.2 Architecture Modules

The software architecture is specified by means of the definition of a set of sub-classes and individuals of the `SystemModule` class. Each class inherited from the *general ontology* is further divided into specialized sub-classes for modules capable of performing a specific type of reasoning. The taxonomic organization of `SystemModules` in the *general ontology* reflects the level of abstraction at which the knowledge is processed, whilst the further decomposition coded in the *domain ontology* reflects the topological and semantic organization of the monitored environment.

Figure 4 shows an example of such organization. Suppose that the target building consists of two rooms, `room_1` and `room_2`, and that each of these rooms contains light sensors. The software architecture includes a class of subsymbolic modules devoted to processing lighting information, called `Light_ssM`, and two instances of this class, namely `Light_ssM_1` and `Light_ssM_2`. These two modules require as input the same type of sensory data, i.e., `LightReading`, and produce as output the same type of qualitative data, `LightLevel`.

5 Autonomic Self-Configuration

The AmI system proposed here is based on the *monitor-analyze-plan-execute* cycle of a typical autonomous system. This paradigm is exploited not only in controlling the environment surrounding the user, but also in dynamically controlling the behavior of the system itself.

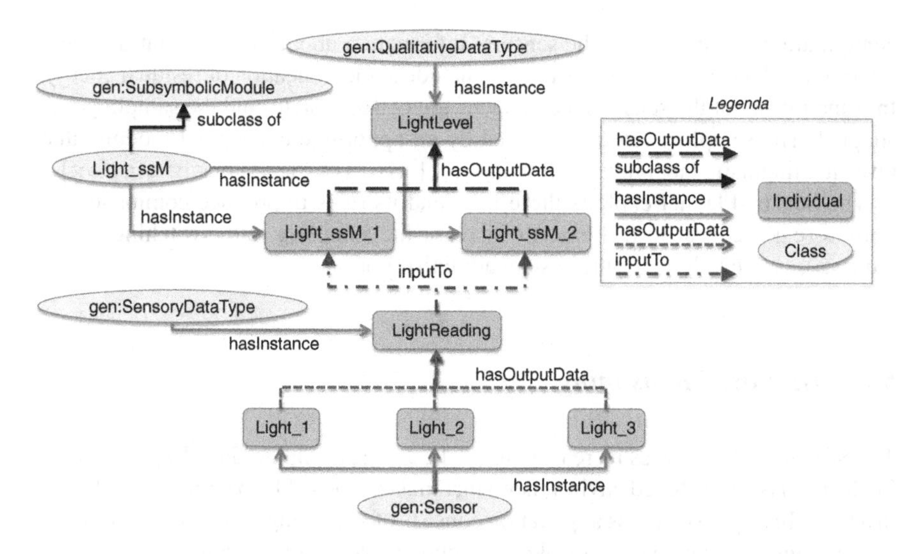

Fig. 4 Example of an AmI system that includes two instances of the `Light_ssM` class; these modules accept as input the same instance of `SensoryDataType` even though the sensory readings come from different sensors; these modules provide, as output, data belonging to the same instance of `QualitativeDataType` class

The application described above for controlling the lighting system is an example of the implementation of such a paradigm for environmental management. The system monitors the environment by gathering sensory readings, analyzes them to obtain a high-level representation of the overall lighting level, plans the sequences of actions to be performed in order to achieve an adequate lighting level, and then executes those actions by means of the actuators available to it.

The implementation of the *Autonomous Computing* paradigm for controlling the system's behavior is embodied in the dynamic configuration of the monitoring activity. The system is able to receive a set of requirements and high-level policies, such as the goal of minimizing the energy consumption of the sensing infrastructure when users are not present in target premises.

Thanks to its capacity for introspection, the system evaluates the quality of its own analysis functionality, and knows which hardware and software elements contribute to inferring a given concept. A planning module, called `SensingPlanner`, exploits this knowledge to optimize the monitoring functionality of the system; this module is based on a set of rules that embodies high-level policies for system management. Information about the system state and inference accuracy flow through different levels of the multi-tier architecture.

In particular, data about sensors' energy consumption are classified by opportune conceptual modules which state whether the consumption is `high`, `medium` or `low`. These linguistic labels are then represented in a symbolic form, by means of the assertion of facts inside the rule engine. Symbolic plans produced by the planning module are provided as input to actuation modules able to translate them into a set of configuration commands for the sensors. Information about the node states is only relevant for devices fed by batteries, and include some indication of residual energy. Information about the sensor state includes the `on/off` state, and the sampling rate adopted. The sampling rate is classified by an opportune conceptual module, that gives it a linguistic label in the following set: {`min`, `low`, `medium`, `high`, `max`}.

The SensingPlanner exploits these facts and its rules to produce commands for sensors (e.g., request to decrease or increase the sampling rate, switching on/off requests), or even alerts for the system administrator.

5.1 Rule-Based Reasoning

The SensingPlanner bases its reasoning on the Jess rule engine. Java Expert System Shell (Jess) is a rule-based environment, that makes it possible to define logical rules in a LISP-like syntax. A Jess application consists of a working memory, a set of rules and an inference engine responsible for applying rules to the working memory.

The working memory consists of a set of facts, which are true statements about the dominion under consideration, and represents the system's knowledge of the world. Facts can be asserted to and retracted from working memory. Each fact has a template (the relation between facts and templates corresponds to the relation between objects and classes in the OOP).

Rules react to changes in working memory, and can exploit auxiliary functions and queries. Each rule is activated when all its antecedents are satisfied, that is when the working memory contains facts matching rule antecedents. Activated rules are fired when the inference engine is executed, thus causing the function execution. Each rule is executed once for any given set of facts in the working memory; a new execution requires that new facts have been asserted.

The templates and rules of the proposed SensingPlanner are described below.

The following templates represent static knowledge:

```
(deftemplate room (slot id))
(deftemplate node (slot id) (slot office)(slot batteryPowered))
(deftemplate nodeSensor (slot nodeId)(slot sensorId))
(deftemplate sensor (slot id) (slot ambientProperty)
(slot energyConsumption) (slot continuousSampling))
(deftemplate user (slot name)(slot office))
(deftemplate ambientProperty (slot id)(slot observable))
(deftemplate affects (slot affectedProperty)
(slot affectingProperty))
```

Facts corresponding to these templates allow the SensingPlanner to know things about the premises being monitored, the users and their offices, and the placement of nodes and sensors. Moreover, these last two templates make it possible to specify which ambient properties can be monitored, and the cause-effect relationship among properties. For example, using these types of facts, the SensingPlanner is able to know that information about the sound level and the workstation activity provides indirect information about a user's presence.

The following set of templates represents facts about the state of the sensory infrastructure and the context. These facts are dynamically updated on the basis of the current states of devices, of the inferred presence of the user in his office or in the building, and of the accuracy of the monitoring activity.

```
(deftemplate nodeState (slot id)(slot batteryLevel))
(deftemplate sensorState (slot id)(slot sampling))
(deftemplate sensingAccuracy (slot room)(slot ambientProperty)
(slot accuracy))
(deftemplate userInOffice (slot user) (slot present))
(deftemplate userInBuilding (slot user) (slot present))
(deftemplate peopleInRoom (slot room) (slot number))
```

Finally, a set of templates define the structure for facts which represent the output of planning. Plans for tuning sensors, indicating a new value for the sampling rate, {`on, off, max, min, up, down`}, correspond to the following template:

```
(deftemplate tuneSensor (slot sensorId) (slot newSampling))
```

Alerts for system administrators can be expressed according to the following templates:

```
(deftemplate insufficient-sensing (slot room)
(slot ambientProperty))
(deftemplate short-sensing-life (slot room)
(slot ambientProperty))
```

An `Insufficient-sensing` fact indicates that, although the sampling rate for available sensors is at its maximum sustainable value, the monitoring is not sufficiently accurate and a structural intervention by the system administrator is probably required. A `short-sensing-life` fact indicates that all sensors involved in monitoring a given ambient property in a given room have low residual energy, and so a system administrator intervention is necessary to guarantee that there will be no interruption in the monitoring functionality.

The main rules of the SensingPlanner module are described below.

The first rule states that, for a given room, if all corresponding users are not present in the entire building, then it is possible to switch all sensors off with the exception of those responsible for monitoring the `UserInOffice` property; for these sensors monitoring is minimized.

```
(defrule stopRoomSensing
  (and
    (room (id ?p))
    (not (and (user (name ?u) (office ?p))
              (or (userInBuilding (user ?u)(present "true"))
                (userInOffice (user ?u)(present "true")))))
  )=>
  (stopRoomSensing ?p)
  (minimizeRoomSensing-ap ?p "UserInOffice"))
```

where minimizeRoomSensing-ap is an auxiliary function which receives as input an ambient property and a room, selects all sensors perceiving this ambient property, directly or indirectly, and sets their sampling rate to the minimum value.

The second rule minimizes monitoring for all ambient properties in a empty room, provided that users normally occupying that room are present in the building.

```
(defrule minimizeRoomSensing
 (and
   (room (id ?p))
   (not (and (user (name ?u) (office ?p))
            (userInOffice (user ?u) (present "true"))))
   (exists (and (user (name ?u) (office ?p))
            (userInBuilding (user ?u)(present "true"))))
 )=>
 (minimizeRoomSensing ?p ))
```

Another two rules deal with increasing and decreasing the sampling rate based on the accuracy of monitoring, inferred by an opportune symbolic module. According to the policy adopted here, an increase in accuracy is considered a goal only if there are people in their offices, so these rules have a lower priority than the previous two. The increaseRoomSensing rule is detailed below.

```
(defrule increaseRoomSensing
 (and
   ?accuracyLowFact <- (sensingAccuracy (room ?p)
                 (ambientProperty ?ap) (accuracy "low"))
   (exists (and (user (name ?u)(office ?p))
              (userInOffice (user ?u)(present "true"))))
 )=>
 (if (increaseRoomSensing-ap ?p ?ap)
   then
   (retract ?accuracyLowFact)
   else
   (assert (insufficient-sensing (ambientProperty ?ap)
                         (room ?p)))))
```

This rule increases the monitoring rate of an ambient property characterized by a low level of accuracy, by means of the increaseRoomSensing-ap auxiliary function. This function tries to increase the sampling rate of sensors, starting from sensors with low energy consumption. If none of sensors has any margin for increasing its sampling rate, the increaseRoomSensing-ap function returns the value false, thus causing the assertion of an insufficient-sensing fact, used to trigger a notification to the system administrator.

6 Conclusions and Future Research

This chapter proposes an ontology-based autonomic system for self-configuring the sensory infrastructure of an ambient intelligence system. The ontology makes it possible to define concepts characterizing the environment, the ambient properties upon which the AmI system reasons, and the structure of the system itself.

The cognitive paradigm adopted is characterized by a flexible scheme easily implementable in new scenarios. The system configuration process can take advantage of an ontological representation of system structure and of its interaction with the environment. Within such a predefined and structured framework of basic knowledge, the designer can easily represent the application domain and specify the configuration of system modules, thus reducing the risk of errors.

Moreover, the knowledge of its own structure and of the current context is exploited by a rule-based planner whose goal is to tune the sampling rate of sensors, in order to maximize the accuracy of the reasoning while minimizing the energy consumption of the system.

The system proposed here may be further expanded with the capability of self-instantiating only on the basis of the high-level description provided by the ontology. This outcome would require the development of a library of parametric modules which cover most of the possible types of processing, besides the possibility of extending the library and the ontology with new functionalities.

Another potential future development concerns communication with final users about the cognitive processing which has occurred, and about the reasons that drive the system to take certain decisions [23, 28]. Such a possibility is enabled by the explicit representation of its knowledge.

Finally, it is possible to imagine a cooperative network of intelligent buildings, able to communicate to each other the knowledge about their structure and about what they have learned, in terms of optimal configurations. Naturally, this type of scenario would require integration with a communication protocol able to identify reliable agents [3] with which it would be opportune to cooperate, with the main goal of protecting user privacy.

Acknowledgments This work has been partially supported by the PO FESR 2007/2013 grant G73F11000130004 funding the SmartBuildings project.

References

1. Atzori, L., Iera, A., Morabito, G.: The internet of things: a survey. Comput. Netw. **54**(15), 2787–2805 (2010)
2. Cook, D., Augusto, J., Jakkula, V.: Ambient intelligence: technologies, applications, and opportunities. Pervasive Mobile Comput. **5**(4), 277–298 (2009)
3. Crapanzano, C., Milazzo, F., De Paola, A., Lo Re, G.: Reputation management for distributed service-oriented architectures. In: 2010 Fourth IEEE International Conference on Self-Adaptive and Self-Organizing Systems Workshop (SASOW), pp. 160–165 (2010)

4. De Paola, A., Farruggia, A., Gaglio, S., Lo Re, G., Ortolani, M.: Exploiting the human factor in a WSN-based system for ambient intelligence. In: International Conference on Complex, Intelligent and Software Intensive Systems, 2009 (CISIS '09), pp. 748–753 (2009)
5. De Paola, A., Gaglio, S., Lo Re, G., Ortolani, M.: An ambient intelligence architecture for extracting knowledge from distributed sensors. In: Proceedings of the 2nd International Conference on Interaction Sciences: Information Technology, Culture and Human, pp. 104–109 (2009)
6. De Paola, A., Gaglio, S., Lo Re, G., Ortolani, M.: Sensor9k: a testbed for designing and experimenting with WSN-based ambient intelligence applications. Pervasive Mobile Comput. **8**(3), 448–466 (2012)
7. Doctor, F., Hagras, H., Callaghan, V.: A fuzzy embedded agent-based approach for realizing ambient intelligence in intelligent inhabited environments. IEEE Trans. Syst. Man Cybern. Part A Syst. Hum. **35**(1), 55–65 (2005)
8. Ducatel, K., Bogdanowicz, M., Scapolo, F., Leijten, J., Burgelman, J.C.: Scenarios for ambient intelligence in 2010. Office for Official Publications of the European, Communities (2001)
9. Farruggia, A., Lo Re, G., Ortolani, M.: Probabilistic anomaly detection for wireless sensor networks. In: AI*IA 2011: Artificial Intelligence Around Man and Beyond, Lecture Notes in Computer Science, vol. 6934, pp. 438–444. Springer, Berlin Heidelberg (2011)
10. Friedman, E.: Jess in action: rule-based systems in Java. Manning Publications Co., Greenwich (2003)
11. Gennari, J., Musen, M., Fergerson, R., Grosso, W., Crubézy, M., Eriksson, H., Noy, N., Tu, S.: The evolution of Protégé: an environment for knowledge-based systems development. Int. J. Hum. Comput. Stud. **58**(1), 89–123 (2003)
12. Gruber, T.: A translation approach to portable ontology specifications. Knowl. Acquisition **5**(2), 199–220 (1993)
13. Gu, T., Wang, X., Pung, H., Zhang, D.: An ontology-based context model in intelligent environments. In: Proceedings of Communication Networks and Distributed Systems Modeling and Simulation Conference, pp. 270–275 (2004)
14. Hagras, H.: Embedding computational intelligence in pervasive spaces. IEEE Pervasive Comput. **6**(3), 85–89 (2007)
15. Hariri, S., Khargharia, B., Chen, H., Yang, J., Zhang, Y., Parashar, M., Liu, H.: The autonomic computing paradigm. Cluster Comput. **9**(1), 5–17 (2006)
16. Herrmann, K., Muhl, G., Geihs, K.: Self management: the solution to complexity or just another problem? IEEE Distrib. Syst. Online **6**(1), 1–17 (2005)
17. Horrocks, I., Patel-Schneider, P., Boley, H., Tabet, S., Grosof, B., Dean, M.: SWRL: a semantic web rule language combining OWL and RuleML. W3C member submission. http://www.w3.org/Submission/2004/SUBM-SWRL-20040521/ (2004)
18. Kephart, J., Chess, D.: The vision of autonomic computing. Computer **36**(1), 41–50 (2003)
19. Klein, M., Schmidt, A., Lauer, R.: Ontology-centred design of an ambient middleware for assisted living: the case of soprano. In: Towards Ambient Intelligence: Methods for Cooperating Ensembles in Ubiquitous Environments (AIM-CU), 30th Annual German Conference on Artificial Intelligence (KI 2007), pp. 1–8 (2007)
20. Kushwaha, N., Kim, M., Kim, D., Cho, W.: An intelligent agent for ubiquitous computing environments: smart home ut-agent. In: Proceedings of the Second IEEE Workshop on Software Technologies for Future Embedded and Ubiquitous Systems, pp. 157–159. IEEE Press, Piscataway, NJ, USA (2004)
21. McGuinness, D., Van Harmelen, F.: OWL web ontology language overview. W3C recommendation. http://www.w3.org/TR/owl-features/ (2004)
22. Mozer, M.: The neural network house: an environment hat adapts to its inhabitants. In: Proceedings of the Intelligent Environments AAAI Spring Symposium, pp. 110–114. AAAI, Palo Alto, CA, USA (1998)
23. Pilato, G., Augello, A., Gaglio, S.: Modular knowledge representation in advisor agents for situation awareness. Int. J. Semant. Comput. **5**(1), 33–53 (2011)

24. Preuveneers, D., Bergh, J., Wagelaar, D., Georges, A., Rigole, P., Clerckx, T., Berbers, Y., Coninx, K., Jonckers, V., Bosschere, K.: Towards an extensible context ontology for ambient intelligence. In: Ambient Intelligence, Lecture Notes in Computer Science, vol. 3295, pp. 148–159. Springer, Berlin (2004)
25. Remagnino, P., Foresti, G.: Ambient intelligence: a new multidisciplinary paradigm. IEEE Trans. Syst. Man Cybern. Part A Syst. Hum. **35**(1), 1–6 (2005)
26. Ribino, P., Oliveri, A., Lo Re, G., Gaglio, S.: A knowledge management system based on ontologies. In: International Conference on New Trends in Information and Service Science, 2009. NISS '09, pp. 1025–1033 (2009)
27. Serrano, J., Serrat, J., Strassner, J., O Foghlu, M.: Facilitating autonomic management for service provisioning using ontology-based functions and semantic control. In: IEEE Network Operations and Management Symposium Workshops, 2008, pp. 77–86 (2008)
28. Sorce, S., Augello, A., Santangelo, A., Gentile, A., Genco, A., Gaglio, S., Pilato, G.: Interacting with augmented environments. IEEE Pervasive Comput. **9**(2), 56–58 (2010)
29. Srivastava, M., Culler, D., Estrin, D.: Overview of sensor networks. Computer **37**(8), 41–49 (2004)
30. Stanfel, Z., Hocenski, Z., Martinovic, G.: A self manageable rule driven enterprise application. In: 29th International Conference on Information Technology Interfaces (ITI 2007), pp. 717–722 (2007)
31. Stankovic, J.: Wireless sensor networks. Computer **41**(10), 92–95 (2008)
32. Tesauro, G.: Reinforcement learning in autonomic computing: a manifesto and case studies. IEEE Internet Comput. **11**(1), 22–30 (2007)

Detection of User Activities in Intelligent Environments

Agnese Augello and Salvatore Gaglio

Abstract Research on Ambient Intelligence (AmI) focuses on the development of smart environments adaptable to the needs and preferences of their inhabitants. For this reason it is important to understand and model user preferences. In this chapter we describe a system to detect user behavior patterns in an intelligent workplace. The system is designed for a workplace equipped in the context of *Sensor9k*, a project carried out at the Department of Computer Science at the University of Palermo (Italy).

1 Introduction

Research in Ambient Intelligence (AmI) focuses on the development of smart environments, generally equipped with wireless sensor networks, allowing the gathering of data about the environment state [1]; such data needs to be processed and analyzed in order to deduce useful information. Ambient Intelligence brings intelligence to our everyday environments making those environments sensitive, and adaptive to us [2]. The definition of appropriate user profiles can allow an AmI system to anticipate their needs, and adapt the environment settings to their preferences [3, 4, 8]. User profiling can also be used to detect significant changes in resident behaviors [2, 5], to customize building energy and comfort management systems [6], or to allow automatic setting of system parameters in order to optimize energy consumption [7].

Most systems perform an explicit profiling or derive users presence and activity by analyzing the sensor data and the use of actuators. In many projects, data mining

A. Augello (✉)
ICAR-CNR, Viale delle Scienze, Edificio 11, Palermo, Italy
e-mail: augello@pa.icar.cnr.it

S. Gaglio
DICGIM-UNIPA, Viale delle Scienze, Edificio 6, Palermo, Italy
e-mail: salvatore.gaglio@unipa.it

S. Gaglio and G. Lo Re (eds.), *Advances onto the Internet of Things*,
Advances in Intelligent Systems and Computing 260, DOI: 10.1007/978-3-319-03992-3_2,

methodologies are used to detect recurrent behaviors in time sequences of events, to find the relationship between variables of interest, or to prefigure the future behavior of some entities. As an example, in MavHome [9] hierarchical models of inhabitant behaviors are learned by means of data-mining techniques aimed to discover periodic and frequent episodes of activity patterns, while in [2] they are used to predict the location, routes and activities of the residents in order to adaptively control home environments .

An important aspect to be considered is the choice of the models used to represent the information acquired in an AmI system. As an example, the authors of [10] propose an ontology to model the domain knowledge of a smart building system. In [11] the problem of modelling interaction events of smart objects in a environment is discussed; ontologies can help developers to infer the possible connections between these objects, enable device interoperability on a semantic level. In [12] there is a discussion on the importance of common sense ontology in a real AmI system; a reasoning on the acquired information can be performed to deal with unforeseen requirements or needs, making decisions out of the prefixed behavioral patterns.

The architecture proposed in this chapter arises from the need to implicitly analyze and profile the users of an intelligent workplace equipped in the context of Sensor9k project [13]. In this work we consider a workplace equipped with a set of sensors aimed at the detection of environmental variables, such as light, temperature and humidity, while there are no sensors providing direct information about the user except for their presence (which is detected by RFId sensors). In our opinion user profiles are implicitly described by the data collected by the sensors located in the workplace rooms, data which explicitly show the consequences of users' actions over the environment state.

The approach proposed her exploits different methodologies of analysis and reasoning on data sensors in order to detect user actions on the environment. In particular the analysis is aimed at the detection of meaningful changes which can be considered as consequences of user actions. The sensory data and the recognized events are arranged in appropriate models in order to highlight the existence of relationships among environmental data or events and the users' presence in the office room. Moreover an analysis and reasoning on the information described in such models can be performed in order to infer users behaviors and preferences.

2 Proposed Approach

The proposed approach consists essentially of three different phases of analysis of the data gathered by the wireless sensor network, regarding the environment state and the user presence (see Fig. 1).

The *pre-processing* analyzes data in order to detect anomalies, remove outliers and replace missing data, the *action detection and modelling* phase analyzes sensor data trends to infer changes which can be ascribed to human actions; appropriate models allow for a better understanding of the extracted information. Finally the *Extraction*

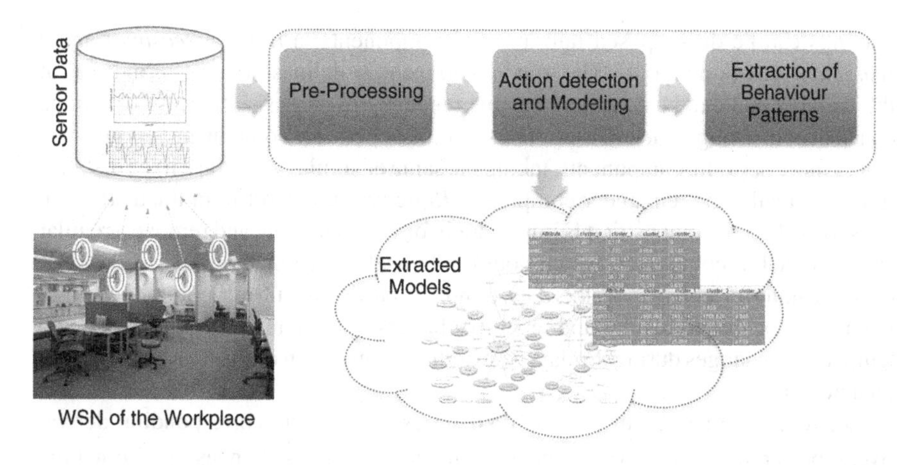

Fig. 1 The proposed approach

of behavior Patterns phase is accomplished to find relationships of interest and to detect similar clusters in the AmI data.

2.1 Pre-Processing

At this phase, the observed variables trends are analyzed to recognize those events that can be ascribed to human intervention. First of all the data collected by the sensors are preprocessed in order to to detect anomalies, remove invalid values and estimate missing values. For example the assumption of a time correlation in the data can be exploited for the estimation of missing values by means of a linear interpolation between preceding and subsequent observations, while the assumption of spatial correlation in the readings of sensors located in small indoor environments can be profitably employed in order to detect outliers, and to replace them with the combination of neighbour sensors readings.

However, for our aim, it is important to analyze differences between sensors belonging to different, but close areas. We assume that variations in inter-areas sensors readings can be due to the use of actuators from users, such as turning on/off the light or changing the settings for the temperature and humidity control systems.

2.2 Action Detection and Modelling

In this phase of analysis we consider the placement of the sensors in areas within the environment and analyze the dynamics of the observed series, i.e., the mechanism by which they evolve over time. In particular, the time series obtained from sensor readings can be decomposed into a set of components: a *trend*, a *seasonal* and a *remainder* component [14]. The trend component T_t defines the long-term trend of the variable and can be considered as the tendency to increase, decrease or remain constant over a long period of time. The seasonal or periodic component S_t is given by one or more periodic components, taking the same or similar values at a fixed distance in time. Finally, the remainder component R_t determines short-term fluctuations in the series. This decomposition is important in order to estimate and remove a regular and predictable component which could hide useful information, and to consider only meaningful changes in the data trend. In particular, in our context, changes in data can be due to regular and repetitive factors, for example to natural light and temperature changes during the day, while other changes can be due to human actions on actuators.

Let R_t the remainder of the analyzed series, and $R_t^{'}$ the corresponding derivative, representing its time variation. The function $R_t^{'}$ shows changes in R_t function. Every change is characterized by a strength and a direction. If the detected change is attributed to a user action, the direction will allow the interpretation of the type of action. For example a positive direction will be considered as a sign of the light being turned on, or in the case we are analyzing temperature data could be interpreted as an increase of temperature settings. Given an experimentally defined threshold, called ϑ, only changes with strength greater than ϑ will be considered.

After the preprocessing of the data we suggest the use of a probabilist model to estimate if the detected changes in sensor measurements can be attributed to users' actions. The choice of using such a model depends on the fact that we are bound to reason about uncertain knowledge. The interest variables and their possible states are modelled as nodes of a dynamic Bayesian network. The nodes are connected with directed links representing the influence famong the nodes where the influence of parent nodes on a variable X_i is quantified by means of conditional probabilities $P(X_i|ParentsValues)$ represented in opportune tables associated to each node. We model two kinds of variables, *UA* variables, representing possible user actions and *SO* variables, representing sensory observations. The model thus built is used to estimate the probability that the event we want to investigate did occur, based on the sensor readings. As an example, Fig. 2 shows the Bayesian network used to estimate the user actions controlling ambient light; the possible action are modelled by the *UserAction t* variable which can take three different states: *on*, corresponding to turning on action, *off*, corresponding to turning off action, and *none* if no action is carried out. We also have four variables of type *SO*: *UserPresence*, *OutdoorLight t +1*, *LightTrend t* and *LightTrend t + 1*. The first can assume the states *in* and *out* depending on users presence in the room. Variable *OutdoorLight t +1* represents the external light at a subsequent time instant and can be *high*, *medium*, or *low*; it is used to better understand how much the light change in the room at a subsequent instant may be due to the external light status or to an user action. The variables *LightTrend t* and *LightTrend t + 1* represent the evolution of light in two successive time instants, and can assume the states *increasing*, *decreasing* and *stable*. The actions are modelled as states of the *UserAction t*, which depends on the state of the light

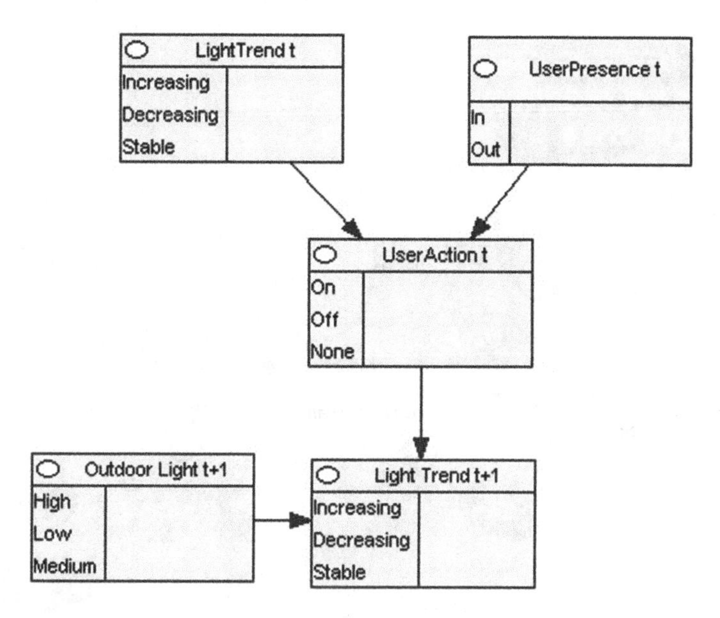

Fig. 2 A probabilistic model to estimate user actions controlling ambient light

(*LightTrend t* variable) and by information on the users presence (*UserPresence t* variable) in workplace. The state of *UserAction t* variable and the state of the outdoor light *OutdoorLight t +1* influence in their turn the state of the light at the next istant (*LightTrend t + 1* variable).

The detected events, which in the particular case are possible user actions on the actuators, can be modelled in an ontology (a portion is shown in Fig. 3).

Each action modelled in the ontology is associated to the room in which the event took place, to the employee who may have executed that action (we have information only about the presence of people in a particular room but we do not know who actually have done that action), to the time in which the action was accomplished, and to other correlated concepts. Into the ontology, other information is also represented, related to the domain knowledge; for example the variables that can be controlled by the user, such as the temperature and light, and the environmental status of the outdoor. The ontology can be used to reason about the acquired information in order to deduce new knowledge. For example, analyzing either the information regarding the state of the temperature inside and outside a room, and the actions performed on the thermostat when in that room is present a specific user, we can acquire the preferences of that user regarding the setting of the temperature.

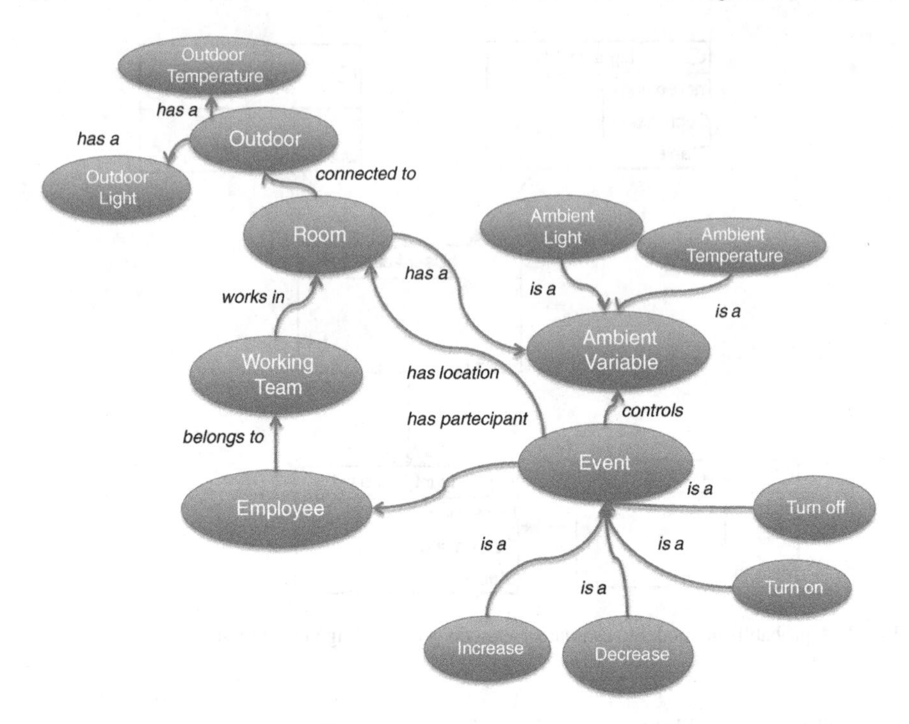

Fig. 3 An ontological formalization of the actions on the intelligent workplace

2.3 Extraction of Behavior Patterns

The sensors data are analyzed to identify relationships among the variables of interest and the users. The measurements recorded by the different sensors for each physical variable in a specific period (such as for instance an entire working day) and the occurrences of events such as user actions in specific instants are represented in a matrix. In particular the rows of the matrix represent the different observations in a given period, while the columns the sample values detected from each sensor during the observations. The number of sensors inside the room can be large, so as to generate several columns. Therefore, we perform a dimensionality reduction to evaluate only the most meaningful informative content. Let **X** indicate the matrix representing a dataset composed by a set of m vectors of length n, each one representing the set of measurements obtained by the n sensor at a specific observation for a specific variable x $X = [\mathbf{x}_1, \ldots \mathbf{x}_n]$.

As an example, let $\mathbf{T} = [\mathbf{T}_1, \ldots \mathbf{T}_n]$ be a $m \times n$ matrix composed of a set of column vectors, each one representing the set of observations regarding the temperature measured by n sensors in an office room.

A dimensionality reduction process, by means of PCA [15] is performed on such set of observations allowing for the projection of original dataset space into a smaller space. PCA extracts a set of orthogonal vectors, called principal components, by

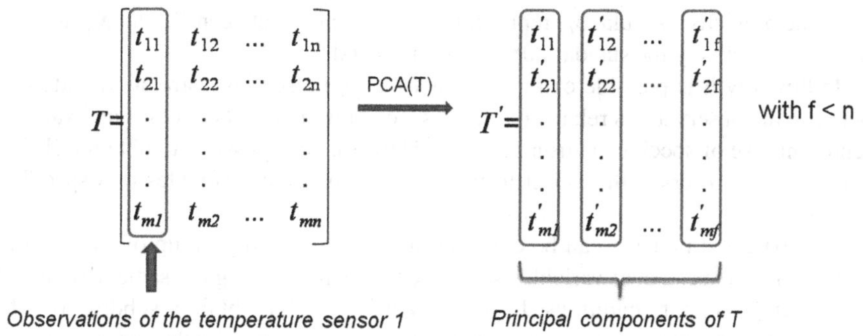

PCA(T)

with f < n

Observations of the temperature sensor 1

Principal components of T

Fig. 4 Principal component analysis performed on temperature dataset

a linear transformation of the original variables, and arranges them according to decreasing variance values. This transformation has the effect to capture the major associational structure in the dataset, removing information which contribute less to the variance of data, and are thus less relevant. It should be highlighted that PCA performs the dimensionality reduction process by a combination of original vectors, while other methods merely select a subset of items from the original dataset [16].

After performing PCA we obtain a $m \times f$ matrix, with $f \leq n, \mathbf{T'} = [\mathbf{T'}_1, \ldots \mathbf{T'}_f]$ (see Fig. 4). The same procedure can be performed over the entire set of observed variables.

The whole set of variables and events observations is then modelled as a matrix $\mathbf{X}(m \times nv)$, where a row $\mathbf{X}_i$ represents an observations at a specific time i and a specific column $\mathbf{X}_j$ represents the entire sample of observations of the j-th variable in the considered period.

In our specific case, matrix $\mathbf{X}$ is given by:

$$\mathbf{X} = [\mathbf{U}_1, \ldots \mathbf{U}_d, \mathbf{T}_1, \ldots \mathbf{T}_f, \mathbf{L}_1 \ldots \mathbf{L}_g]$$

composed of a set of vectors, each one representing the set of observations of a specific variable: the set $\mathbf{U} = \{\mathbf{U}_j\}_{j=1 \ldots d}$ represents observations about the presence of d users in the considered period, the sets $\mathbf{T} = \{\mathbf{T}_j\}_{j=1 \ldots f}$ and $L = \{\mathbf{L}_j\}_{j=1 \ldots g}$ represent observations about temperature and light exposure respectively, related to the f and g variables obtained after the application of PCA on temperature and light matrices as described in the previous section.

We therefore compute a correlation matrix $\mathbf{C}(nv \times nv)$ in order to highlight the relationships among the variables, where the i, j-th element of $\mathbf{C}$ is given by the correlation coefficient c_{ij} between the i-th and the j-th variable, as given by:

$$c_{ij} = Corr(\mathbf{X}_i, \mathbf{X}_j) = \frac{\sigma_{ij}}{\sigma_i \sigma_j}$$

In the previous formula, σ_{ij} represents the covariance between $\mathbf{X}_i$ and $\mathbf{X}_j$ and σ_i and σ_j respectively the standard deviation of $\mathbf{X}_i$ and $\mathbf{X}_j$.

In this way it is possible extract sub-matrices, representing correlation patterns between the observations related to the presence of users in office rooms and values representative of specific environment variables. It is thus possible to obtain a characterization of users with respect to values of the observed variables in a specific period.

The extracted patterns can be clustered in order to identify group of users with similar preferences about variables setting, or users performing the same actions in similar environment conditions. In this way it is possible obtain a subdivision of users' behavior patterns into a set of similar profiles . In particular the data extracted from the correlation matrix **C** are classified by means of a K-means [17] algorithm, setting experimentally the number of clusters to be obtained and the function to evaluate distances between data points and cluster centers. An iterative process is then performed, during which k data in the dataset are randomly chosen to constitute the first centroids of the clusters. The metric distance allows to assign the remaining data to the cluster on the strength of their closeness with the centers of clusters, then, new centers are detected evaluating the average of each cluster. The process ends when the obtained result satisfies a predetermined criterion of termination.

3 Experimental Results

In this section we report some experimental results obtained testing the proposed approach on a dataset of the *Sensor9k* project [18, 19]:. The dataset contains a set of measurements obtained from a testbed for WSN-based Ambient Intelligence applications [20–22], built in a workplace of the University of Palermo. The office rooms of the workplace have been equipped with sensor nodes monitoring indoor and outdoor physical quantities such as relative humidity, temperature, and light exposure; additionally, RFId sensors allow for detecting the employees' presence in the workplace through the use of personal badges. To test the proposed approach we have considered a subset of the sensors used in [18], in particular the experiments have been conducted analyzing data measured from MTS300 sensor nodes, where the analyzed variables are light, and temperature, the access of the employees in the workplace and outdoor measurements of light and temperature. In particular we have data regarding two office rooms, *Room1* and *Room2*. The former is an office room, used by two employees *User1* and *User2*, whereas the latter in a common area. The two rooms share similar exposition (thus similar trends for the considered variables), and are connected by a door.

We have analyzed data measured from two sensors per room. We will indicate light and temperature measurements collected by the two sensors in *Room1* as *Light101*, *Light102*, *Temperature101* and *Temperature102*, respectively; analogously, *Light201*, *Light202*, *Temperature201* and *Temperature202* will be the

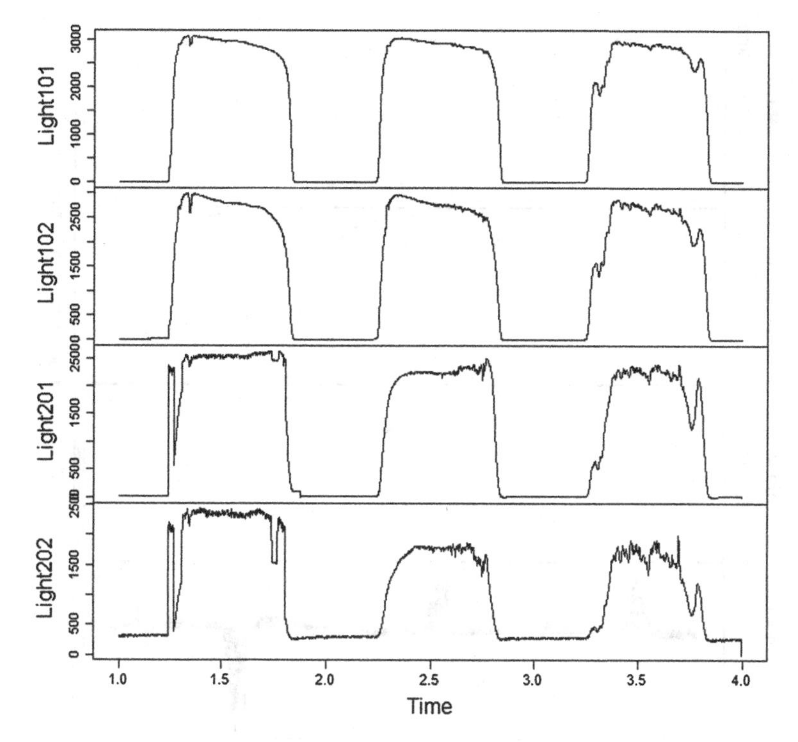

Fig. 5 Time series for *Light101*, *Light102*, *Light201* and *Light201*

measurements related to *Room2*. Figure 5 shows the time series of the all four light measurements in a period of three days.

The two topmost plots clearly show the similarity in the trends of sensors located within the same room, and the same consideration holds for the two plots at the bottom; if we consider the two central plots, we can also identify significant similarities, although not as striking as in the previous cases; we argue that the differences in measurements for sensors in different rooms are partially due to different placements, and mainly to the effects caused by a different use of the actuators by the users.

Figures 6 and 7 show the decomposition of two light time series belonging to the two rooms, i.e. *Light101* and *Light201*.

Each figure shows the original series, and its seasonal, trend and remainder components. The plots show how the seasonal component is related to day-night cycles, while the trend is related to level of brightness of days.

Figure 8 shows the analysis of the remainder component of the *Light101* series and the corresponding relevant variations computed as derivative of the function, which presumably correspond to actions on part of some user. The plot shows only instantaneous changes since we are looking for actions on artificial light settings (a different analysis would be performed for the detection of actions on temperature settings, because temperature takes longer to stabilize).

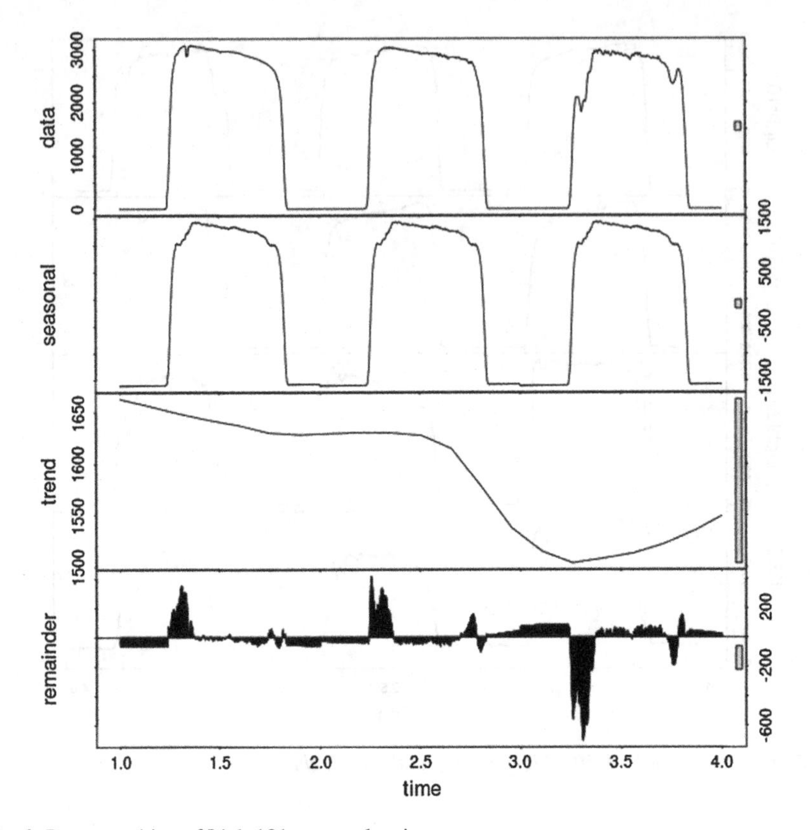

Fig. 6 Decomposition of Light101 temporal series

Figure 9 shows a reasoning process to disambiguate one of the detected variations. In the example, evidence coming from sensor observations is set, and the probability of user action states is evaluated. In particular, the user was in the room, the external light was high and the internal light had an increasing change in two subsequent instants. The result of the reasoning process is that the user action could be a *turning on* action, with a probability of 0.32, a *turning off* action with a probability of 0.11, and with a probability of 0.57 the increasing will be due to the increasing of the external light (independent of the user action).

The dataset used to validate our approach so far contains information about the presence of only two users. For this reason, we conducted a proof-of-concept experiment involving a set of time slots in a working day, also considering information about the users' presence and environmental conditions. Figure 10 shows the results obtained by a k-means clustering on one-day matrix observations, related to the presence of the two users *User1* and *User2*, and the measurements of light and temperature.

The Table shows that *User2* had more significant influence on the measured quantities as it was more present; this is especially evident for the last two clusters,

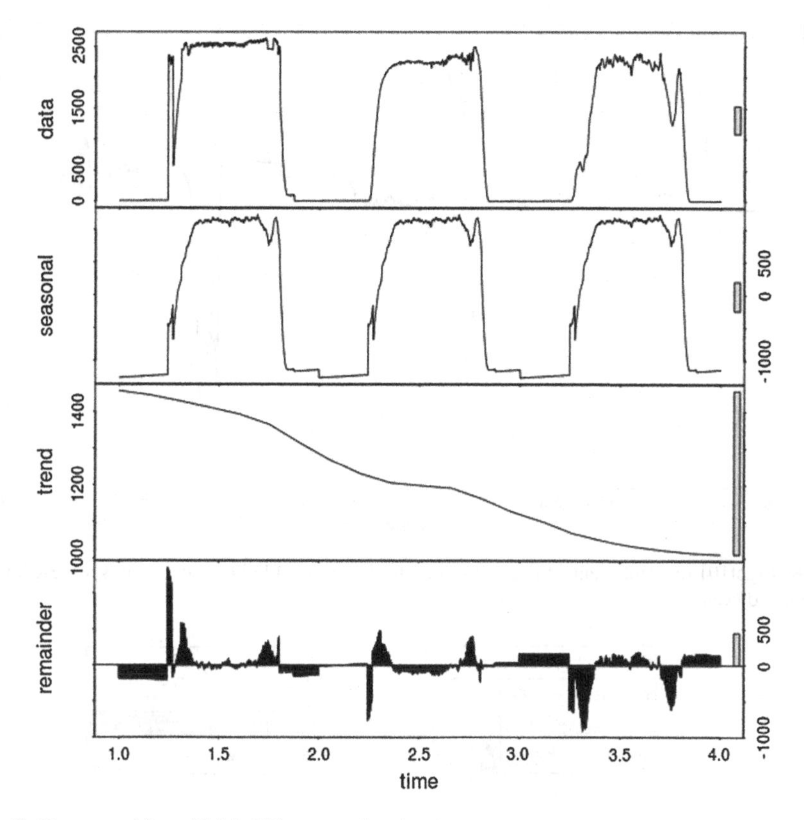

Fig. 7 Decomposition of Light201 temporal series

regarding the later time of the day, where the influence of *User2* may be easily singled out; the significant differences in the numerical values captured by cluster 3 are easily explainable by considering that this cluster contained data measured during nighttime.

4 Conclusion

In this chapter we have described an approach aimed at implicitly detecting behavior patterns of users working in an intelligent environment, equipped in the context of *Sensor9k* project. The approach consists in different modules for the extraction of useful information regarding user actions and habits. Such information can be used for different purposes, for example to adapt the environment settings to users preferences, or to monitor the energy consumption in the workplace.

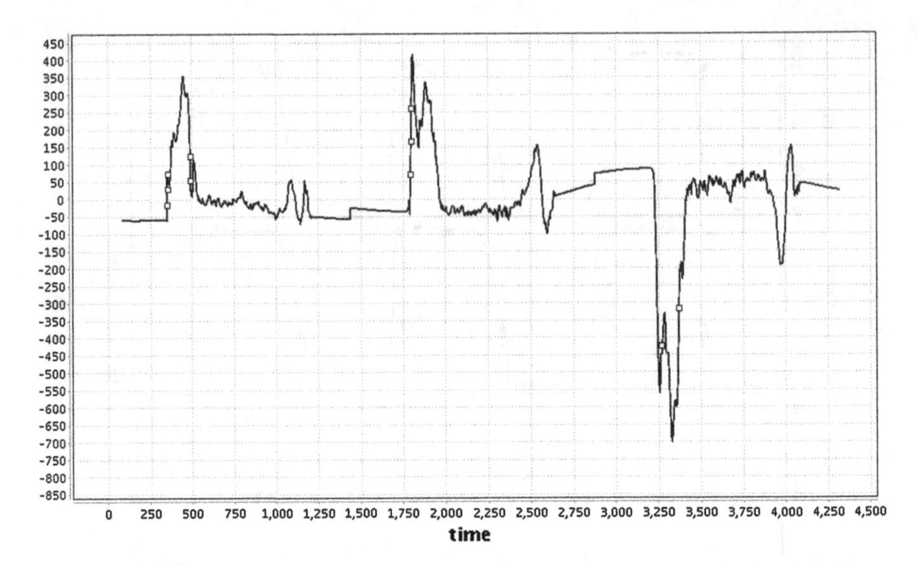

Fig. 8 Light101 relevant events: the curve represents the trend of light, while squares represent the recognized events

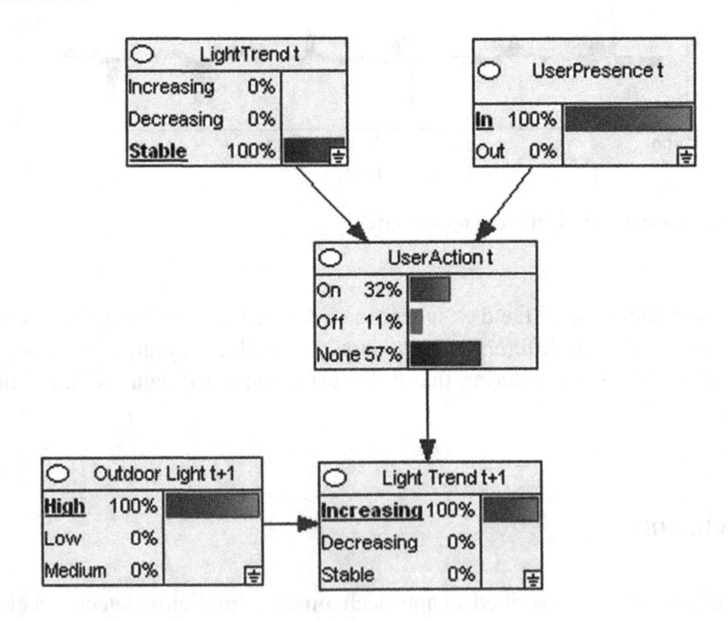

Fig. 9 Probabilist reasoning on a recognized event

Acknowledgments This work has been partially supported by the PO FESR 2007/2013 grant G73F11000130004 funding the SmartBuildings project.

Attribute	cluster_0	cluster_1	cluster_2	cluster_3
user1	0.307	0.175	0	0
user2	0.631	0.535	0.659	0.148
Light102	2686.862	2402.147	1508.620	9.898
Light101	2603.906	2246.683	1306.187	7.833
Temperature101	25.577	25.729	25.614	9.395
Temperature102	26.273	25.969	26.399	9.609

Fig. 10 Relation between users presence and temperatures and lights values in a day

References

1. Akyildiz, I.F., Su, W., Sankarasubramaniam, Y., Cayirci, E.: A survey on sensor networks. IEEE Commun. Mag. **40**(8), 102–114 (2002)
2. Cook, D.J., Augusto, J.C., Jakkula, V.R.: Ambient intelligence: Technologies, applications, and opportunities. Pervasive Mob. Comput. **5**(4), 277–298, (2009). ISSN 1574–1192, doi:10.1016/j.pmcj.2009.04.001
3. Khalili, A., Wu, C., Aghajan, H.: Autonomous learning of user's preference of music and light services in smart home applications. In: Behavior Monitoring and Interpretation Workshop at German AI Conference, September (2009)
4. Mozer, M.C.: Lessons from an adaptive home. In Cook, D.J., Das, S.K. (eds.) Smart Environments: Technology, Protocols, and Applications, pp. 273–298. Wiley, UK (2004)
5. Akhlaghinia, M.J., Lotfi, A., Langensiepen, C., Sherkat, N.: Occupant behaviour prediction in ambient intelligence computing environment. In: Special Issue on Uncertainty-Based Technologies for Ambient Intelligence Systems, vol. 2, Issue number 2, May (2008)
6. Dong, B., Andrew, B.: Sensor-based occupancy behavioral pattern recognition for energy and comfort management in intelligent buildings. In: Proceedings of Building Simulation '2009, an IBPSA Conference. Glasgow, U.K. (2009)
7. Barbato, A., Borsani, L.., Capone, A., Melzi, S.: Saving, home energy, through a user profiling system based on wireless sensors. In: ACM Buildsys, : (in conjunction with SenSys 2009), Berkeley, CA. Nov. 3, (2009)
8. Doctor, F., Hagras, H., Callaghan, V.: A fuzzy embedded agent-based approach for realizing ambient intelligence in intelligent inhabited environments. IEEE Trans. Syst. Man Cybern. Part A **35**(1), 55–65 (2005)
9. Youngblood, G.M.: Automating inhabitant interactions in home and workplace environments through data-driven generation of hierarchical partially-observable Markov decision processes. Ph.D. thesis, The University of Texas at Arlington, (2005)
10. Stavropoulos, T.G., Vrakas, D., Vlachava, D., Bassiliades, N.: BOnSAI: a smart building ontology for ambient intelligence. In: Proceedings of the 2nd International Conference on Web Intelligence, Mining and Semantics (WIMS '12). ACM, NY, USA (2012)
11. Niezen, G: Ontologies for interaction: enabling serendipitous interoperability in smart environments. JAISE **5**(1), 135–137 (2013)
12. Santofimia, M.J., Moya, F., Villanueva, F.J., Villa, D., Lpez, J. C.: How intelligent are ambient intelligence systems? Int. J. Ambient Comput. Intell. (IJACI) **2**(1), 66–72 (2010). doi:10.4018/jaci.2010010106
13. De Paola, A., Farruggia, A., Gaglio, S., Lo Re, G., Ortolani, M.: Exploiting the human factor in a WSN-based system for ambient intelligence. In: CISIS pp. 748–753 (2009)
14. Brockwell, P.J., Davis, R.A.: Time Series: Theory and Methods. Springer, NY (1998). ISBN: 038797429
15. Jolliffe, I.T.: Principal Component Analysis, p. 487, Springer, NY (1986). doi:10.1007/b98835 ISBN 978-0-387-95442-4

16. Han, J., Kamber, M.: Data mining: concepts and techniques, 2nd edn. In: Jim G. Series (ed.) The Morgan Kaufmann Series in Data Management Systems. Morgan Kaufmann Publishers, San Francisco (March 2006). ISBN 1-55860-901-6
17. MacQueen, J.B.: Some methods for classification and analysis of multivariate observations. In: Proceedings of 5-th Berkeley Symposium on Mathematical Statistics and Probability. vol. 1, pp. 281–29. University of California Press, Berkeley (1967)
18. De Paola, A., Gaglio, S., Lo Re, G., Ortolani, M.: Sensor9k : a testbed for designing and experimenting with WSN-based ambient intelligence applications. Pervasive Mob. Comput. **8**(3), 448–466 (2012)
19. Augello, A., Ortolani, M., Lo Re, G., Gaglio, S.: Sensor mining for user behavior profiling in intelligent environments. Advances in Distributed Agent-Based Retrieval Tools 143–158 (2011)
20. De Paola, A., La Cascia, M., Lo Re, G., Morana, M., Ortolani, M.: User detection through multi-sensor fusion in an Am I scenario. In: Proceedings of the 15th International Conference on, Information Fusion, pp. 2502–2509 (2012)
21. De Paola, A., La Cascia, M., Lo Re, G., Morana, M., Ortolani, M.: Mimicking biological mechanisms for sensory information fusion. J. Biologically Inspired Cogn. Architectures **3**, pp. 27–38 (2013)
22. Cottone, P., Lo Re, G., Maida, G., Morana, M.: Motion sensors for activity recognition in an ambient-intelligence scenario. In Proocedings of the 5th International Workshop on Smart Environments and Ambient, Intelligence, pp. 646–651 (2013)

An AMI System for User Daily Routine Recognition and Prediction

Salvatore Gaglio and Gloria Martorella

Abstract Ambient Intelligence (AmI) defines a scenario involving people living in a smart environment enriched by pervasive sensory devices with the goal of assisting them in a proactive way to satisfy their needs. In a home scenario, an AmI system controls the environment according to a user's lifestyle and daily routine. To achieve this goal, one fundamental task is to recognize the user's activities in order to generate his daily activities profile. In this chapter, we present a simple AMI system for a home scenario to recognize and predict users' activities. With this predictive capability, it is possible to anticipate their actions and improve their quality of life. Our approach uses a Hidden Markov Model (HMM) to recognize activities and deal with the intrinsic uncertainty of sensory information. The concepts of this domain have been formally defined to allow a higher-level system to enrich its knowledge base.

1 Introduction

Ambient Intelligence defines a scenario in which an intelligent system supports users in their everyday life, assisting them in a proactive way. Many invisible embedded interconnected devices are deployed in the environment, thereby creating a pervasive infrastructure able to gather useful information about users and their requirements. In a home environment, an intelligent system perceives users and environmental conditions through sensors pervasively deployed in the environment and acts on this information to reach its goals, as proposed in [1, 8]. Several objects can be equipped with sensors to gather information about their use and to act independently, without human intervention. An AmI architecture exploiting heterogeneous sensory and actuator devices was proposed in [10]. Appliances are examples of this kind of object. Recent applications include the use of smart homes to provide a higher level of

S. Gaglio (✉) · G. Martorella
DICGIM, University of Palermo, Viale delle Scienze, Edificio 6, 90128 Palermo, Italy
e-mail: salvatore.gaglio@unipa.it

S. Gaglio and G. Lo Re (eds.), *Advances onto the Internet of Things*,
Advances in Intelligent Systems and Computing 260, DOI: 10.1007/978-3-319-03992-3_3,

comfort to disabled people. In this chapter, we focus on this intelligent home scenario. Our system exploits sensory information to recognize activities performed by the user. We use a Hidden Markov model (HMM) since it is good at modeling a scenario where observable events are used to infer unobservable phenomena. In addition, a HMM makes it possible to handle uncertainty in user behavior and sensory models. For this reason, HMMs have been successfully employed in activity recognition applications. We focus on activity recognition, because it represents an important functionality in designing an AmI system capable of reacting to context switches. We also design a profiler to predict future user actions in order to allow the system to respond according to the user's requirements. The remainder of the chapter is organized as follows. Section 2 discusses related research, while Sect. 3 describes the design and implementation of our system, including the formal definition of the concepts of the domain under consideration. The experimental results are reported in Sect. 4 and finally, conclusions are discussed in Sect. 5.

2 Related Works

Activity recognition is a growing research area and many different approaches have been proposed. Many papers in the literature suggest vision-based video processing techniques to recognize activities. However, the use of a camera could represent a privacy violation, because people generally do not agree to be monitored by cameras, especially at their own homes. A survey of vision-based approaches for activity recognition can be found in [19, 20]. Other works [16, 18] suggest performing activity recognition by using wearable sensors such as accelerometers and gyroscopes or temperature and heart rate monitors. These sensors allow a system to recognize activities involving body movements or specific body postures. However, wearable sensors often restrict the mobility of a patient, thereby causing the user to undergo an excessive degree of intrusiveness. The use of non-intrusive sensors is therefore often preferred, as proposed in [6]. This implies handling noisy data with low correlation to the phenomena observed. For this reason, data fusion techniques exploiting HMMs can be often required. HMMs have frequently been used for activity recognition [5, 13, 15]. Several papers dealing with activity recognition use ontologies to model entities in a formal way, by describing properties and relationships between these concepts. In [4], the authors propose an ontology-based method for activity recognition in a home scenario, modeling user profile as a set of activities. Each activity contains properties to describe user preferences. The authors of [21] aim to perform ontology-based activity recognition by modeling user behavior in a smart home and introducing concepts like context, posture, location, activity, object and sensor. Another approach can be found in [2], which describes an ontology-based system for discovering and monitoring patterns of activities of daily living (ADL). Other ontology-based approaches for activity recognition can be found in [3]. Predicting users' activities is a fundamental task to anticipate user behaviors, because anticipating their actions could improve their comfort and quality of life. User activities are thus analyzed and processed in order to build a user behavior schema. Therefore, prediction algorithms are required. The authors of [7] present two prediction algorithms for user mobility management at home. The authors of [12] propose a prediction algorithm based on pattern matching. In [14], the authors provide an overview of activities prediction algorithms, analyzing requirements and evaluating performance using three different datasets. User profiles can also be useful in detecting behavior anomalies, as in [22]. Unlike the previously described systems, our model exploits context information, such as time, with such information being integrated into sensory events to infer activities. Furthermore, an adaptive mechanism allows the system to interact with the user and adapt its behavior at runtime according to the current percentage of user interventions. In addition, prediction is dynamically adjusted in order to improve the prediction capability of the system.

3 The Proposed System

Activity recognition represents a fundamental task required to create smart environments. An AmI system may need to know the activity currently being performed by the user in order to adequately control environmental conditions. We model activity recognition through a HMM where the observable symbols are sensory readings. Hidden states represent activities, since they are not directly observed, but can be inferred through sensory events. The learning of model parameters is performed using a supervised learning method requiring a training set composed of labeled data. The number of occurrences of state transitions and sensory events in the sample data are used to compute the HMM parameters. The inference process allows the system to identify hidden phenomena on the basis of observed events. For our purpose, we use the Viterbi algorithm [11] to infer activities from a sequence of sensory readings. On the basis of the HMM parameters, this inference algorithm gives as output the most likely sequence of hidden states generating the observed sequence of sensory events. Moreover, the Viterbi algorithm employs dynamic programming techniques to reduce the computational cost of the inference phase. An analogue approach was described in [5], but unlike [5], our HMM model includes an observable node representing time. Time information is the fulcrum of our system design, since daily routine consists of activities frequently occurring within the same time intervals. Figure 1 shows the HMM adopted here in two sequential time steps. The time node is discrete and it can assume 24 possible values corresponding to the 24 h in a day.

3.1 Adaptive Activity Recognition

Having the model parameters trained using the supervised learning approach, the Viterbi algorithm gives as output the most probable sequence to have generated the observed sensory readings. Moreover, it furnishes as output a belief about the

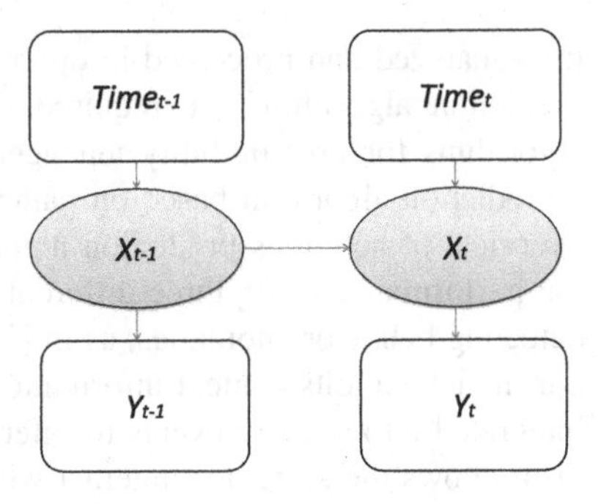

Fig. 1 The HMM model in two sequential time steps. X_t is the activity performed by the user at time t, while Y_t is the sensory event gathered at time t

current state, that is, a probability distribution for the latest state. We employ the belief about the latest state to perform activity recognition in an adaptive manner. When the system is not sure about the activities it has just recognized, it asks the user, and the feedback gathered is used to update HMM parameters. Since entropy is a measure of the uncertainty of a probability distribution, we consider the uncertainty of the system as the entropy value of the latest state probability distribution. Entropy is generally related to the a priori probability distribution of a random variable, but we extend this concept to the a posteriori belief, as described in [9].

$$uncertainty(X_t) = -\sum_{x_t} Bel(x_t) \log Bel(x_t) \,. \tag{1}$$

Given the probability distribution on the last inferred state, the system computes its uncertainty. If the uncertainty value exceeds a given threshold, the system questions the user and updates the HMM parameters, otherwise it just updates the HMM parameters without involving the user. If the system is uncertain, it computes the current error as the current percentage of user interventions. If this value exceeds a given threshold, either the HMM parameters must be incorrect or the user must be changing his routine. To deal with this issue, there is another phase for gradually updating the HMM parameters, called “on line learning”, which is described in the next section. Otherwise, the system continues the adaptive phase, a flow diagram of which is shown in Fig. 2.

3.2 On-line Learning

A training set is not available in many practical contexts. In order to overcome this limitation, we propose an “on-line learning” mechanism that only requires a default initial value for model parameters and that gradually updates HMM parameters. For example, the initial default parameters could be provided by a recognizer temporarily installed in the monitored premises and based on intrusive but very precise sensors, such as cameras. We will refer to this further recognizer as the “camera-

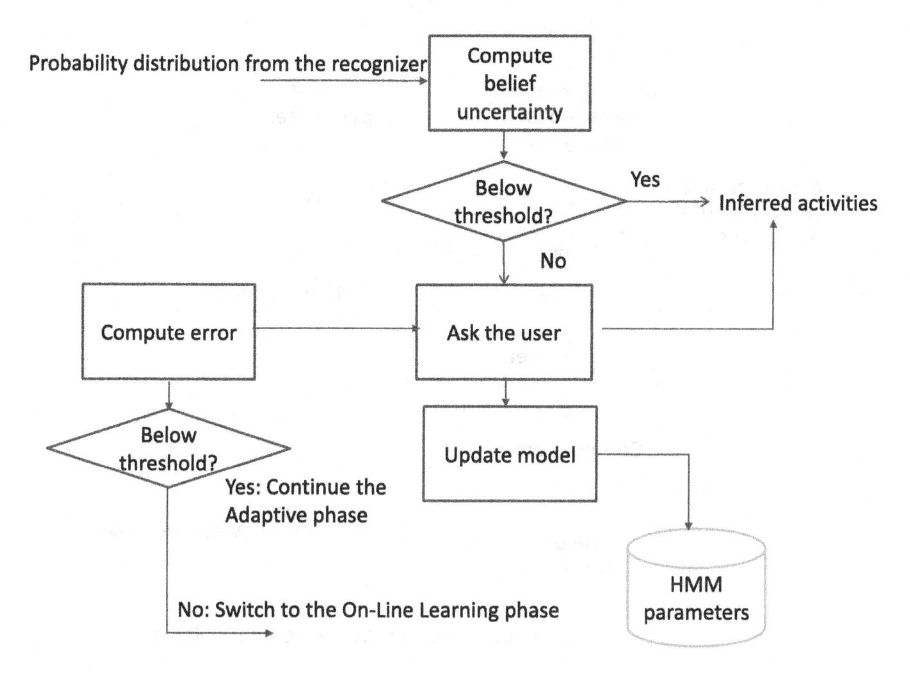

Fig. 2 Flow diagram of the Adaptive Phase. This phase provides the system a mechanism for interacting with the user and for updating the HMM parameters only if necessary. The system switches to the On-Line Learning phase according to the current percentage of user intervention

based recognizer" because it is based on camera sensors, while we will refer to our recognizer as the "sensor-based recognizer" because we are using a sensor-based approach for activity recognition. Figure 3 shows the flow diagram of the On-Line Learning phase. Although our sensor-based recognizer has these partial default parameters, it performs inferences together with the camera-based recognizer, and the activities inferred from both recognizers are then compared. If the output activities are not equal, it first questions the user and then updates the parameters, otherwise it avoids asking the user just updating the HMM parameters. This phase is thus computationally onerous, because two recognizers are running at the same time. In order to decrease this computational effort, and to preserve user privacy as much as possible, once the percentage of user intervention falls below a certain threshold, the system switches to the previously described adaptive phase.

3.3 The Proposed Architecture

We propose a classical layered architecture (see Fig. 4) where the lower level is the hardware level composed of many sensors deployed in the environment. The middleware layer handles the sensor heterogeneity, gathering data from sensors and inserting

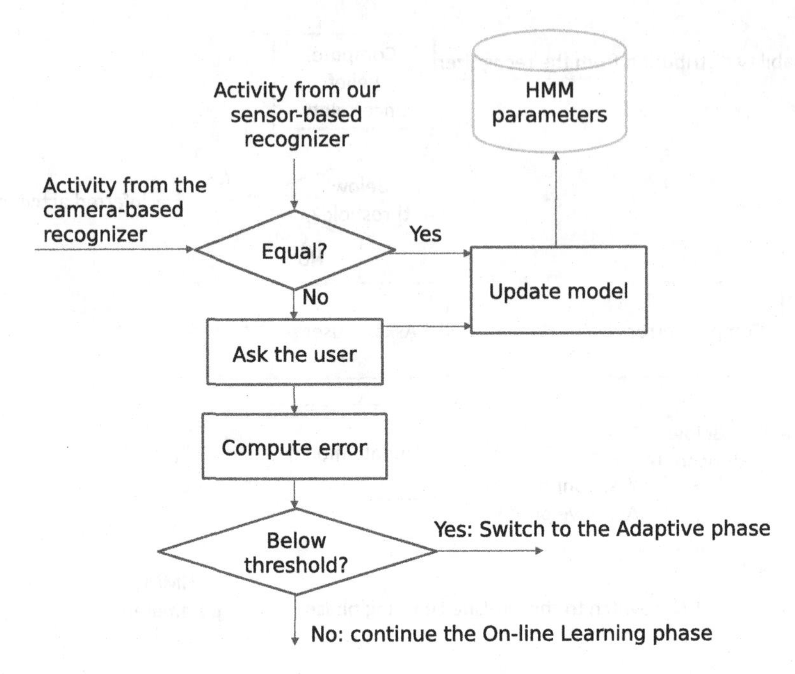

Fig. 3 Flow diagram of the on-line learning phase. This is the initial phase which allows the sensor-based recognizer to work in unknown scenarios. User feedback is required in order to update the parameters with greater certainty. The system switches to the Adaptive phase once the current percentage of error falls below a certain threshold

them in a database. Higher level modules include our sensor-based recognizer, the profiler, the camera-based recognizer and a module for interacting with the user. Our sensor-based recognizer performs activity recognition on the basis of cheap noisy sensors deployed in the environment, such as movement and temperature sensors, while the camera-based recognizer uses data gathered by expensive precise sensors such as cameras. The system employs a software module able to communicate with the other system modules. This central module acts as a coordinator sending trigger or sleep signals to the other system components. It activates the Adaptive or On-line learning phase, depending on the current percentage of error. The coordinator considers how many times the system has questioned the user. If it exceeds a given threshold, it switches to the learning phase, otherwise to the adaptive one. Separating the on-line learning phase from the adaptive one avoids the overhead due to continuous employment of the camera-based recognizer. The profile module handles user profile generation using the HMM parameters, properly updated by the coordinator. Finally a module for interacting with the user is required to perform the adaptive and on-line phases and give the system the necessary user feedback. Moreover, the proposed architecture provides a memory area containing the conditional probability tables, which are accessible both to the recognizer and the profiler modules.

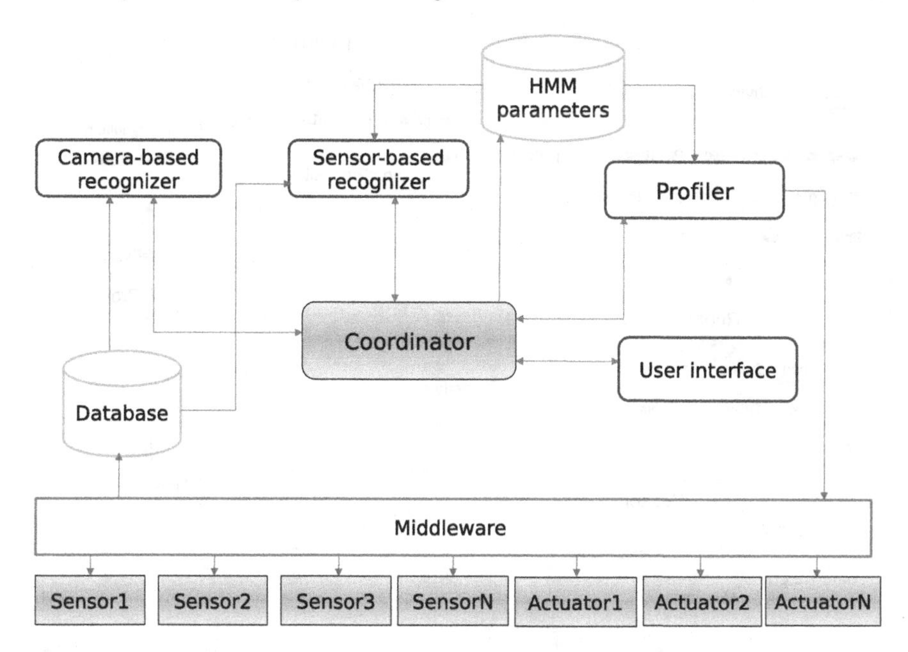

Fig. 4 The proposed architecture. The coordinator is the main component performing the adaptive or on-line learning phases, depending on the current error percentage. The *red lines* indicate the interactions between the coordinator and the other modules in the system

3.4 User Activity Profiling

Profiling therefore represents another important task in building an AmI system able to predict activities and thus to satisfy user's requirements. In a realistic scenario, a software module could generate the sequence of users' future activities as an input for actuators in order to anticipate their actions. Once the daily activity profile has been generated, new sensory events will occur, and the system may verify whether the inferred belief was correct. It may eventually correct its error and thus profile again. In the profile design, we take an hour as the prediction interval. Using a supervised learning, the system computes the state transition model containing the transition probability to an activity in the next hour $P(X_t|X_{t-1})$. The profiling module proposed adopts a simplified state transition model in which it is assumed that a user can only perform one activity within any given 1 h time frame. The prediction of the next most probable activity is performed according to the following equation:

$$P(X_{t+1}|e_{1:t})=\sum_{X_t} P(X_{t+1}|X_t, e_{1:t}) * P(X_t|e_{1:t})=\sum_{X_t} P(X_{t+1}|X_t) * P(X_t|e_{1:t}). \tag{2}$$

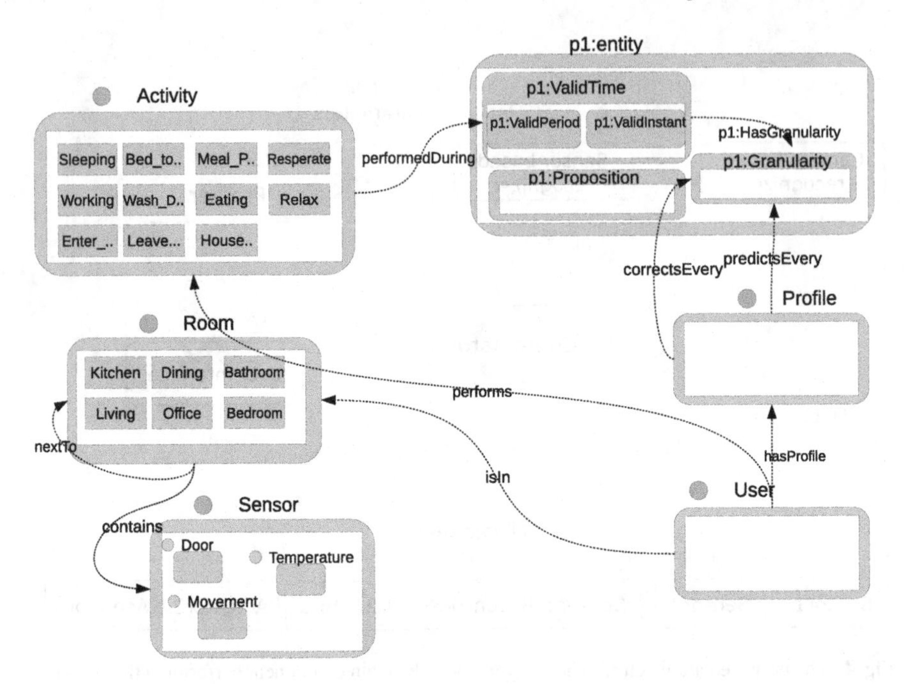

Fig. 5 Significant concepts in our domain include Sensor, Activity, Room, Profile, User and Time. Time information is very precious in this context, since activity occurs within a time interval and profile running relies on correction and prediction times

3.5 System Ontology

Ontologies allow us to formally represent domain concepts, together with their relationships and properties. Ontologies play an important role when automatic interaction between the system and other modules of the AmI system is required. In order to understand the information exchanged, they must adopt the same conceptualization, that is, the same representation of concepts within the domain. In our scenario, the significant entities are Activity, Room, User, Profile, Sensor and Time, as shown in Fig. 5. Time representation is very important in our domain, since activities and events occur at a single instant or over an interval of time. Hence, we have used the time representation described in [17]. In a home environment, the user performs an activity for a certain period of time. He can be in a certain room and has his own generated profile. A profile is based on the prediction and the correction times. Each room contains sensors, each of which has an ID and an instant value. If another system has the same representation of these concepts, interoperability is easily ensured. Moreover, this formal concept representation could allow a higher-level AmI system to enrich its knowledge base.

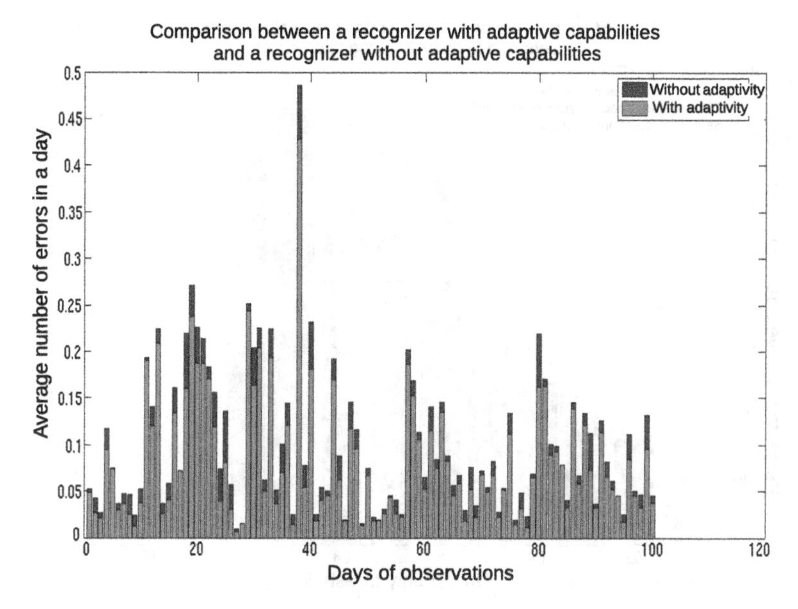

Fig. 6 Comparison between a recognizer with the adaptive capability and a recognizer without this functionality. The adaptive recognizer is more precise for all of the days tested

4 Experimental Results

This section describes the results of the experimental evaluation undertaken to analyze the performance of the proposed system. The experimental evaluation was performed by exploiting the sensory dataset of the WSU CASAS Smart Home Project [5] as a basis for generating a synthetic dataset containing simulated interaction with the user. The ARUBA dataset was divided into three parts and any observations with missing labels were removed. Two parts were used as a training set for the HMM, while the remaining one was used for verification. The system is able to detect activities with a minimum duration of 10 s and it updates the parameters whenever the recognizer uncertainty exceeds a threshold fixed at 20 %. The experimental evaluation revealed that adaptive behavior makes the proposed recognizer more precise, as shown in Fig. 6.

The uncertainty threshold also affects the average number of recognition errors (see Fig. 7). By increasing this threshold, average recognition errors also go up because the system will question the user less frequently and will consequently correct the inference errors less frequently (see Fig. 8). The effect of the uncertainty threshold variation on both the average number of recognition errors and on user intervention is shown in Fig. 9.

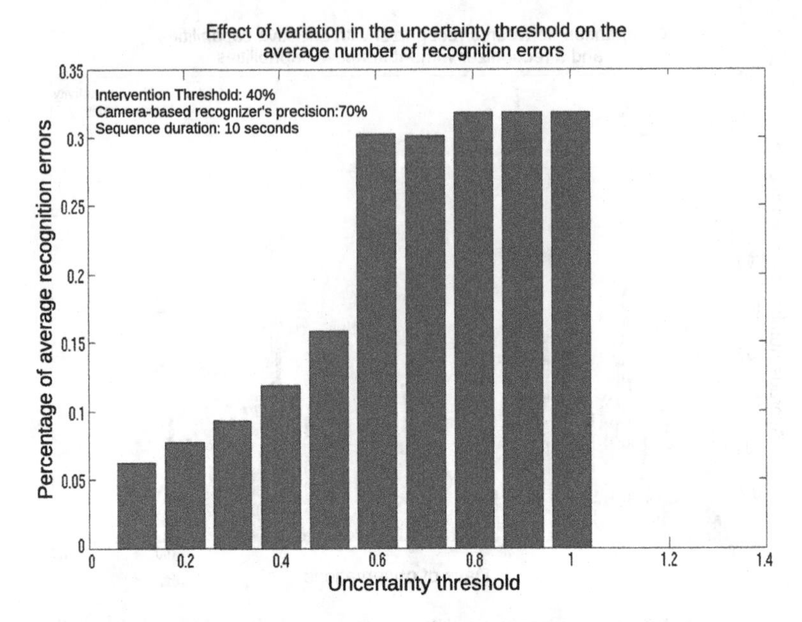

Fig. 7 Effect of variation in the uncertainty threshold on the average number of recognition errors

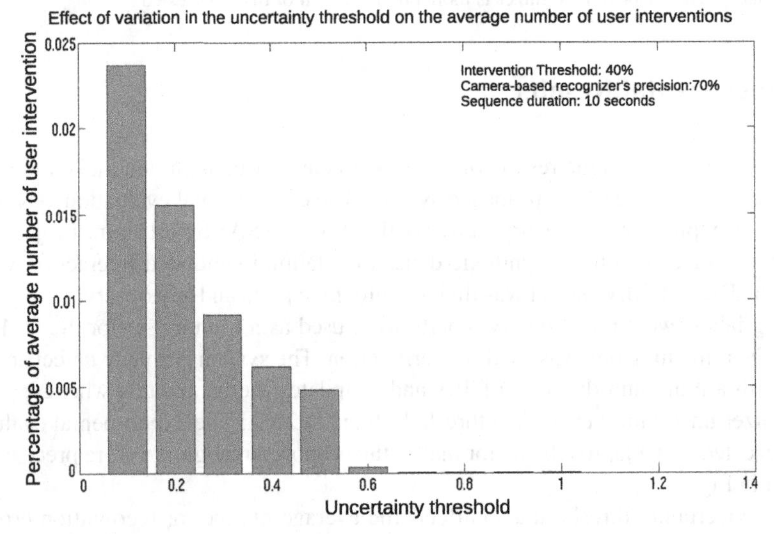

Fig. 8 Effect of variation in the uncertainty threshold on the average number of user interventions

5 Conclusions

The main goal of an AmI system in a home scenario is to assist users in their daily life and to improve their quality of life. For this purpose, we have presented a simple AmI system to recognize and predict activity performed by users in a home sce-

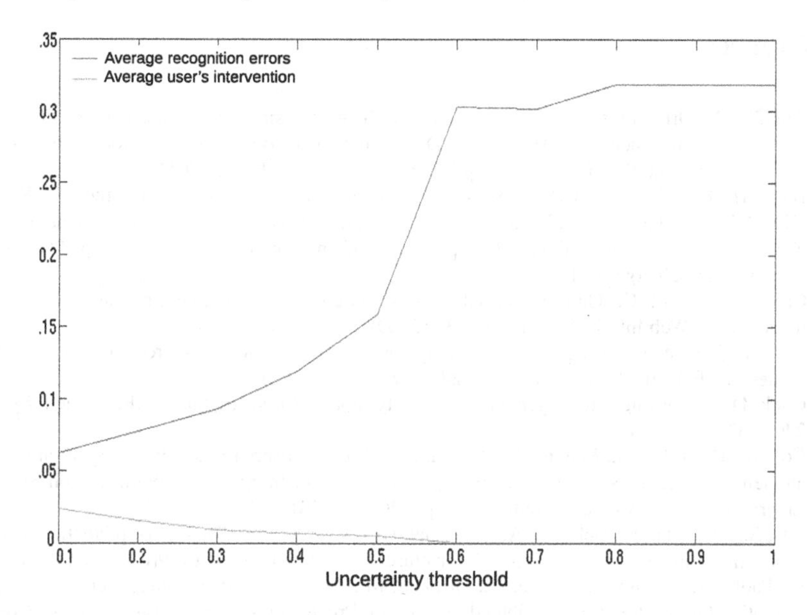

Fig. 9 Effect of variation in the uncertainty threshold on both average number of recognition errors and user interventions

nario. Activity recognition represents a fundamental ability, allowing the system to understand how the user behaves and acts at home. Hence, we focused on the design and implementation of a recognizer based on HMMs. The probabilistic approach makes it possible to describe phenomena characterized by a non-negligible level of uncertainty, such as user behavior or the sensor model. Our recognizer exploits time information to infer the most probable sequence of activities performed. Moreover, we enriched the recognizer by providing it with the ability to learn from its errors by stimulating user interventions and thus updating the HMM parameters. The experimental results reported here show an improvement in performance attributable to this additional functionality. The "on-line" learning mechanism makes it possible to work in unknown scenarios by using an additional recognizer to partially fill the conditional probability tables representing the HMM parameters. The activity recognizer was used as constituent part of a profiler able to anticipate user activities and to monitor their daily routines. We also provided a formal representation of the concepts of our domain, describing all the entities going to make up our system. It is worth noting that the module for interacting with the user was not really implemented. This interaction was only simulated. We now intend to develop our user interface further. We are planning to conduct more tests to simulate the contemporary use of the recognizer and profiler, and also want to explore the effects of time granularity variation on the profiler.

Acknowledgments This work has been partially supported by the PON R&C grant MI01_00091 funding the SeNSori project.

References

1. Augello, A., Ortolani, M., Lo Re, G., Gaglio, S.: Sensor mining for user behavior profiling in intelligent environments. In: Advances in Distributed Agent-Based Retrieval Tools, Studies in Computational Intelligence, vol. 361, pp. 143–158. Springer, Berlin (2011)
2. Bae, I.H., Kim, H.: An ontology-based ADL recognition method for smart homes. In: Kim, T.H., Adeli, H., Fang, W.C., Vasilakos, T., Stoica, A., Patrikakis, C.Z., Zhao, G., Villalba, J., Xiao, Y. (eds.) Communications in Computer and Information Science, vol. 266, pp. 371–380. Springer, Heidelberg (2011)
3. Chen, L., Nugent, C.: Ontology-based activity recognition in intelligent pervasive environments. Int. J. Web Inf. Syst. **5**(4), 410–430 (2009)
4. Chen, L., Nugent, C., Wang, H.: A knowledge-driven approach to activity recognition in smart homes. IEEE Trans. Knowl. Data Eng. **24**(6), 961–974 (2012)
5. Cook, D.J.: Learning setting-generalized activity models for smart spaces. IEEE Intell. Syst. **27**(1), 32–38 (2012)
6. Cottone, P., Lo Re, G., Maida, G., Morana, M.: Motion sensors for activity recognition in an ambient-intelligence scenario. In: Proceedings of the 5th International Workshop on Smart Environments and Ambient Intelligence, pp. 646–651 (2013)
7. Das, S., Cook, D., Battacharya, A., Heierman E.O., I., Lin, T.Y.: The role of prediction algorithms in the mavhome smart home architecture. Wirel. Commun. IEEE **9**(6), 77–84 (2002)
8. De Paola, A., Gaglio, S., Lo Re, G., Ortolani, M.: An ambient intelligence architecture for extracting knowledge from distributed sensors. In: Proceedings of the 2nd International Conference on Interaction Sciences: Information Technology, Culture and Human, pp. 104–109 (2009)
9. De Paola, A., Gaglio, S., Lo Re, G., Ortolani, M.: Multi-sensor fusion through adaptive bayesian networks. In: AI*IA 2011: Artificial Intelligence Around Man and Beyond, Lecture Notes in Computer Science, vol. 6934, pp. 360–371. Springer, Berlin (2011)
10. De Paola, A., Gaglio, S., Lo Re, G., Ortolani, M.: Sensor9k: A testbed for designing and experimenting with wsn-based ambient intelligence applications. Pervasive Mob. Comput. **8**(3), 448–466 (2012)
11. Forney, G.D.J.: The viterbi algorithm. Proc. IEEE **61**(3), 268–278 (1973)
12. Gorniak, P., Poole, D.: Predicting future user actions by observing unmodified applications. In: Proceedings of the Seventeenth National Conference on Artificial Intelligence and Twelfth Conference on Innovative Applications of Artificial Intelligence, pp. 217–222. AAAI Press (2000)
13. Han, C.W., Kang, S.J., Kim, N.S.: Implementation of HMM-based human activity recognition using single triaxial accelerometer. IEICE Trans. Fundam. Electron. Commun. Comput. Sci. **93**(7), 1379–1383 (2010)
14. Hartmann, M., Schreiber, D.: Prediction algorithms for user actions. In: I. Brunkhorst, D. Krause, W. Sitou (eds.) 15th Workshop on Adaptivity and User Modeling in Interactive Systems (2007)
15. Kim, E., Helal, S., Cook, D.: Human activity recognition and pattern discovery. Pervasive Comput. IEEE **9**(1), 48–53 (2010)
16. Lee, S.W., Mase, K.: Activity and location recognition using wearable sensors. Pervasive Comput. IEEE **1**(3), 24–32 (2002)
17. OConnor, M., Das, A.: A method for representing and querying temporal information in owl. In: A. Fred, J. Filipe, H. Gamboa (eds.) Biomedical Engineering Systems and Technologies, Communications in Computer and Information Science, vol. 127, pp. 97–110. Springer, Berlin (2011)
18. Parkka, J., Ermes, M., Korpipaa, P., Mantyjarvi, J., Peltola, J., Korhonen, I.: Activity classification using realistic data from wearable sensors. IEEE Trans. Inf. Technol. Biomed. **10**(1), 119–128 (2006)
19. Poppe, R.: A survey on vision-based human action recognition. Image Vis. Comput. **28**(6), 976–990 (2010)

20. Weinland, D., Ronfard, R., Boyer, E.: A survey of vision-based methods for action representation, segmentation and recognition. Comput. Vis. Image Underst. **115**(2), 224–241 (2011)
21. Wongpatikaseree, K., Ikeda, M., Buranarach, M., Supnithi, T., Lim, A., Tan, Y.: Activity recognition using context-aware infrastructure ontology in smart home domain. In: 2012 Seventh International Conference on Knowledge, Information and Creativity Support Systems (KICSS), pp. 50–57 (2012)
22. Yin, J., Yang, Q., Pan, J.: Sensor-based abnormal human-activity detection. IEEE Trans. Knowl. Data Eng. **20**(8), 1082–1090 (2008)

A Fuzzy Adaptive Controller for an Ambient Intelligence Scenario

Alessandra De Paola, Giuseppe Lo Re and Antonio Pellegrino

Abstract The definition of effective energy saving strategies capable of satisfying users' requirements for environmental wellness is a complex task that requires the definition of well-tuned optimization algorithms. Sensory information depends on the environments observed, hence the model adopted to describe it should be adaptive and dynamic. This chapter presents a methodology for the tuning of a fuzzy controller capable of minimizing energy consumption while maximizing the users comfort in an Ambient Intelligence Scenario. A meta-heuristic search algorithm produces different sets of fuzzy rules depending on the needs of the system. An ontology has been developed to describe the configurations of environments and user requirements, thus enabling automatic reconfiguration of the whole system.

1 Introduction

Energy saving in buildings is mandatory nowadays if we are to safeguard the planet. This may be especially difficult with building complexes in which the energy management policy is different for each structure. In order to cope with such heterogeneity, it is necessary to adapt the energy-saving strategy to each single environment under consideration. In [4], a general architecture is described for AmI systems devoted to energy saving, work which is specialized for detecting user presence in [3].

The optimization of energy consumption must also take into account the preferences of the users living in the buildings.

Users who occupy the premises being monitored perform various activities during which they may express different preferences for temperature, humidity and lighting

A. De Paola (✉) · G. Lo Re · A. Pellegrino
DICGIM—University of Palermo, Viale delle Scienze, ed 6, 90128 Palermo, Italy
e-mail: alessandra.depaola@unipa.it

G. Lo Re
e-mail: giuseppe.lore@unipa.it

S. Gaglio and G. Lo Re (eds.), *Advances onto the Internet of Things*,
Advances in Intelligent Systems and Computing 260, DOI: 10.1007/978-3-319-03992-3_4,

levels. Environmental sensors monitor the instantaneous usage of energy and the values of the environmental phenomena.

Although accurate and continuous monitoring performed by distributed sensors allows us to collect the basic information, such simple activity does not make it possible to infer the preferences of users or the ideal environmental values directly. In other words, this goal requires a more complex reasoning procedure which is the subject of this chapter.

We propose the adoption of an intelligent system capable of determining the actions that actuators need to perform in order to achieve a given goal. Such a goal is typically represented by the minimization of energy consumption, or the accomplishment of user preferences or a weighted combination of both. That is, depending on the optimization policy, the system may assign more weight to energy saving or, to user wellbeing.

The behavior of actuators is represented by fuzzy rules. A controller exploits these rules that depend on the sensory information collected in the environment and on the specific goals to be achieved, in order to select the best actions to be performed by the actuators.

In order to adapt the fuzzy controller to the specific contexts, a planner generates the optimal fuzzy rules through the TabuSearch [6] algorithm. TabuSearch is a heuristic search methodology that explores the space of all possible solutions for a problem in order to identify the optimal one. The search is accomplished through the usage of admissible moves and the exclusion of tabu moves, the latter being defined as those moves which generate a loop, since they return to visit configurations which have already been examined.

Rules depend on the environment, the different preferences expressed by users and the energy consumption objectives set. The resulting system is dynamic and capable of adapting itself to the specific conditions.

In our approach, each fuzzy rule is composed of a set of antecedents and a set of consequent. The former regard both the current environmental state, in terms of temperature, humidity and lighting level, and the estimate of the activity currently being performed by the user. The consequent represent the actions to be performed in order to achieve system goals. The optimization procedure looks for the optimal set of fuzzy rules, driven by a fitness function that evaluates the energetic cost of each configuration and the level of user satisfaction, with a mechanism similar to that adopted in [1].

The remainder of the chapter is structured as follows. Section 2 describes some related work illustrating the optimization of fuzzy controllers by means of artificial intelligence techniques. The environmental measures that will be taken into account and the ontological system for representing the processing of sensory data at a symbolic level are described in Sect. 3, whilst Sect. 4 illustrates the architecture proposed here with our conclusions.

2 Related Work

In our system, the planning agent generates the actions that modify the environment according to certain global objectives. In particular, we decided to adopt a Fuzzy approach.

As described in [10], Fuzzy logic is inspired by real life, since it assigns a membership degree for any given property to a specific object or class of objects. Fuzzy logic uses specific membership functions and ad-hoc operators to model and manage information through a reasoning process that is similar to human reasoning [13].

The fuzzy rules are the fundamentals of the environment controller, because getting the correct output of the actions depends on the appropriate use of these rules. The selection of the rules is delegated to a higher level planner whose task is that of finding the best rules for the correct operation of the entire system.

A large variety of approaches for fuzzy rule generation is reported in the literature, ranging from the use of neural networks, to clustering techniques, and the use of evolutionary and heuristic search algorithms.

The authors of [15] suggest using neural networks to learn fuzzy rules. A neural network simulates the behavior of the fuzzy rules inside the controller. In order to be able to consider all the rules, it is necessary to design a neural network with lots of nodes representing the preconditions and the results of each rule. All nodes of the neural network are connected through weighted links. The weights are modified depending on desired output through a back propagation of the errors. The huge number of nodes required by large Fuzzy systems makes such methodology unsuitable because of its excessive computational load.

A clustering technique for fuzzy rule generation is adopted by Shi and Mizumoto [17]. Clustering methods are usually used to classify large sets of data according to predetermined features, and in this chapter the cluster centroids represent the parameters of the membership functions. However, when there are lots of input variables and the membership functions are complex, the method is particularly difficult to implement. A similar technique is used in [2], in which Chang and Chen initially determine the fuzzy rules using the C-Means algorithm, and the inferences are calculated through an interpolation scheme by a reassignment of the weights.

As described in [12], rules can be learned through a reinforcement algorithm such as Q-Learning. This technique creates a table where each configuration is rewarded or penalized according to the approach of the target system. The configurations that achieve the highest scores will be selected as the best set of rules for the system. The authors of [14, 16], adopt a hierarchical approach to learn fuzzy rules in medical applications, for reducing the memory consumption.

In [7] a genetic algorithm, a heuristic search methodology, is adopted in order to determine to best membership functions that optimize the behavior of a fuzzy controller designed to manage packet queues on routers of the Internet [8, 9].

the Tabu Search [6] algorithm falls into the same category of heuristic methodologies. The high level symbols, representing membership functions, are coded as binary strings on which actions are carried out to generate new configurations.

Table 1 Ideal temperatures according to Italian labor laws

Type of work	Min temperature (°C)	Max temperature (°C)
Manual	12	18
Light work	18	22
Sedentary	20	26

3 Sensory Subsystem

The basic layer of our system is constituted by the sensory infrastructure that pervades the monitored premises with the aim of collecting the measurements of physical phenomena. In our architecture the sensory subsystem provides the input to the upper process responsible for the abstraction of sensory data collected within the buildings.

Such raw data are merged and processed in order to extract high-level aggregated information. The planning of the actions to be performed in order to achieve the optimal environmental conditions is a process that works at a higher level of abstraction, and requires symbolic information rather than raw data.

That is, through a process of reasoning applied to symbolic information, it is possible to determine what actions to perform to achieve the goals set by the system.

3.1 Environmental Variables

In the scenario observed in this chapter, our system takes into account environmental measurements regarding temperature, humidity and lighting. All such variables may be interpreted subjectively since perceived temperature and lighting preferences may vary heavily from one user to another.

The relationship that exists between humidity and temperature is an important factor to consider, because the sensation of hot or cold is influenced by the degree of humidity in the air. For example the perceived temperature in the summer is more pleasant if there is low rather than high humidity.

Bioclimatic indices, such as the PET (Physiological Equivalent Temperature) and HI (Heat Index), evaluate many physiological conditions related to human perception of the environmental situation on the basis of the correlation of various parameters [11].

These weather parameters determine the apparent temperature. HI as a combination of temperature and humidity, whilst PET considers different values, such as the heat balance of the human body, skin temperatures, activity, basic metabolism, the heat resistance of clothing, air temperature and humidity.

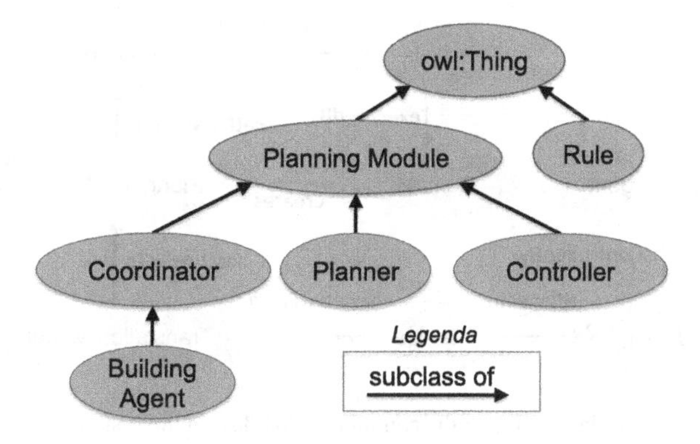

Fig. 1 Taxonomy of classes defined in the proposed ontology

PET shows how the perception of temperature depends on humidity and other factors, but in sedentary activities such as office work, the management of temperature and humidity is simpler because there aren't any complex parameters (Table 1).

3.2 Ontological Representation of Domain

An Ontology in computer science is a formalism used to represent knowledge expressed as classes of objects and the relationships between them. We have developed an ontology representing the information flow at different levels of abstraction, from raw sensory data up to a symbolic representation of the context.

In particular, the ontology describes the abstraction flow of data processing with a correspondence where the different system components are the objects of the formalism and the exchange of data between the various modules is represented through appropriate relations, as shown in Fig. 1.

More specifically, the elements in the ontology represent the main modules of the planning system and reproduce the communication flow and the connections with other subsystems.

The main class of the ontology is the PlanningModule, which generates the Coordinator, Planner and Controller classes. As described in the following section, in which the architecture of the planning system is reported, these objects communicate with each other to determine the right actions to perform in order to achieve the final objective.

In particular, the Coordinator exploits the Building class that describes the maximum and current usage of energy, while the Rule class is used both by the Planner, whose purpose is to generate the fuzzy rules and by the Controller class, which integrates and manages these rules. Finally, the Device class describes all the sensor and actuator devices used in the monitored environment. Communication flow among

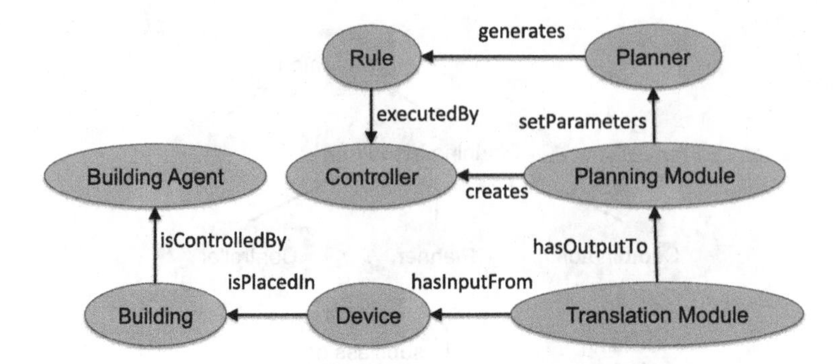

Fig. 2 Relations among classes coding the communication flow in the system

these devices and information processing are described respectively by DataType and Execution Modules.

The Translation Module exploits the ontology's description in order to appropriately connect one subsystem to another. In particular, the Translation Module exploits knowledge about which devices are deployed in the controlled environment in order to opportunely define the actions available for the Controller (Fig. 2).

In this sense, the definition of the Controller, and consequently of the Planner, is parametric with respect to the knowledge of the actuator infrastructure coded into the ontology. Similarly, input variables to the fuzzy controller are dynamically linked to the outputs of the cognitive modules responsible for processing raw sensory data and producing high-level symbolic representations of the current context.

4 The Proposed System

According to the architecture described in [5], a BuildingAgent is implemented for each considered environment. Such Agent collects environmental data through the sensor devices spread over all of the spaces being monitored and sends the aggregated information to the AmI Box. The Controller, Planner, and Planner Coordinator modules constitute the intelligent subsystem and the reasoning engine of the overall architecture, as shown in Fig. 3.

A single instance of the Planner Coordinator is responsible for the whole system, whilst the Planner is replicated for each building and several Controllers are instantiated one for each single environment. All these modules operate on symbolic information after the processing of sensorial data through their collection, analysis and modeling by other system modules and services.

The role of the Coordinator is to evaluate different energy consumption requests from the different buildings and then decide how to set the opportune goals for each planner. Such goals are represented as the trade-off between energy consumptions and user preferences, and are expressed by a weight that assumes values in the [0,

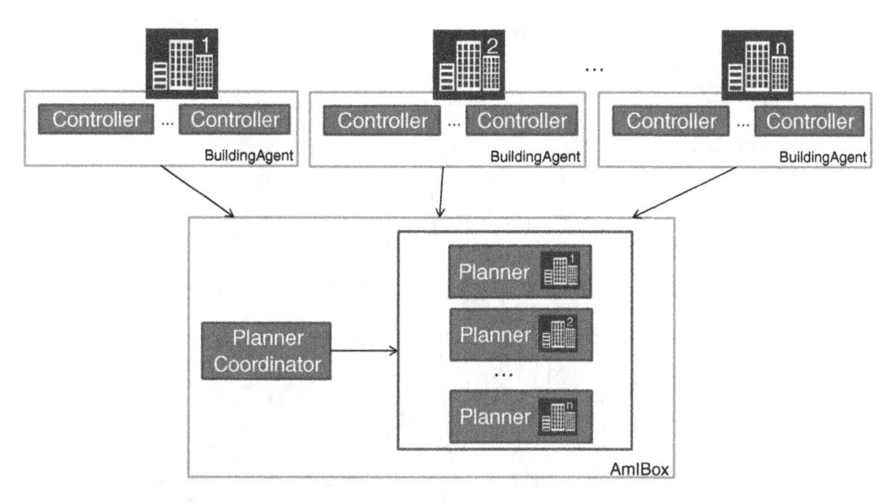

Fig. 3 Deployment diagram of the proposed system

1] interval, where 0 represents a preference for user welfare, whilst 1 the will to minimize energy consumption.

Whenever rules are generated, they are transferred to the controllers. Thus in our solution, we designed two different ad-hoc controllers, the first being responsible for the regulation of temperature and humidity and the second for the management of lighting.

4.1 Fuzzy Controllers

Fuzzy logic differs from classical Boolean logic in that it can attribute to each proposition a degree of truth between 0 and 1. Properties are represented by functions called membership functions. Fuzzy logic is the basis of our Controller, which constitutes the reactive component of our system, and implements mapping between the actual sensory values and the actions to be performed.

The Controller is hosted in the BuildingAgent software module and, from a physical point of view, depends on the actual number of sensors and actuators deployed in the environment, and on the list of actions they make possible. As described above, the knowledge of available sensor and actuator devices, and of their capabilities, is coded into the ontology. When the software system is installed, the TranslatorModule is responsible for translating this knowledge by setting opportune parameters in the PlanningModule.

Moreover, the TranslatorModule is responsible for storing this infrastructural knowledge for a specific building in a sub-module of the BuildingAgent, called TopologyManager, which is responsible for providing this information whenever it

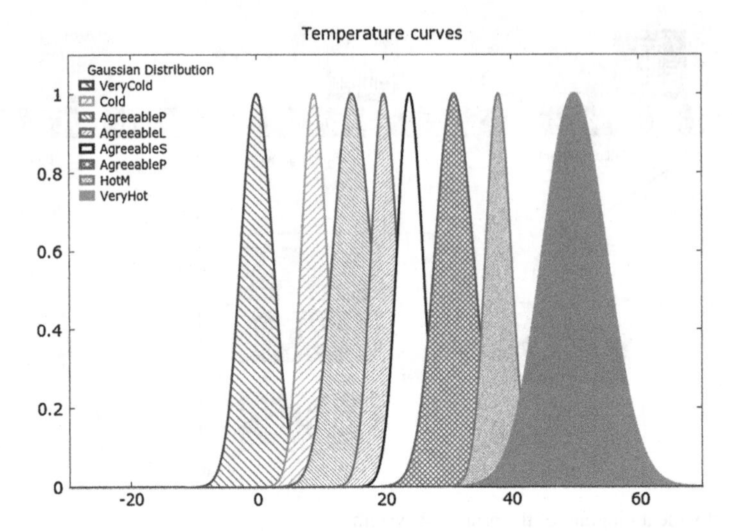

Fig. 4 Membership functions for the qualitative evaluation of temperature

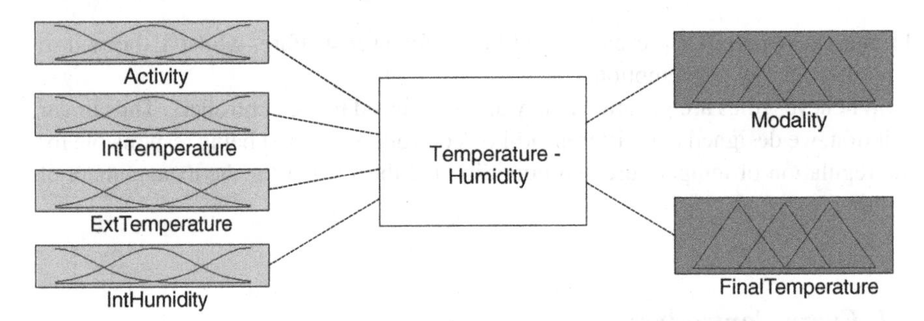

Fig. 5 Scheme of the fuzzy controller devoted to managing temperature and humidity

is required rules. In particular, when the Planner optimizes the Controllers rules it exploits this information in dynamically setting the input and output of the Controller.

Temperature, humidity and lighting were considered controlling variables. Since there is a strict relation between the first two variables, there is only one controller in our system for temperature and humidity and another specifically devoted to lighting.

Figure 4 provides an example of a membership function in which the temperature values are assigned to eight different Gaussians, in such a way that a value greater than 45° Celsius is translated in the VeryHot concept and a temperature lower than 5° Celsius is described as VeryCold. The two controllers have different internal structures. The temperature-humidity controller always takes the same number of input values, while the output values depend entirely on the number of actuators The number of input values also depend on the available sensors in the lighting controller.

For the sake of simplicity, we describe a deployment here, shown in Fig. 5, with only one actuator and where the inputs variables considered are:

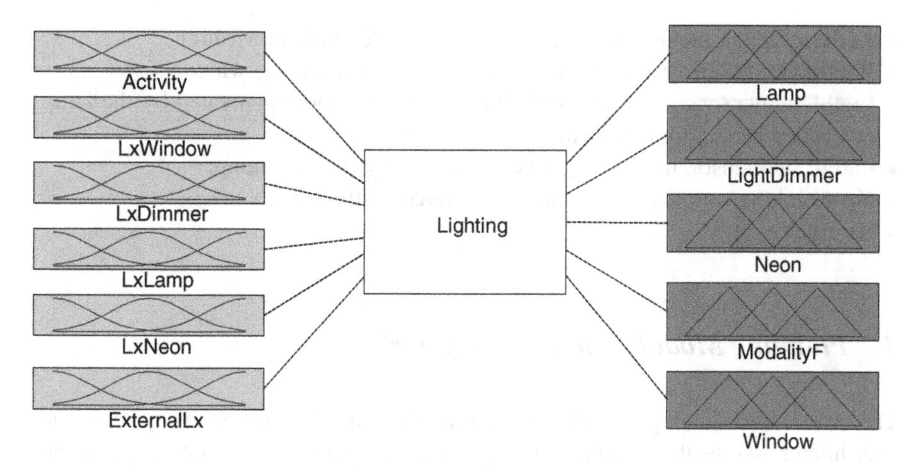

Fig. 6 Scheme of the fuzzy controller devoted to managing lighting

- Activity: such variable represents the activity performed,
- Internal Temperature,
- External Temperature,
- Internal Humidity.

The outputs represent the actions to be performed by the actuator, as follows:

- ModeT: this variable is the modality of the actuator and its possible values are: Dry, Cold, Hot, Off,
- FinalTemperature: is the final temperature value.

The structure of the lighting controller is more complex, since it has dynamic inputs and outputs. In the scenario considered here, four actuators and five sensors control and monitor lighting, as shown in Fig. 6.

There are two types of actuators, an "on-off" one, such as the lamp placed on the desk, or the neon installed on the roof, and those which can regulate intensity, such as the lamp with a dimmer switch and the window with an adjustable percentage of opening. In particular, we have the following outputs:

- Lamp: with the two command values: on or off;
- Neon: with the two command values: on or off;
- LightDimmer: with has various intensity values expressed in percentages from 0 % (off) to 100 % (maximum intensity);
- Window: which has various values for the level of opening, expressed in percentages from 0 % (off) to 100 % (maximum intensity).

Four of the available light sensors are placed near the actuators and the fifth is placed in the exterior of the window to monitor the available level of natural lighting. These inputs are used together with the estimation of the activity currently being performed by the user, produced by one of the system modules. Thus, Controller inputs are as follows:

- Activity, which includes user activities such as PC Work or Meeting;
- WindowLightSensor, the lighting level monitored beside the window;
- LigthDimmerSensor, which the lighting level monitored generated by the lamp controlled by the dimmer switch;
- LampLightSensor, the lighting level generated by the desk lamp;
- NeonLightSensor, the lighting level generated by the Neon;
- ExternalLight.

4.2 Planning Module and Search Algorithm

The task of the planning module is to achieve the objectives set by the system. The coordinator assigns the weights to user preferences and to energy consumption and then, on the basis of this information, it determines a single value in the range [0, 1] that will be used by the planner to generate the fuzzy rules.

Rule Generation is a quite difficult task, since their correctness and suitability characterize the intelligence of the system as a whole. TabuSearch (TS), a meta heuristic search algorithm, is exploited by the planner to generate various rules, depending on the needs of the system.

Values known as cost, fitness, or energy value, control the search process, looking for optimal solutions and avoiding local minima. A number of random configurations are used to generate the right actions within a larger space of configurations. A configuration is a point in this space and represents mapping between sensory values and actions.

The input of the algorithm is a vector which codes the values representing the membership functions (MF) and the output is constituted by the set of fuzzy rules. The input values of MF are coded in binary form, whilst gray code [6] is adopted for the output MFs.

The gray code makes it possible to move from the current configuration to an adjacent one by the complementation of a single bit. The switching of a bit represents a move, and the TS algorithm guarantees that the same configuration cannot be reached after further moves, by distinguishing admissible moves from inadmissible ones (tabu moves).

To this end, TS exploits several lists in order to store the sets of admissible and tabu moves and the best configurations.

The lists used in the algorithm are:

- Tabu Moves: initially this is an empty set, but during the search contains the moves that lead to configurations that have already been visited.
- Admissible moves: initially this includes all possible moves that can be performed on a configuration. Subsequently the tabu moves are removed from this list.
- Best Configurations: this list contains the configurations with the best fitness values.

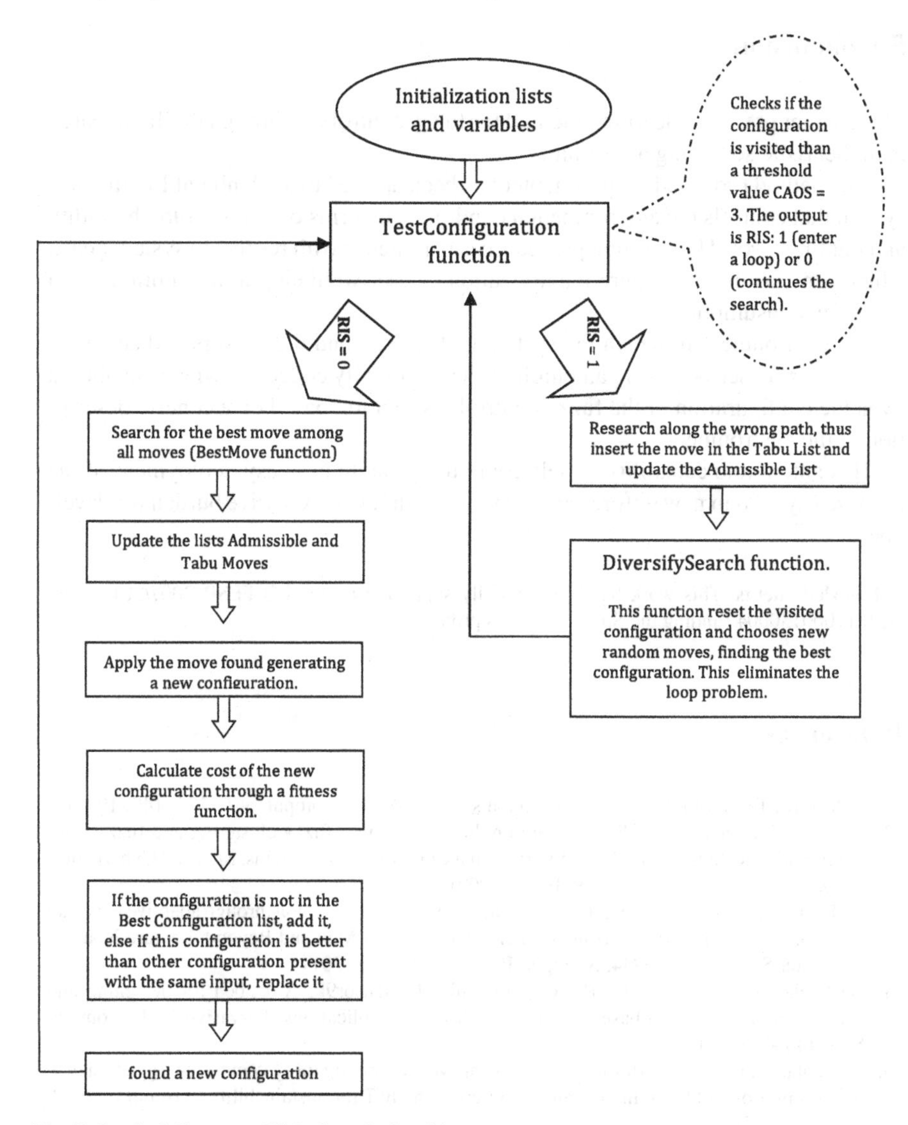

Fig. 7 Logical diagram of TabuSearch algorithm

By definition a move is admissible if it does not repeat moves executed in the last T steps, where T is called Period. Such values change during the search process and indicate the number of tabu moves.

The Fig. 7 shows a logical diagram of the TS algorithm.

5 Conclusion

This chapter presents a learning method for fuzzy controllers through the TabuSearch meta-heuristic searching algorithm.

The method proposed in this chapter has been applied to an Ambient Intelligence system that controls the environmental conditions in terms of temperature, humidity and lighting level. The learning procedure is parametric with respect to system goals, which can vary from the pure maximization of user wellbeing to the minimization of energy consumption.

The environment in which the system is deployed and the sensors and actuators available are described using an ontology. This formally coded knowledge enables a dynamic configuration of the fuzzy controller structure, based on a concise description of the environment.

This autonomic capability of self-configuring facilitates easy deployment of the proposed system in new unforeseen scenarios without an excessive burden for developers.

Acknowledgments This work has been partially supported by the PO FESR 2007/2013 grant G73F11000130004 funding the SmartBuildings project.

References

1. Battiti, R., Tecchiolli, G.: The reactive tabu search. ORSA j. comput. **6**(2), 126–140 (1994)
2. Chang, Y.C., Chen, S.M.: Temperature prediction based on fuzzy clustering and fuzzy rules interpolation techniques. In: IEEE International Conference on Systems, Man and Cybernetics, 2009, SMC 2009, pp. 3444–3449. IEEE (2009)
3. De Paola, A., Gaglio, S., Lo Re, G., Ortolani, M.: Multi-sensor fusion through adaptive bayesian networks. In: AI*IA 2011: Artificial Intelligence Around Man and Beyond, Lecture Notes in Computer Science, vol. 6934, Springer, Berlin Heidelberg (2011)
4. De Paola, A., Gaglio, S.: Lo Re, G., Ortolani, M.: Sensor9k: A testbed for designing and experimenting with wsn-based ambient intelligence applications. Pervasive Mob. Comput. **8**(3), 448–466 (2012)
5. De Paola, A., Lo Re, G., Morana, M., Ortolani, M.: An intelligent system for energy efficiency in a complex of buildings. In: Sustainable Internet and ICT for Sustainability (SustainIT), 2012, pp. 1–5 (2012)
6. Denna, M., Mauri, G., Zanaboni, A.M.: Learning fuzzy rules with tabu search-an application to control. IEEE Trans. Fuzzy Syst. **7**(3), 295–318 (1999)
7. Di Fatta, G., Hoffmann, F., Lo Re, G., Urso, A.: A genetic algorithm for the design of a fuzzy controller for active queue management. IEEE Trans. Syst. Man Cybern. Part C Appl. Rev. **33**(3), 313–324 (2003)
8. Di Fatta, G.: Lo Re, G., Urso, A.: A fuzzy approach for the network congestion problem. In: Computational Science—ICCS 2002, Lecture Notes in Computer Science, vol. 2329, pp. 286–295. Springer, Berlin Heidelberg (2002)
9. Di Fatta, G.: Lo Re, G., Urso, A.: Parallel genetic algorithms for the tuning of a fuzzy AQM controller. In: Computational Science and Its Applications (ICCSA 2003), Lecture Notes in Computer Science, vol. 2667, pp. 417–426. Springer, Berlin Heidelberg (2003)

10. Dubois, D., Prade, H.: Fuzzy Sets and Systems: Theory and Applications. Academic Press, New York (1997)
11. Höppe, P.: The physiological equivalent temperature a universal index for the biometeorological assessment of the thermal environment. Int. J. Biometeorol. **43**, 71–75 (1999). DOI:10.1007/s004840050118
12. Hosoya, Y., Umano, M.: Dynamic fuzzy q-learning with facility of tuning and removing fuzzy rules. In: Fuzzy Systems (FUZZ-IEEE), 2012 IEEE International Conference on, pp. 1–8. IEEE (2012)
13. Kaufmann, A., Gupta, M.M.: Introduction to Fuzzy Arithmetic: Theory and Applications. Van Nostrand Reinhold Company, New York (1985)
14. Lhotska, L., Macek, J., Peri, D.: Evaluation of ecg: comparison of decision tree and fuzzy rules induction. In: European Meetings on Cybernetics and Systems Research (EMCSR), pp. 713–718 (2004)
15. Nauck, D., Kruse, R.: A fuzzy neural network learning fuzzy control rules and membership functions by fuzzy error backpropagation. In: IEEE International Conference on Neural Networks, 1993, pp. 1022–1027. IEEE (1993)
16. Navara, M., Peri, D.: Automatic generation of fuzzy rules and its applications in medical diagnosis. In: Proceedings of the 10th International Conference on Information Processing and Management of Uncertainty, pp. 657–663 (2004)
17. Shi, Y., Mizumoto, M.: An improvement of neuro-fuzzy learning algorithm for tuning fuzzy rules. Fuzzy Sets Syst. **118**(2), 339–350 (2001)

Design of an Adaptive Bayesian System for Sensor Data Fusion

Alessandra De Paola and Luca Gagliano

Abstract Many artificial intelligent systems exploit a wide set of sensor devices to monitor the environment. When the sensors employed are low-cost, off-the-shelf devices, such as Wireless Sensor Networks (WSN), the data gathered through the sensory infrastructure may be affected by noise, and thus only partially correlated to the phenomenon of interest. One way of overcoming these limitations might be to adopt a high-level method to perform multi-sensor data fusion. Bayesian Networks (BNs) represent a suitable tool for performing refined artificial reasoning on heterogeneous sensory data, and for dealing with the intrinsic uncertainty of such data. However, the configuration of the sensory infrastructure can significantly affect the performance of the whole system, both in terms of the uncertainty of the inferred knowledge and in term of the hardware performance of the sensory infrastructure itself. This chapter proposes an adaptive Bayesian System whose goal is to infer an environment feature, such as activities performed by the user, by exploiting a wide set of sensory devices characterized by limited energy resources. The system proposed here is able to adaptively configure the sensory infrastructure so as to simultaneously maximize the inference accuracy and the network lifetime by means of a multi-objective optimization.

1 Motivations and Related Work

Artificial intelligence systems often adopt a sensory infrastructure characterized by elevated device heterogeneity both in terms of the energy consumption profile and the type of measurements collected. One of the application scenarios in artificial intelligence, where this feature is more evident is Ambient Intelligence, characterized by the adoption of pervasive and ubiquitous sensors for monitoring relevant ambient

A. De Paola (✉) · L. Gagliano
Viale delle Scienze, University of Palermo, ed 6, 90128 Palermo, Italy
e-mail: alessandra.depaola@unipa.it

S. Gaglio and G. Lo Re (eds.), *Advances onto the Internet of Things*,
Advances in Intelligent Systems and Computing 260, DOI: 10.1007/978-3-319-03992-3_5,

features. Data fusion, by enabling high-level context information to be obtained from raw sensory data, may offer a solution to the need to cope with such heterogeneity, and to manage data that may be only partially correlated with the phenomenon of interest [1, 2]. Considerable attention has been devoted to context information, such as user presence in monitored areas [3–5] or current user activities [6, 7].

When dealing with multi-sensor data fusion, one of the most relevant issues is the management of the non-negligible level of uncertainity and noise in data gathered by low-cost devices. To deal with this problem, several papers in the literature have suggested adopting a probabilistic approach, such as Naive Bayes classifiers, Hidden Markov Models (HMMs) and Conditional Random Fields (CRF), as described in [7], which compares the performance of these three approaches in different type of datasets, adopting a semi-supervised learning scheme. In [8] a distributed and adaptive Bayesian network is proposed for the detection of data anomalies in WSN data.

In line with state of the art research, we propose a Bayesian adaptive system devoted to inferring user activity through a set of low-cost sensors, embedded into a Wireless Sensor Network (WSN) [9], as preliminarly described in [10] for detecting user presence [11]. A WSN comprises a huge set of wireless sensor nodes, pervasively deployed in the environment and capable of performing on-board computations. These devices are characterized by limited, non-renewable, energy resources. This latter feature makes the maximization of the network lifetime a crucial goal, with the proviso that its achievement should not excessively sacrifice the inference accuracy. The proposed system aims to dynamically find the best trade-off between these two contrasting goals, maximizing both the WSN lifetime and the quality of the information gathered. This problem dealt with by minimizing both the uncertainty of inferred knowledge and the energy consumption of the sensory infrastructure.

In order to solve our multi-objective problem, two objective functions have to be formally defined. For the uncertainty function we used the classic definition provided in [12]. The definition of an objective function for representing energy consumption of sensor nodes is a rather more complex problem. Several papers have dealt with the issue of minimizing the energy consumption of a WSN [13–15]. In [14] the author describes PAMAS, a MAC layer protocol which reduces the cost of routing packets over the shortest-hop routing. In [13] the authors propose a node cost model for their clustering-based protocol that utilizes randomized rotation of local cluster basestations (cluster heads) to distribute the energy load among sensor nodes. In [15], the authors offer an analysis of the power consumption model for the communication module of a generic WSN node. To the best of our knowledge, most of the research in the literature deals with the problem of maximizing the WSN lifetime either at MAC level or at routing level. In contrast, our system manages the entire sensory infrastructure at a higher level, making our approach independent from low-level details.

This chapter is structured as follows. Section 2.1 provides a general description of the system proposed here, in terms of the concepts involved and the relations among them. Section 2 provides a formal definition of the Bayesian Network (BN) adopted, and of the quality indices exploited to evaluate system performance.

The self-configuration problem through which the system is able to adapt its sensory infrastructure is described in Sect. 2.6. Section 3 details the results of the experimental evaluation of the proposed system, and finally, Sect. 4 states our conclusions and proposes some future developments of our work.

2 Proposed System

We propose the adoption of an AmI system whose sensory infrastructure is based on Wireless Sensor Networks composed of off-the-self and low-cost devices. This feature makes it possible to maintain a low intrusiveness for the users and for the monitored premises, but implies that the signals gathered are, in general, only partially correlated with the feature of interest.

To overcome this problem, the adopted system exploits Bayesian network (BN) as a framework for performing multi-sensor data fusion. In particular, the BN aims to detect the activity performed by the user in the monitored premises.

With a view to evaluating the behavior of the current sensory infrastructure, we defined two *quality indices*, expressing the actual energy consumption of sensory devices and the quality of the gathered information. These quality indices are continuously monitored in order to detect anomalous situations, and whenever one of them goes over a given threshold an *alarm* is triggered. In such cases, the system then reconfigures the sensory infrastructure.

A meta-level for self-configuration is implemented over the BN, as shown in Fig. 1. Such high-level component try to achieve the best trade-off between the degree of confidence of the Bayesian network and the energy consumption of the sensory infrastructure; a plan is produced stating which sensory devices have to be activated or de-activated.

2.1 Conceptual Representation

We formally modeled the concepts characterizing our domain and the relationships between them through an ontology. This formalism allows us to understand the structure and the behavior of our system better, and to support the automatic interaction with other AmI components. The proposed ontology also makes it is possible to describe the components of our system, namely the *sensory infrastructure*, the *inference engine* and the *optimization module*. The relationships among these components are showed in Fig. 2: the optimization module changes the configuration of the sensory infrastructure in order to find the best trade-off between energy consumption and quality of the information obtained, thus affecting the accuracy of the inference engine.

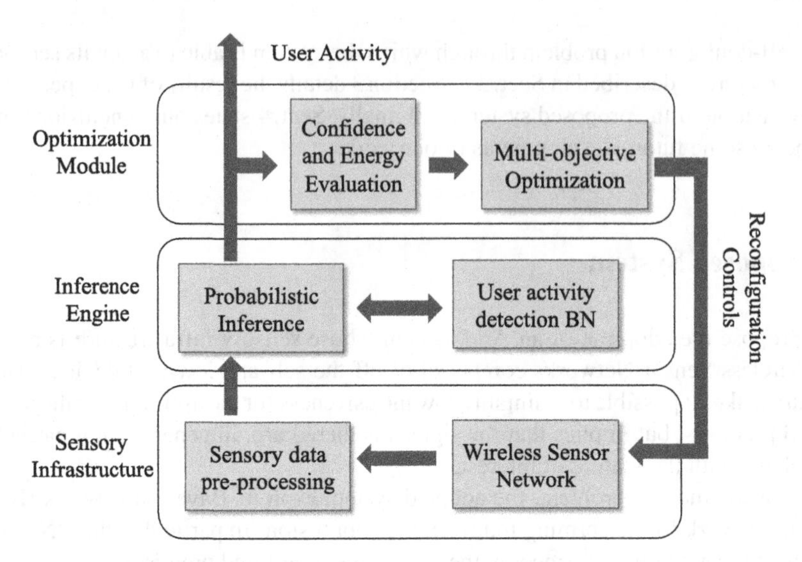

Fig. 1 Block diagram for the proposed system

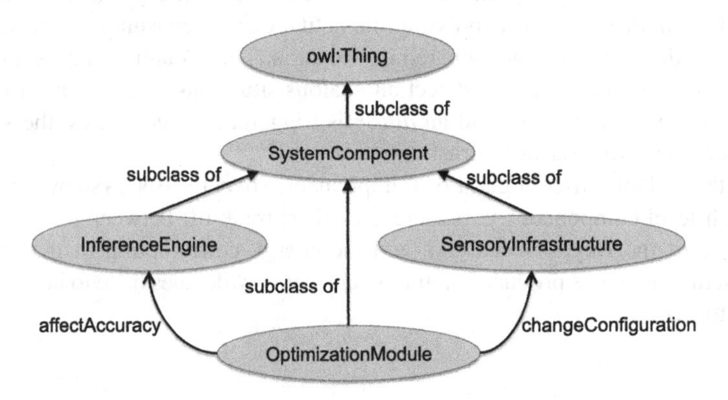

Fig. 2 Taxonomy of system components and their relationships, as described in the proposed ontology

The role of the optimization module is represented by the concepts and relationships depicted in Fig. 3. At each time step, the optimization module observes the inference accuracy characterizing the inference engine and the power consumption caused by the sensory infrastructure. These two indices are verified against two fixed thresholds, and whenever one index exceeds its threshold, an alarm is fired, thus triggering the reconfiguration of the sensory infrastructure. The formal definition of such indices is provided in the following section.

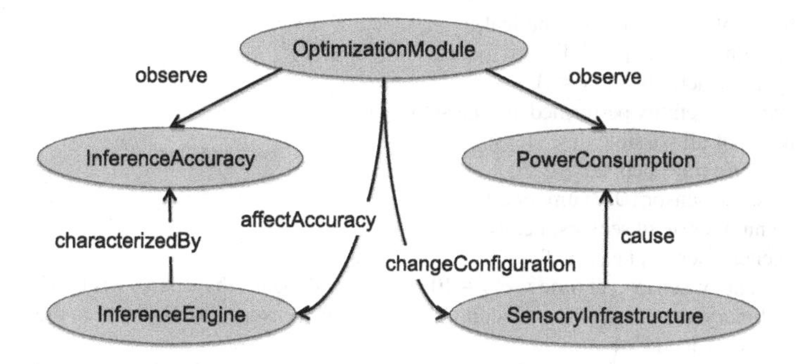

Fig. 3 Description of concepts involved in the functioning of the optimization module

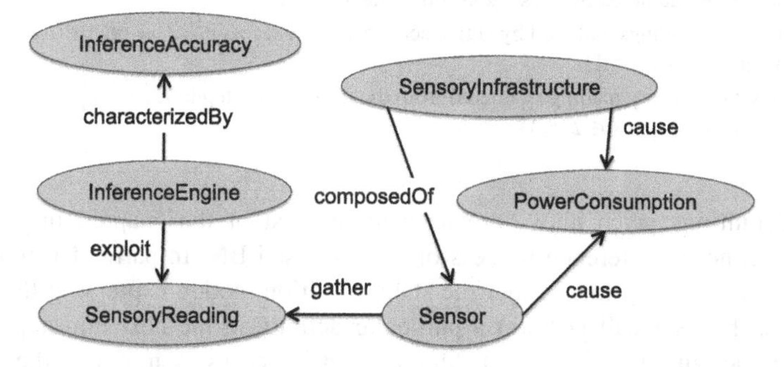

Fig. 4 The ontology proposed represents the indirect dependency between the status of the sensory infrastructure and both energy consumption and inference accuracy

As demonstrated in Fig. 4, the system knows that the sensory infrastructure is composed of several sensors, and that each of these sensors consumes energy and contributes to the energy consumption of the whole sensory infrastructure. Switching a sensor on or off affects not only such consumption, but also the set of sensory readings gathered in a given time step. Because the inference engine uses as input the sensory readings gathered, each change in the state of the sensory infrastructure indirectly affects the accuracy of the inference process.

2.2 Basic Definitions

Before describing the structure of the BN, we provide some formal definitions, which are required to formally state both the structure of the Bayesian system and the multi-objective problem.

$\mathcal{X}$: the set of activity IDs (numerical);
$n_{\mathcal{X}}$: the number of all possible activities, i.e., $n_{\mathcal{X}} = \sharp(\mathcal{X})$;
x : a generic activity, i.e. $x \in \mathcal{X}$;
x_t : a generic activity performed at time step t, i.e. $x_t \in \mathcal{X}$;
$\mathcal{T}$: the set of all possible time steps;
t : a generic time step, i.e. $t \in \mathcal{T}$;
$\mathcal{S}$: the set of sensor IDs (numerical);
$n_{\mathcal{S}}$: the number of all sensors, i.e. $n_{\mathcal{S}} = \sharp(\mathcal{S})$;
s : a generic sensor, i.e., $s \in \mathcal{S}$;
$c_{s,t}$: the state of sensor s at time t; $c_{s,t} \in \{0, 1\}$, where 0 means that sensor s is OFF;
$\mathbf{c}_t$: the binary vector encoding the configuration of the sensory infrastructure at the time step t, i.e. $\mathbf{c}_t \in \{0, 1\}^{n_{\mathcal{S}}}$;
$\mathcal{I}(\mathbf{c}_t)$: the subset of sensors ON in the configuration $\mathbf{c}_t$, i.e., $\mathcal{I}(\mathbf{c}_t) = \{s \in \mathcal{S} \mid c_{s,t} = 1\}$;
$\mathcal{E}$: the set of numerical IDs, one for each possible value of sensory readings;
e_t^s : the reading gathered by sensor s at time t, i.e. $e_t^s \in \mathcal{E}$;
$e_t^{\mathcal{I}(\mathbf{c}_t)}$: the set of readings gathered by active sensors at time t, i.e., $e_t^{\mathcal{I}(\mathbf{c}_t)} = \{e_t^s \mid s \in \mathcal{I}(\mathbf{c}_t)\}$ (ordered by sensor ID);
$e_{1:t}^{\mathcal{I}(\mathbf{c}_k)}$: the set of sensory readings gathered from the initial time step to t, i.e., $e_{1:t}^{\mathcal{I}(\mathbf{c}_k)} = \{e_k^s \mid 1 \leq k \leq t\ ,\ s \in \mathcal{I}(\mathbf{c}_k)\}$.

The definitions given above are used in the rest of the chapter, in order to formally define the inference process of the proposed BN. In particular, to define the BN, it is necessary to consider the state transition model, expressing the probability that the user will perform a particular activity in the next timestep, given the current activity, i.e., $p(x_t|x_{t-1})$. Moreover, it is necessary to define the sensor model, expressing the probability that a specific set of sensor readings is gathered by the sensory infrastructure, given a specific activity performed by the user, i.e., $p(e_t^{\mathcal{I}(\mathbf{c}_t)}|x_t)$. The state of the *sensory infrastructure* is fully specified by the binary vector $\mathbf{c}_t = (c_{1,t}, c_{2,t}, \ldots, c_{n_{\mathcal{S}},t})$, if we assume that the location of each device does not change over the time. It is worth noting the relevance of $\mathcal{I}(\mathbf{c}_t)$, which can be seen as an operator which, given a sensory infrastructure, returns the set of active sensors at time t, thus making it possible to indicate which sensors really contribute to inferring context knowledge.

2.3 Inference Engine

Given the structure of the Bayesian network shown in Fig. 5, the probabilistic state transition model, i.e., $p(x_t|x_{t-1})$, and the probabilistic sensor model, i.e., $p(e_t^{\mathcal{I}(\mathbf{c}_t)}|x_t)$, fully define the Bayesian network. The Bayesian network allows the inference engine to build its own *belief* about the activity currently being performed by the user, taking as input the whole observation set, as follows:

$$Bel(x_t; \mathbf{c}_t) = p(x_t|e_{1:t}^{\mathcal{I}(\mathbf{c}_k)}), \tag{1}$$

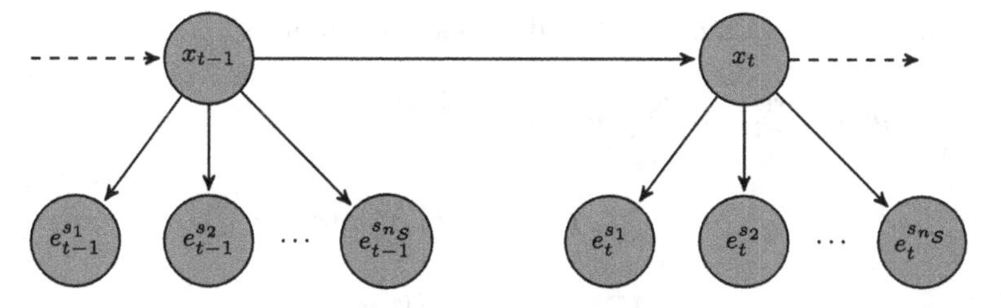

Fig. 5 Structure of the Bayesian Network for detecting user activity

This belief is evaluated over a state $x_t \in \mathcal{X}$, and it is parametric with respect to the configuration of the sensory infrastructure $\mathbf{c}_t$. The evaluation of such belief requires knowledge both of the evolution of sensory infrastructure over the time, i.e. $\mathbf{c}_1, \mathbf{c}_2, \ldots, \mathbf{c}_t$, and of the whole set of sensory readings gathered over the time in question. Equation 1 can be expressed as a recursive equation thanks to the assumption of independence between the different measures given a state value, and to the validity of the Markov assumption [16].

Indeed, by using the Bayes rule, it is possible to derive the following equation:

$$\begin{aligned} Bel(x_t; \mathbf{c}_t) &= p(x_t | e_{1:t}^{\mathcal{I}(\mathbf{c}_k)}) = p(x_t | e_t^{\mathcal{I}(\mathbf{c}_t)}, e_{1:t-1}^{\mathcal{I}(\mathbf{c}_k)}) = \\ &= \eta \times p(e_t^{\mathcal{I}(\mathbf{c}_t)} | x_t, e_{1:t-1}^{\mathcal{I}(\mathbf{c}_k)}) \times p(x_t | e_{1:t-1}^{\mathcal{I}(\mathbf{c}_k)}), \end{aligned} \tag{2}$$

where η is a normalizing factor.

The Markov assumption makes it possible to neglect the sensory readings gathered up to $t-1$, when the knowledge of the state $x_{t.1}$ is given, thus the following equation holds:

$$p(e_t^{\mathcal{I}(\mathbf{c}_t)} | x_t, e_{1:t-1}^{\mathcal{I}(\mathbf{c}_k)}) = p(e_t^{\mathcal{I}(\mathbf{c}_t)} | x_t). \tag{3}$$

The assumption of measures independence, given the state x_t, allows factorization as follows:

$$p(e_t^{\mathcal{I}(\mathbf{c}_t)} | x_t) = \prod_{s \in \mathcal{I}(\mathbf{c}_t)} p(e_t^s | x_t). \tag{4}$$

Consequently, the belief can be expressed through the following equation:

$$Bel(x_t; \mathbf{c}_t) = \eta \prod_{s \in \mathcal{I}(\mathbf{c}_t)} p(e_t^s | x_t) p(x_t | e_{1:t-1}^{\mathcal{I}(\mathbf{c}_k)}). \tag{5}$$

The last term in Eq. 5 can be further decomposed as follows:

$$\begin{aligned} p(x_t|e_{1:t-1}^{\mathcal{I}(\mathbf{c}_k)}) &= \sum_{x_{t-1}\in\mathcal{X}} p(x_t, x_{t-1}|e_{1:t-1}^{\mathcal{I}(\mathbf{c}_k)}) \\ &= \gamma \sum_{x_{t-1}\in\mathcal{X}} p(x_t|x_{t-1}, e_{1:t-1}^{\mathcal{I}(\mathbf{c}_k)})p(x_{t-1}|e_{1:t-1}^{\mathcal{I}(\mathbf{c}_k)}) \\ &= \gamma \sum_{x_{t-1}\in\mathcal{X}} p(x_t|x_{t-1}, e_{1:t-1}^{\mathcal{I}(\mathbf{c}_k)})Bel(x_{t-1}; \mathbf{c}_{t-1}), \end{aligned} \tag{6}$$

where γ is a normalizing factor.

The substitution of equation (6) in equation (5) and a further application of the Markov assumption lead to the following recursive definition of the belief:

$$Bel(x_t; \mathbf{c}_t) = \eta \prod_{s\in\mathcal{I}(\mathbf{c}_t)} p(e_t^s|x_t) \sum_{x_{t-1}\in\mathcal{X}} p(x_t|x_{t-1})Bel(x_{t-1}; \mathbf{c}_{t-1}), \tag{7}$$

where γ is integrated in the normalization factor η. It is worth noting that such expression of the belief is directly reflected in the graphical representation of the proposed BN shown in Fig. 5.

2.4 Uncertainty Index

We define the uncertainty index at the timestep t, on the basis of the classical definition of *entropy* for the *a priori* probability distribution of a random variable:

$$U(\mathbf{c}_t) = -\sum_{x_t\in\mathcal{X}} Bel(x_t; \mathbf{c}_t) \log_2(Bel(x_t; \mathbf{c}_t)). \tag{8}$$

By varying the configuration $\mathbf{c}_t$ of the sensory infrastructure, it is possible to decrease belief uncertainty and thus to improve the information inferred at next timestep. This index makes it is possible, at least, to predict a better configuration of the sensory infrastructure and to obtain a lower degree of uncertainty for the inferred knowledge.

2.5 Power Consumption Index

Generally, sensor nodes are able to monitor their own residual energy. If $E_s(t)$ indicates the quantity of residual energy of node s at the timestep t, typically associated with its battery charge, the residual energy for the entire sensory infrastructure can be expressed as follows:

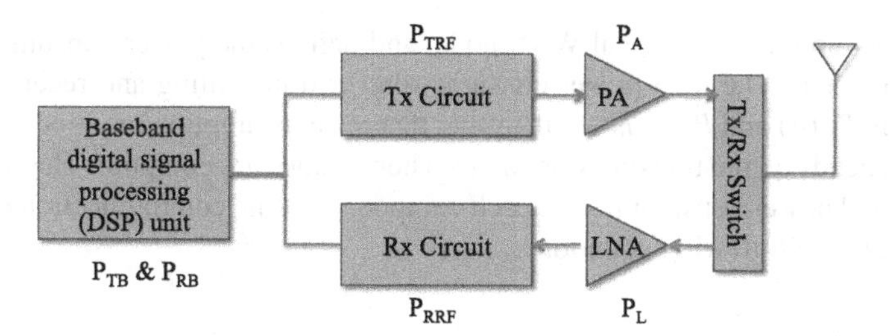

P_{TB} & P_{RB}: Power consumption in baseband DSP circuit for transmitting or receiving (mW)

P_{TRF} & P_{RRF}: Power consumption in front-end circuit for transmitting or receiving (mW)

P_A: Power consumption of power amplifier (PA) for transmitting (mW)

P_L: Power consumption of low-noise amplifier (LNA) for receiving (mW)

Fig. 6 Structure of the communication module of a WSN sensor node [15]

$$E(t) = \sum_{s=1}^{n_S} E_s(t). \tag{9}$$

In what follows we will omit an explicit indication of the dependency of E on t. By supposing that E is differentiable for small timesteps, the following approximation of the energy variation with a first order differential equation holds:

$$dE = \sum_{s=1}^{n_S} dE_s. \tag{10}$$

By dividing both members by dt, it is possible to obtain the following expression:

$$\frac{dE}{dt} = \sum_{s=1}^{n_S} \frac{dE_s}{dt} \Rightarrow P = \sum_{s=1}^{n_S} P_s, \tag{11}$$

where $P = P(t)$ is the total power consumption of the sensory infrastructure and $P_s = P_s(t)$ is the power consumption of the sensor s at t.

Obviously, the power consumption depends heavily on the configuration of the sensory infrastructure, thus we express the power consumption as a parametric function, as follows:

$$P(\mathbf{c}_t) = \sum_{s \in \mathcal{I}(\mathbf{c}_t)} P_s. \tag{12}$$

In the literature, there is a considerable body of work on the form of P_s for a single WSN node. By adopting one of these models, it is possible to compute the power consumption of the whole sensory infrastructure. In this chapter, the model presented in [15] is adopted. Figure 6 illustrates the internal structure of the communication module of a typical WSN node, and defines the power consumption of each component. The total power consumption for transmitting and receiving, are denoted by $P_T(d)$ and P_R; it is worth noting that the consumption required for transmitting depends on the transmission range. These values are computed Based on the structure and power consumption of each component of the communication module, according to the following equations:

$$\begin{aligned} P_T(d) &= P_{TB} + P_{TRF} + P_A(d) = P_{T0} + P_A(d), \\ P_R &= P_{RB} + P_{RRF} + P_L = P_{R0}. \end{aligned} \tag{13}$$

In Eq. (13) the term $P_A(d)$ represents the power consumption of the amplifier, and it is the only term depending on the transmission range. Other terms can be modeled as constant values: P_{T0} for the constant part of the power consumption of the transmitting circuit, and P_{R0} for the power consumption of the receiving circuit. $P_A(d)$ depends on several physical features, like antenna and propagation medium features. For example, by supposing that signals propagate in free space, i.e. in a vacuum without obstacles, the term $P_A(d)$ can be expressed as follows:

$$P_A(d) = \frac{P_R}{G_T G_R}\left(\frac{4\pi d}{\lambda}\right)^2, \tag{14}$$

where G_T and G_R are the gains for the transmitting antenna and for the receiving antenna respectively, P_R is the power required by the receiving antenna, λ is the wavelength adopted, and d is the distance between antennas. Equation (14) is the well-known *Friis Formula* [17] and summarizes the features of the medium and physical characteristics of the device. There are more general versions of such equations, which take into account the non vacuum space, namely both the presence of obstacles and different media [17]. Equation (14) shows heavy interdependence between transmission power, device features and the environment in which the sensory infrastructure is deployed.

2.6 Self-Configuration Behavior

The self-configuration capability of the proposed system allows it to find the optimal configuration of the sensory infrastructure autonomously, based on the uncertainty of the inference engine and on the energy consumption of the sensor nodes. In order to quantify these contrasting goals we propose to exploit the uncertainty index $U(\mathbf{c}_t)$, described in Sect. 2.4 , and the power consumption index $P(\mathbf{c}_t)$, described in Sect. 2.5.

The configuration problem is a multi-objective problem with two objective functions to be minimized:

$$\begin{cases} f_1(\mathbf{c}_t) = U(\mathbf{c}_t) \\ f_2(\mathbf{c}_t) = P(\mathbf{c}_t). \end{cases} \tag{15}$$

In order to avoid drastic changes in the sensory infrastructure, the configuration is forced to change at most the status of a single sensor at each time step. Formally, this dynamic constraint is expressed as follows:

$$\mathbf{c}_{t+1} \in \Gamma(\mathbf{c}_t), \tag{16}$$

where $\Gamma(\mathbf{c}_t)$ defines the region containing possible configurations of the sensory infrastructure, given the current one. This set of configuration is obtained from $\mathbf{c}_t$, by switching on or off only one sensor. Formally, it is defined as follows:

$$\Gamma(\mathbf{c}_t) = \left\{ \hat{\mathbf{c}}_t \; : \; \sum_{i=1}^{n_S} \left| c_{i,t} - \hat{c}_{i,t} \right| \leq 1 \right\} \tag{17}$$

In order to solve the multi-objective problem the multi-objective problem defined in Eq. (15), we chose to look for the Pareto optimal solutions, as proposed in [18] in the context of multi-objective genetic algorithms.

The pseudocode for the self-configuration algorithm is shown in Algorithm 1. The algorithm proposed here consists of three parts: (i) the delimitation of the admissible region, according to Eq. 17, (ii) the identification of the Pareto optimal solutions, and (iii) the selection from the admissible and Pareto-optimal solutions, of the one that improves the index which triggered the alarm.

2.7 System Overview

The overall behavior of the proposed system is described by the pseudocode in Algorithm 2. Two main parts are identifiable: the belief update and the self-configuration. Belief update is performed according to the classical equations of a Bayesian filter, as described in Sect. 2.3. This involves verifying whether the current sensory configuration triggers some alarms. Then, if necessary, self-configuration is performed, as described in Sect. 2.6.

3 Experimental Evaluation

3.1 Experimental Setting

In order to evaluate the performance of the proposed system we used a synthetic dataset built on the basis of the WSU CASAS Datasets [7], which consists of rows, as follows:

```
<day, time, sensor_name, sensor_measure, activity,
                    label>.
```

Each term in a row is expressed according to the following BNF grammar:

Algorithm 1 Pseudo-code of the self-configuration algorithm.

```
1: function RECONFIGURE_INFRASTRUCTURE(c_t)
2:   Γ(c_t) ← c_t
3:   for i = 1 → n_S do                                   ▷ Find the eligible region
4:     Γ(c_t) ← Γ(c_t) ∪ {c_{1:i−1,t} c̄_{i,t} c_{i:n_S,t}}
5:   end for
6:   F ← Ø
7:   for all q ∈ Γ(c_t) do                                ▷ Find the non dominated front
8:     n_q ← 0
9:     for all p ∈ Γ(c_t) do
10:      if U(p) < U(q) ∧ P(p) < P(q) then
11:        n_q ← n_q + 1
12:      end if
13:    end for
14:    if n_q == 0 then
15:      F ← F ∪ {q}
16:    end if
17:  end for
18:  if alarm is about "Uncertainty" then                 ▷ Select a solution
19:    Ω ← {q ∈ F | U(q) ≤ U(c_t)}
20:    if Ω == Ø then
21:      ĉ_t ← arg min_{q∈F} U(q)
22:    else
23:      ĉ_t ← arg max_{q∈Ω} U(q)
24:    end if
25:  else
26:    Ω ← {q ∈ F | P(q) ≤ P(c_t)}
27:    if Ω == Ø then
28:      ĉ_t ← arg min_{q∈F} P(q)
29:    else
30:      ĉ_t ← arg max_{q∈ıOmega} P(q)
31:    end if
32:  end if
33:  return ĉ_t
34: end function
```

```
day → yy-mm-dd
time → hh:mm:ss
sensor_name → M0[01-31] | D001 | D002 | D004
sensor_measure → ON | OFF | OPEN | CLOSE
activity → activity_label | ϵ
label → begin | end | ϵ
```

It is worth noting that our synthetic dataset only contains readings of movement sensors and sensors about the state of doors, whereas temperature readings present in the original DB have been discarded because of the low correlation between this physical phenomenon and the activity performed by the user.

On the basis of the dataset adopted, it is possible to properly define the sets $\mathcal{X}$, $\mathcal{S}$, $\mathcal{T}$ and $\mathcal{E}$ as required in Sect. 2.2. In the case under consideration, the definition of $\mathcal{X}$

Algorithm 2 Main System Pseudocode

```
1: function UPDATE_BELIEF(Bel(x_{t-1}; c_{t-1}), c_t, p(x_t|x_{t-1}), p(e_t^{I(c_t)}|x_t))
2:     for all x_t ∈ X do
3:         Bel(x_t; c_t) = η ∏_{s∈I(c_t)} p(e_t^s|x_t) Σ_{x_{t-1}∈X} p(x_t|x_{t-1}) Bel(x_{t-1}; c_{t-1})
4:     end for
5:     P(c_t) ← compute the power consumption of c_t
6:     U(c_t) ← compute the uncertainty of information with c_t
7:     c_{t+1} ← c_t
8:     if P(c_t) > P_th || U(c_t) > U_th then
9:
10:        c_{t+1} ← RECONFIGURE_INFRASTRUCTURE(c_t)
11:     end if
12:     return (Bel(x_t; c_t), c_{t+1})
13: end function
```

simply requires distinct activity labels to be considered, and for each of them to be associated with a unique numerical ID. An analogous procedure involving sensors is required to define $\mathcal{S}$. In order to define $\mathcal{T}$ we considered the number of seconds in a 24 h day and then we divided them into interval of 30 s. Finally, we assigned a unique numerical ID to each interval. In order to define set $\mathcal{E}$, a preprocessing of the original DB was required. Let us suppose that the DB contains two distinct rows (denominated row_i and row_j, where $i < j$), associated to the same sensor s, and that the `label` is `ON` for row_i and `OFF` for row_j. If t_1 and t_2 are the value of the `time` field of row_1 and row_2 respectively, then our DB has to contain an entry for each $t \in [t_1, t_2]$ indicating that the sensor s is active, i.e., $e_t^s = 1$.

3.2 Experimental Results

The original DB contains some unclassified sensory readings and the authors of [7] adopted a semi-supervised approach [19] to deal with this lack of information. To fulfill the same purpose, we used the Expectation Maximization (EM) algorithm. In order to evaluate the performance of our system we adopted the cross validation method dividing our DB into ten parts.

We compared the performance of three different systems. The first system is obtained by deactivating self-configuring behavior and favors minimization of the uncertainty index, thus setting all sensors permanently to on. The second system is obtained also by deactivating self-configuring behavior, but it favors minimization of the power consumption index, thus setting only a minimal subset of sensors to on; this set is fixed and it consists of 10 of the 34 sensors available. The third system is obtained by activating the self-configuring behavior.

The performance of these three systems are compared in Fig. 7. Figure 7 shows the trend of the uncertainty index during a given day, with Fig. 7 showing the trend of the power consumption index during the same day. As expected, with the first

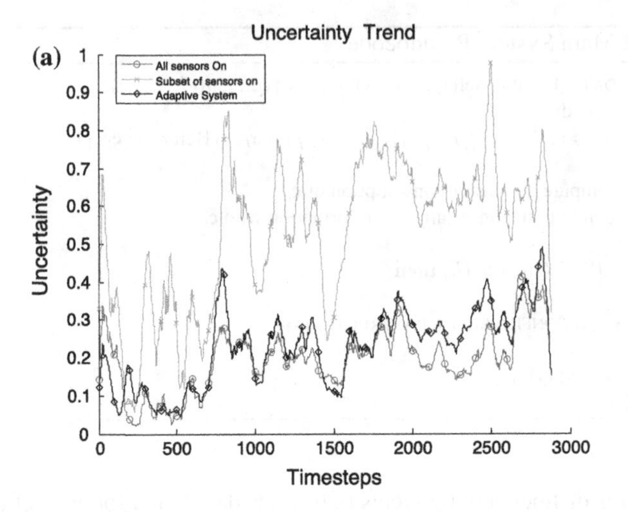

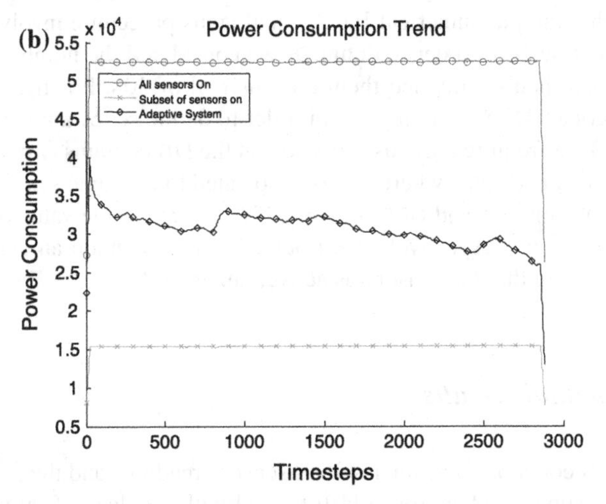

Fig. 7 Comparison of the trend of the uncertainty index and of the power consumption index during a given day for the system proposed with the two base-line systems considered here

base-line system, when all sensors are on, it is possible to obtain the lowest level of uncertainty, but the maximum level of power consumption. In contrast, the second base-line system, with a fixed and limited set of on sensors, is characterized by the highest level of uncertainty and the minimum level of power consumption. The proposed adaptive system, able to self-configure the sensory infrastructure, shows an uncertainty level close to that of the first base-line system, with a significant reduction in power consumption. Table 1 and Table 2 summarize the mean accuracy for all of the tests in the cross validation phase.

Table 1 Mean accuracy, for all of the tests considered in the cross, of the proposed *Adaptive System* compared with the two base-line systems considered (*All Sensors On* and *Subset of Sensors On*)

All sensors on (%)	Subset of sensors on (%)	Adaptive system (%)
78.03	69.88	75.20
82.69	75.27	76.42
78.42	32.85	71.70
70.87	47.22	70.64
56.91	33.08	63.07
52.89	32.50	59.97
56.54	36.34	62.06
54.59	32.62	63.85
69.25	34.63	72.46
78.65	41.60	71.08

Table 2 Overal mean accuracy of the proposed *Adaptive System* compared with the two base-line systems considered (*All Sensors On* and *Subset of Sensors On*)

All sensors on (%)	Subset of sensors on (%)	Adaptive system (%)
67.92	43.60	68.65

4 Conclusions

This chapter describes formal and practical details of the design and implementation of an adaptive Bayesian system for performing multi-sensor data fusion in an Ambient Intelligence scenario. The adaptivity consists of dynamic self-configuration of the underlying sensor network, with the aim of finding the best trade-off between the uncertainty of the inferred knowledge and the power consumption of sensory devices.

The proposed system has been evaluated on a synthetic dataset based on a well-known dataset for Smart Homes, available in the literature. The experimental results show a clear energy saving as compared with a static approach where all sensor nodes are always on, at the cost of a small reduction in inference accuracy. On the other hand, the capability of dynamically selecting which sensors to hold on was found to produce a clear advantage in terms of inference accuracy over a static approach in which only a fixed subset of sensor nodes are on.

Acknowledgments This work is partially supported by the PO FESR 2007/2013 grant G73F11000130004 funding the SmartBuildings project and by the PON R&C grant MI01_00091 funding the SeNSori project

References

1. De Paola, A., La Cascia, M., Lo Re, G., Morana, M., Ortolani, M.: Mimicking biological mechanisms for sensory information fusion. Biol Inspired Cogn. Archit. **3**(0), 27–38 (2013)
2. De Paola, A., Farruggia, A., Gaglio, S., Lo Re, G., Ortolani, M.: Exploiting the Human factor in a WSN-based system for ambient iIntelligence. In: International Conference on Complex, Intelligent and Software Intensive Systems, 2009 (CISIS '09), pp. 748–753 (2009)
3. De Paola, A., La Cascia, M., Lo Re, G., Morana, M., Ortolani, M.: User detection through multi-sensor fusion in an Am I scenario. In: Proceedings of the 15th International Conference on Information Fusion (FUSION 2012), pp. 2502–2509 (2012)
4. Pirttikangas, S., Tobe, Y., Thepvilojanapong, N.: Smart environments for occupancy sensing and services. In: Handbook of Ambient Intelligence and Smart Environments, pp. 825–849. Springer, Berlin (2010)
5. Bernardin, K., Ekenel, H., Stiefelhagen, R.: Multimodal identity tracking in a smart room. Personal Ubiquitous Comput. **13**(1), 25–31 (2009)
6. Li, N., Yan, B., Chen, G., Govindaswamy, P., Wang, J.: Design and implementation of a sensor-based wireless camera system for continuous monitoring in assistive environments. Personal Ubiquitous Comput. **14**(6), 499–510 (2010)
7. Cook, D.: Learning setting-generalized activity models for smart spaces. IEEE Intell. Syst. **27**(1), 32–38 (2012)
8. Lo Re, G., Milazzo, F., Ortolani, M.: A distributed bayesian approach to fault detection in sensor networks. In: Proceedings of the 2012 IEEE Global Communications Conference (GLOBE-COM), pp. 634–639 (2012)
9. Akyildiz, I., Su, W., Sankarasubramaniam, Y., Cayirci, E.: A survey on sensor networks. IEEE Commun. Mag. **40**(8), 102–114 (2002)
10. De Paola, A., Gaglio, S.: Lo Re, G., Ortolani, M.: Sensor9k: A testbed for designing and experimenting with WSN-based ambient intelligence applications. Pervasive Mobile Comput. **8**(3), 448–466 (2012)
11. De Paola, A., Gaglio, S.: Lo Re, G., Ortolani, M.: Multi-sensor fusion through adaptive bayesian networks. In: AI × IA 2011: artificial Intelligence Around Man and Beyond, vol. 6934, pp. 360–371. Lecture Notes in Computer Science. Springer, Berlin (2011)
12. Shannon, C.: A Mathematical theory of communication. SIGMOBILE Mob. Comput. Commun. Rev. **5**(1), 3–55 (2001)
13. Heinzelman, W., Chandrakasan, A., Balakrishnan, H.: Energy-efficient communication protocol for wireless microsensor networks. In: Proceedings of the 33rd Annual Hawaii International Conference on System Sciences, pp. 1–10 (2000)
14. Singh, S., Woo, M., Raghavendra, C.: Power-aware routing in mobile ad hoc networks. In: Proceedings of the 4th annual ACM/IEEE International Conference on Mobile Computing and Networking, pp. 181–190 (1998)
15. Wang, Q., Hempstead, M., Yang, W.: A Realistic power consumption model for wireless sensor network devices. In: Proceedings of the 3rd Annual IEEE Communications Society on Sensor and Ad Hoc Communications and, Networks, 1, pp. 286–295 (2006)
16. Koller, D., Friedman, N.: Probabilistic Graphical Models: principles and Techniques. The MIT Press, Cambridge (2009)
17. Verdone, R., Dardari, D., Mazzini, G., Conti, A.: Wireless Sensor and Actuator Networks: technologies, Analysis and Design. Academic Press, San Diego (2008)
18. Deb, K., Pratap, A., Agarwal, S., Meyarivan, T.: A fast and elitist multiobjective genetic algorithm: NSGA-II. IEEE Trans. Evol. Comput. **6**(2), 182–197 (2002)
19. Witten, I., Frank, E.: Data Mining: Practical Machine Learning Tools and Techniques. Morgan Kaufmann, San Francisco (2005)

A Heterogeneous Sensor and Actuator Network Architecture for Ambient Intelligence

Enrico Daidone, Orazio Farruggia and Marco Morana

Abstract One of the most important characteristics of a typical ambient intelligence scenario is the presence of a number of sensors and actuators that capture information about user preferences and activities. Such nodes, i.e., sensors and actuators, are often based on different technologies so that types of networks which are typically different coexist in a real system, for example, in a home or a building. In this chapter we present a heterogeneous sensor and actuator network architecture designed to separate network management issues from higher, intelligent layers. The effectiveness of the solution proposed here was evaluated using an experimental scenario involving the monitoring of an office environment.

1 Introduction

In recent years, numerous standards for the design of wireless and wired networks for home and building automation have been proposed. Some of the most well-known solutions are Konnex Dali, Powerline Communication, ZigBee and Insteon. Regardless of the standard considered, wired networks are not as flexible as wireless ones and usually require more infrastructural support [1, 11].

In real Ambient intelligence (AmI) systems, it is necessary to combine different networks in order to gather different types of information [7]. However, each network has its own centralized management systems, so a number of issues may arise,

E. Daidone · O. Farruggia · M. Morana (✉)
University of Palermo, Viale delle Scienze, ed 6, 90128 Palermo, Italy
e-mail: marco.morana@unipa.it

O. Farruggia
e-mail: orazio.farruggia@unipa.it

E. Daidone
e-mail: enrico.daidone@unipa.it

S. Gaglio and G. Lo Re (eds.), *Advances onto the Internet of Things*,
Advances in Intelligent Systems and Computing 260, DOI: 10.1007/978-3-319-03992-3_6,

e.g., duplication of infrastructures, difficulties in expanding the existing systems or integrating the sensor networks into elaboration systems.

Nowadays, even though some attempts have been made, (e.g., the *IEEE* has tried to standardize the lowest network level by promoting the standard IEEE1451 [10]), the adoption of a single common standard is still a utopian dream. New solutions therefore need to be developed to integrate the multitude of sensor networks present on the market.

A possible solution might be the one used by Geographic Information Systems (GIS) that standardizes the way both the networks and the interaction methods are described. Specifically, this approach defines a Service Oriented Architecture that is implemented through web services. Authors in [14] propose a method of integrating low-cost and resource-constrained heterogeneous devices and a large-scale placement of servers and wireless sensor networks. The solution proposed consists of a middleware to integrate heterogeneous sensors and devices, and an abstraction layer for AmI applications.

One important aspect in any AmI scenario is the use of context-aware technologies to characterize user preferences according to environmental conditions. In this case, context information can be gathered by distributed sensors deployed in the environment. Authors in [3, 13] underline the difficulties involved in integrating different types of devices into a single network. Gatani et al. [9] presents a data gathering approach in wireless sensor networks that takes into account the need for energy saving.

A distributed telemonitoring system is proposed in [5]. The system is based on a Service-Oriented Architecture (SOA), which allows heterogeneous wireless sensor networks to communicate in a distributed way without time and location restrictions. An SOA approach was chosen because such architectures are asynchronous and independent from the context. This system can run on multiple wireless devices independently of their micro-controller or the programming language they use. It works in a distributed way so that the application code does not have to reside almost entirely on the central node.

Wireless sensor networks are also used in robotics for coordinating the operations of multiple robots in an indoor environment. The framework presented in [2], based on a hybrid wireless network, grants robots to enhance their perceptive capabilities and to exchange information in order to achieve a global common goal.

De Paola et al. [6] describes the implementation of a testbed providing the hardware and software tools for the development and management of AmI applications based on wireless sensor and actuator networks, whose main goal is energy saving for global sustainability.

This chapter describes a heterogeneous sensor and actuator network architecture designed to separate network management issues from higher, intelligent layers.

The remainder of the chapter is organized into three sections. Section 2 gives a description of the architecture proposed here, whilst Sect. 3 outlines the experimental scenario adopted, with Sect. 4 reporting our conclusions and suggestions for future developments.

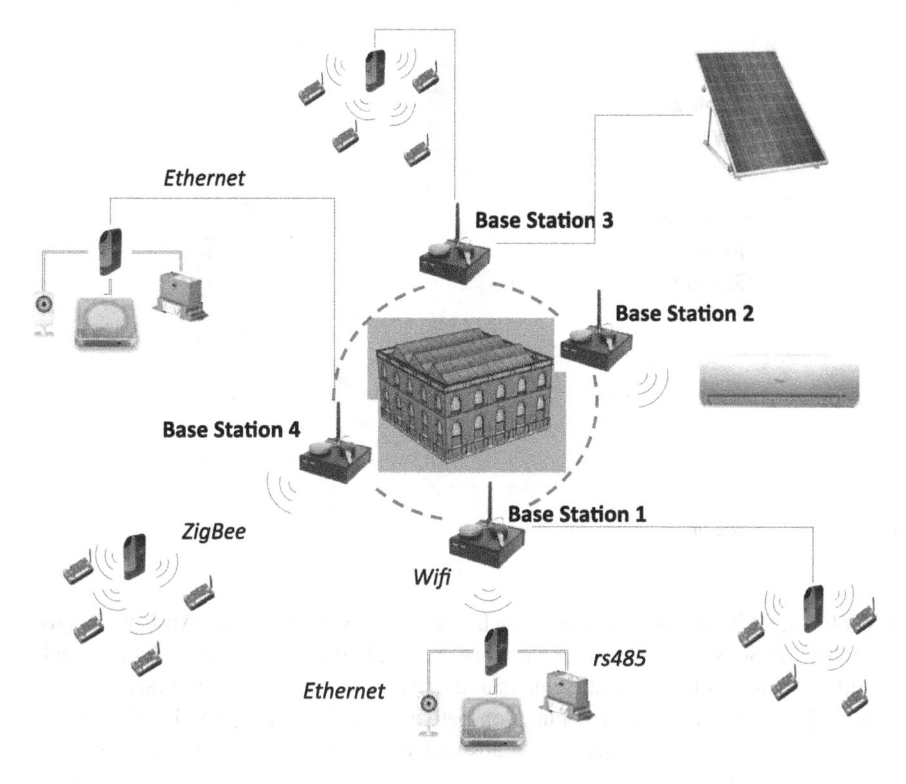

Fig. 1 Example of a deployment with four base stations and different networks

2 System Overview

AmI systems collect information from different sources, so it is usually necessary to interface them with different types of networks, each with its own characteristics. The development of an appropriate layer between the various physical networks and the AmI system thus allows network management procedures to be dealt with by higher layers.

In this section we present a novel, flexible and scalable architecture adaptable to different types of networks. Scalability is guaranteed by using a hierarchical network, where some nodes, called *base stations*, are responsible for managing one or more subnets. The number of base stations is limited only by the higher levels' capacity to manage these devices, whilst the number of subnets managed by a base station depends on the processing capacity of the base station itself. This very flexible physical organization of the network can be adapted to different goals. Figure 1 demonstrates a typical network deployment with several base stations.

The architecture proposed here (Fig. 2) consists of the four elements discussed below:

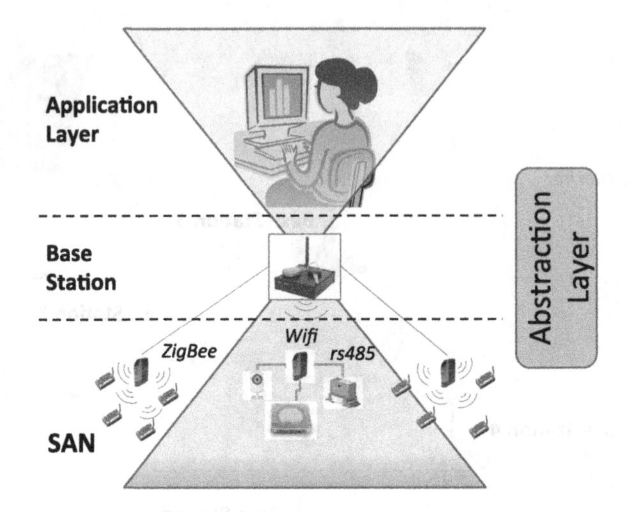

Fig. 2 The reference architecture proposed here

- **Building SAN Agent:** this allows the higher-level system, i.e., the AmI system, to control the Sensor and Actuator Network (SAN). In particular, it provides a virtual uniform representation of all sensors and actuators managed by the system.
- **Base Station:** the point of connection between the *Building SAN Agent* and the *SAN*. Each base station provides an abstract view of the physical sensors and the networks it manages.
- **Abstraction Layer:** this logic layer is distributed between the *Building SAN Agent* and *Base Station*. Specifically, it defines the models used to obtain an abstract representation of the *SAN* and the methods employed to access relevant information.
- **Sensor and Actuator Network:** the sensor and actuator networks include any possible network type, from classic home and building automation systems to the most innovative WSN.

A description of the four proposed layers is given below.

2.1 Building SAN Agent

The Building SAN Agent coordinates and manages the various base stations deployed in the system. It provides a homogeneous, independent high level view of all the SANs. The role of the Building SAN Agent is particularly important in complex buildings where different SANs are managed by different base stations.

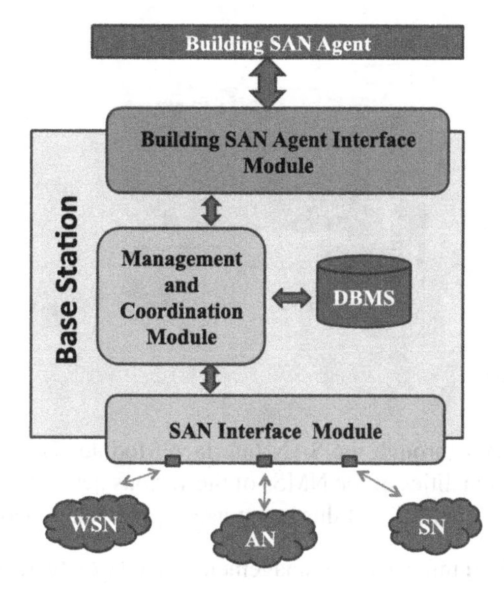

Fig. 3 The Base station architecture

2.2 *The Base Station*

The Base Station is principally responsible for managing the overall network. It communicates upwards with the Building SAN Agent and downward with the sensor and actuator networks. Since the base station serves as the hub of the sensors network, it must have better computational capabilities than conventional sensor nodes. Furthermore, the base station is equipped with a number of communication interfaces sufficient to enable it to exchange information with the various types of sensor networks to be managed.

The base station also provides various functionalities for the communication with the building SAN agent and for the management of the network. Some of the communication functionalities include registration, commands reception, reporting new hot-plug sensors, and data and control information transmission. The management of the physical network involves forwarding commands from the upper level to the SAN, collecting data from sensor networks and implementation of the centralized functionalities of the Network Management System (NMS) for each subnet.

Figure 3 provides a block diagram that illustrates the base station architecture. From a logical point of view, the base station can be divided into four macro components.

The **Building SAN Agent Interface Module** implements the interfaces to the Building SAN Agent in accordance with the specifications defined in the Abstraction Layer. It provides homogeneous access to the systems managed by the Base Station.

The **Management and Coordination Module** is responsible for managing the SAN and translating the Abstraction Layer commands in a way that allows them

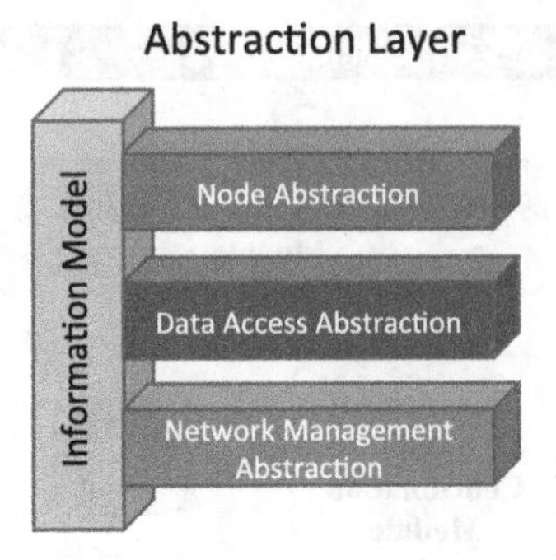

Fig. 4 Abstraction layer

to be used by the SAN through the SAN Interface Module. This module includes the centralized functionalities of the NMSs of the various networks connected to the base station. The functionalities of this component can be classified as follows:

- **Sensor registration:** information management related to actuators, sensors, nodes and networks;
- **Command dispatch:** command collection management;
- **Event Management:** a set of functions that manages the events;
- **Fault management:** a set of methods to detect, isolate, and correct network faults;
- **Data deliverable:** forwarding data from sensor networks to the higher levels;
- **Configuration management:** configuration management of the networks;
- **Energy management:** a system component to monitor the energy status of each node;
- **Topology management:** a set of functions for activation and/or deactivation of the nodes.

The main role of the **SAN Interface Module** is to interface the base station with the SAN, for this reason its specifications depend on the types of sensor networks deployed in the system. In particular, this module translates requests from the management module to the SAN and forwards information received from sensor networks to the management and coordination module. Moreover, the SAN Interface Module is able to extend the basic functionalities provided by the various sensor networks in order to implement additional features to be exported to the management and coordination module.

The **DBMS** facilitates effcient management of information related to the network and the devices deployed.

2.3 Abstraction Layer

The Abstraction Layer is a logic layer between the Building SAN Agent and the Base Station. The main goal of this component is to allow the separation of physical networks from higher level components. This objective is achieved by defining an abstract view and a uniform access interface to the different types of networks. To be specific, we used a communication paradigm called *Data Centric*, which allows application layers to be separated from issues related to network management and we defined a unique communication interface to higher levels, in order to make them independent from the peculiarities of each network. A block diagram of the Abstraction Layer is given in Fig. 4.

The *Node Abstraction* layer allows the nodes to present their characteristics and functionalities to the higher levels in order to provide a homogeneous mechanism for the discovery of the various sensors deployed. The description of the networks is based on *SensorML* (OpenGIS Sensor Model Language Encoding Standard) which provides a standard model and an XML encoding of the measurement process. In particular, some discovery models (Net-Centric, Location-Centric and Sensor-Centric) have been developed to guarantee access to the SAN in a standardised way.

The *Data Access Abstraction* layer provides an interface to collect sensor network data and transmit it to the upper levels. In particular, the following data delivery models are provided: *Continuous*, in which information is transmitted at regular time intervals; *Event-triggered*, with information being transmitted only when a particular event occurrs; and *Query-triggered*, in which information is only transmitted on request.

The *Network Management Abstraction* layer provides an interface towards the functionalities related to network management system, e.g., network configuration and node management.

The *Information Model* defines a conceptual model for representing data and describing the sensors and the commands accepted by the network. The context (sensors, actuators and physical phenomena) is modeled by means of an ontology based on the *Ontology Web Language* (OWL), so that it is always possible to maintain an updated representation of the environment.

The concepts involved in the ontology are shown in Fig. 5. In particular, the main concepts are *Device*, *Command*, *NetworkSAN*, *ProcessModel*, *Location*, *Time*, *Phonomena*, *Observation*, and *OperationMode*. *Command* can be SensorCommand, BaseStationCommand, *NodeCommand*, *NetCommand* or *ActuatorCommand* while the *Device* may be *Actuator*, *Gateway*, *ActuatorNode*, *SensorNode*, *BaseStation* or *Sensor*. The most important properties of objects modeled are shown in Fig. 6.

2.4 Sensor and Actuator Network

The Sensor and Actuator Network represents both the passive and active part of the system since it is able to collect information about the environment and users and also act on the environment itself in order to satisfy users' needs. One of the most important advantages of adopting the architecture proposed is that it is possible to integrate existing wireless and wired networks simply by defining the communication module between such networks and the base station.

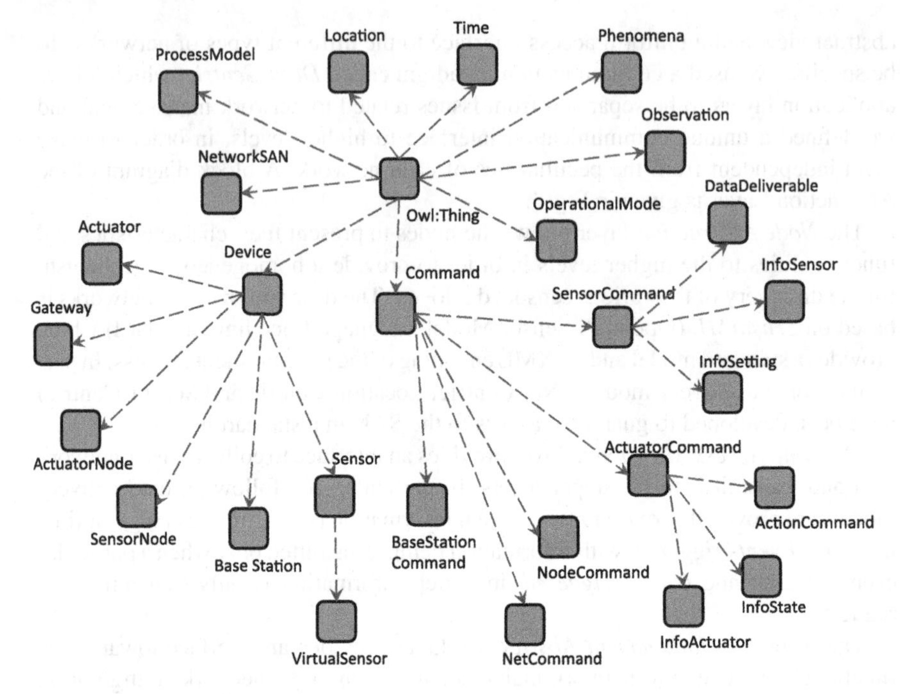

Fig. 5 Taxonomy of classes

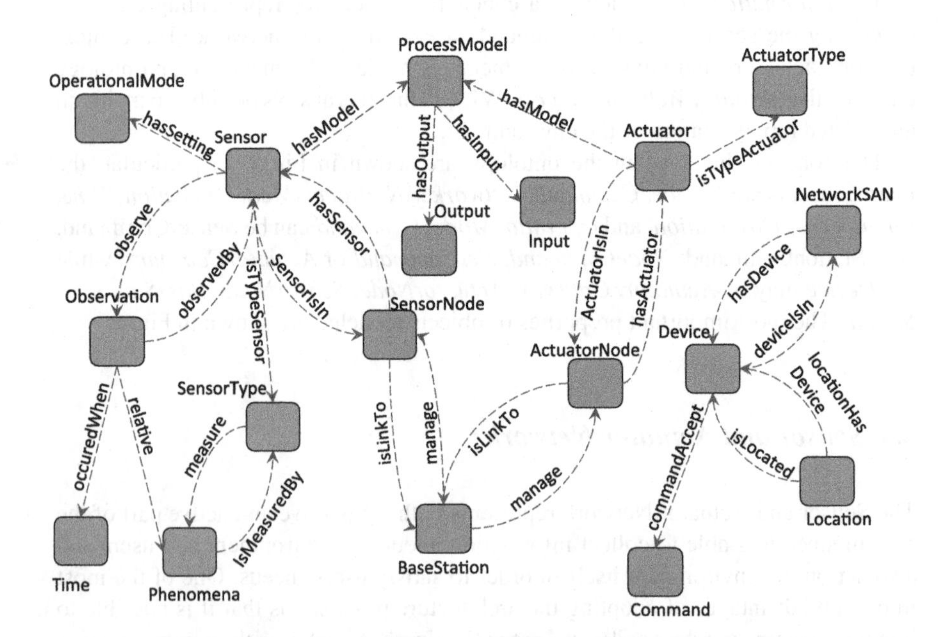

Fig. 6 Properties of the classes

3 Experimental Scenario

The architecture proposed here has been adopted for a prototypal AmI system developed at our department [8]. In particular, the experimental scenario involves monitoring an office environment by means of a number of both wireless and wired nodes. The main characteristics of the sensors and actuators deployed are summarized in Tables 1 and 2 respectively.

The wireless part of the sensor network is responsible for monitoring certain important environmental parameters, including temperature, humidity, lighting conditions, CO2, noise level and HVAC settings. As discussed in the preceding section, the monitoring system is able to perform both continuous and event sampling. We used Crossbow IRIS sensor nodes equipped with MTS300 and MTS400 crossbow sensor boards, and with two sensor boards designed ad-hoc: one to monitor the level of CO2 and one based on the IR sensor IR38DM to intercept the commands sent to the HVAC.

In order to simplify network management, it is possible to logically organize sets of nodes into *groups* so that a message can be sent directly to a *group*. Such organization makes it possible to optimize the number of messages forwarded in the network and improve the energy efficiency of the nodes. The network functionalities include setting the data rate for each physical quantity to be acquired, setting group membership of a node, setting the rate for transferring the network configuration data, querying the nodes to capture individual physical quantity, querying the nodes to ascertain their status, setting the data collection mode (on event or periodic) and activating/deactivating a node or group.

The wired network is responsible for monitoring the energy consumption and the status of given ad-hoc actuators. In the scenario under consideration, we focused on the monitoring of the energy consumption of both the office as a whole and of certain devices such as lights, HVAC and electrical sockets. The measurements are performed by means of an RS-485 digital transducer managed by a master node equipped with a programmed micro-controller. The master node also handles the motion and reads the state of the rolling shutters and the office curtains. Moreover, it also controls the relay for lighting management and door opening.

The most appropriate hardware platform for the base station was chosen while trying to achieve a good compromise between energy consumption and processing capacity. Thus we chose a miniature fanless PC, i.e., FitPC2 [4], based on an Intel Atom processor with SSD storage that guarantees a power consumption of only 8 Watts.

Finally, in order to guarantee a high degree of pervasiveness, the interactions between the user and the AmI system are managed by a gesture recognition module based on a motion sensor device [12].

Table 1 List of the sensors used in the deployed case study

Measure	Sensor	Network	Characteristics
Temperature and Relative humidity	Sensirion SHT11	WSN	Temperature range: −40 C to +123.8 C Temperature accuracy: +/−0.5 C @ 25C Humidity range: 0–100 % RH Absolute RH accuracy: +/−3.0 % RH Low power consumption
Temperature and Pressure	Intersema MS5534AM	WSN	Temperature range: −10 C to +60 C Temperature accuracy: +/−0.8 C @ 25 C Pressure range: 400–1100 mBar Pressure accuracy: +/−1, 5 % at 25 C Low power consumption
Temperature	Panasonic ERT-J1VR103J	WSN	Range: −40 C to +125 C Accuracy: +/−2 %
Light	TAOS TSL2550D	WSN	Range: 0–1847 lux Spectral responsivity: 400–1000 nm
Air conditioning sniffer	IR receiver base on chip IR38DM	WSN	Developed ad hoc
CO2	SenseAir K33LP	WSN	CO2 range: 0–5000 ppm CO2 accuracy: +/−30 ppm Low power consumption
Voltage, current, power factor, active power, reactive power, active energy, reactive energy	CE-AJ12-34BS3-1.0	Wired	Accuracy: 0.5 %
Curtain sensor Rolling shutter sensor Light on/off	–	Wired	Developed ad hoc
RFID Reader	LabID KITNLO	Wired	Supported protocols: ISO 15693 ISO 14443 A ISO 14443 B - ST SRI family
Proximity reader ISO	LabID RFID Reader RWBLUE	Wired	Supported protocols: ISO 15693 ISO 14443 A ISO 14443 B - ST SRI family

Table 2 List of the actuators used in the deployed case study

Action	Actuator	Network
Air conditioning setting	IguanaWorks IR transceiver	Wired
Curtain up/down	Curtain motor	Wired
Rolling shutter up/down	Rolling shutter motor	Wired
Light on/off	Relay	Wired
Door open	Electric lock	Wired

4 Conclusions

In this chapter we presented a new heterogeneous sensor and actuator network architecture to be used in ambient intelligence scenarios where the presence of pervasive devices with different technologies is required to ensure the appropriate control and monitoring of an area. In particular, the architecture proposed aims to make ubiquitous observations of buildings easier in order to understand the users' preferences efficiently.

The three-tier architecture proposed is based on an abstraction layer to ensure the independence of the monitoring and actuating infrastructure from the upper level system, thereby guaranteeing scalability and adaptability in different contexts.

Moreover, the construction of a real prototype of the monitoring and controlling system at the University of Palermo allowed us to constantly verify the design choices we made in order to adopt the most robust and energy efficient solutions.

Acknowledgments This work has been partially supported by the PO FESR 2007/2013 grant G73F11000130004 funding the SmartBuildings project.

References

1. Ahmed, A., Ali, J., Raza, A., Abbas, G.: Wired vs wireless deployment support for wireless sensor networks. In: TENCON 2006. 2006 IEEE Region 10 Conference, pp. 1–3 (2006). doi:10.1109/TENCON.2006.343679
2. Chella, A., Lo Re, G., Macaluso, I., Ortolani, M., Peri, D.: Multi-robot interacting through wireless sensor networks. Lect. Notes Comput. Sci. (including subseries Lecture Notes in Artificial Intelligence and Lecture Notes in Bioinformatics) **4733**, 789–796 (2007)
3. Cho, J., Shim, Y., Kwon, T., Choi, Y.: Sarif: A novel framework for integrating wireless sensor and rfid networks. IEEE Wirel. Commun. **14**(6), 50–56 (2007). doi:10.1109/MWC.2007.4407227
4. CompuLab:Fitpc2 specifications. http://fit-pc.com/web/products/fit-pc2/fit-pc2-2i-specifications/fit-pc2-specifications/
5. Corchado, J., Bajo, J., Tapia, D., Abraham, A.: Using heterogeneous wireless sensor networks in a telemonitoring system for healthcare. IEEE Trans. Inf. Technol. Biomed. **14**(2), 234–240 (2010). doi:10.1109/TITB.2009.2034369

6. De Paola, A., Gaglio, S., Lo Re, G., Ortolani, M.: Sensor9k : A testbed for designing and experimenting with wsn-based ambient intelligence applications. Pervasive Mob. Comput. **8**(3), 448–466 (2012). http://dx.doi.org/10.1016/j.pmcj.2011.02.006
7. De Paola, A., La Cascia, M., Lo Re, G., Morana, M., Ortolani, M.: User detection through multi-sensor fusion in an ami scenario. In: Information Fusion (FUSION), 2012, pp. 2502–2509 (2012)
8. De Paola, A., Lo Re, G., Morana, M., Ortolani, M.: An intelligent system for energy efficiency in a complex of buildings. In: Sustainable Internet and ICT for Sustainability (SustainIT), 2012, pp. 1–5 (2012)
9. Gatani, L., Lo Re, G., Ortolani, M.: Robust and efficient data gathering for wireless sensor networks. In: Proceedings of the Annual Hawaii International Conference on System Sciences 9, 235a (2006)
10. Lee, K.: IEEE 1451: A standard in support of smart transducer networking. In: Proceedings of the 17th IEEE Instrumentation and Measurement Technology Conference, 2000. IMTC 2000, vol. 2, pp. 525–528 (2000). doi:10.1109/IMTC.2000.848791
11. Lee, M.H., Yoe, H.: Comparative analysis and design of wired and wireless integrated networks for wireless sensor networks. In: 5th ACIS International Conference on Software Engineering Research, Management Applications, 2007. SERA 2007, pp. 518–522 (2007). doi:10.1109/SERA.2007.65
12. Lo Re, G., Morana, M., Ortolani, M.: Improving user experience via motion sensors in an ambient intelligence scenario. In: Pervasive and Embedded Computing and Communication Systems (PECCS), 2013, pp. 29–34 (2013)
13. Marin-Perianu, M., Meratnia, N., Havinga, P., de Souza, L., Muller, J., Spiess, P., Haller, S., Riedel, T., Decker, C., Stromberg, G.: Decentralized enterprise systems: a multiplatform wireless sensor network approach. IEEE Wirel. Commun. **14**(6), 57–66 (2007). doi:10.1109/MWC.2007.4407228
14. Stavropoulos, T.G., Gottis, K., Vrakas, D., Vlahavas, I.: awesome: A web service middleware for ambient intelligence. Expert Syst. Appl. **40**(11), 4380–4392 (2013). http://dx.doi.org/10.1016/j.eswa.2013.01.061. URL http://www.sciencedirect.com/science/article/pii/S0957417413000936

Short-Term Sensory Data Prediction in Ambient Intelligence Scenarios

Enrico Daidone and Fabrizio Milazzo

Abstract Predicting data is a crucial ability for resource-constrained devices like the nodes of a Wireless Sensor Network. In the context of Ambient Intelligence scenarios, in particular, short-term sensory data prediction becomes a key enabler for more difficult tasks such as prolonging network lifetime, reducing the amount of communication required and improving user-environment interaction. In this chapter we propose a software module designed for clustered wireless sensor networks, able to predict various environmental quantities, namely temperature, humidity and light. The software module is supported by an ontology that describes the topology of the AmI scenario and the effects of the actuators on the environment. We applied our module to real data gathered from a public office at our department and obtained significant results in terms of prediction error even in presence of environmental actuators.

1 Introduction

Ambient Intelligence (AmI) is an emergent field of AI aimed at developing smart, distributed pervasive systems able to support human-environment interaction [8]. The basic infrastructure of an AmI system is made up of sensors, actuators and reasoners [2, 4, 7, 9]. The sensory components monitor the environment by measuring physical phenomena like temperature, humidity and light, but also by acquiring digital images and sounds, detecting user presence and so on. The actuators are those elements able to affect the environment according to the users' needs. The reasoners are able to learn, recognize and infer users' needs as well as to predict environmental phenomena.

E. Daidone · F. Milazzo (✉)
University of Palermo, DICGIM-Viale delle Scienze, Edificio 6, 90128 Palermo, Italy
e-mail: fabrizio.milazzo@unipa.it

S. Gaglio and G. Lo Re (eds.), *Advances onto the Internet of Things*,
Advances in Intelligent Systems and Computing 260, DOI: 10.1007/978-3-319-03992-3_7,

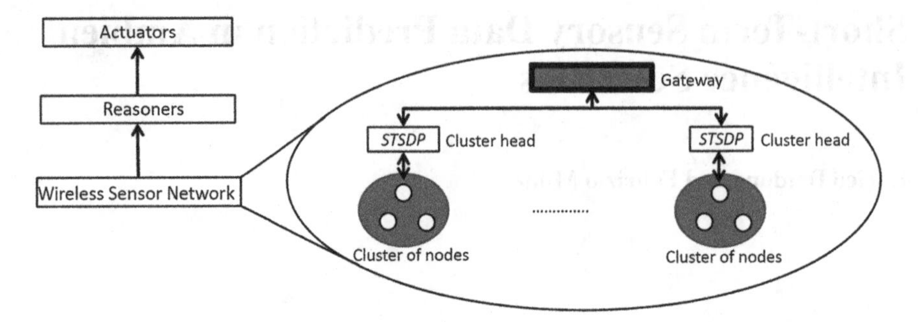

Fig. 1 Reference architecture of the AmI system and placement of STSDP software module

Over the last few years, AmI designers have begun to implement sensory infrastructure by using the so-called Wireless Sensor Network (WSN) technology, that is a network made up of sensor nodes able to sense physical phenomena and to perform small on-board computations. The growing use of such technology has been triggered by the presence of consolidated communication protocols like IEEE 802.15.4 and Wireless HART and a lightweight operating system like TinyOS. Nevertheless, wireless sensor nodes are limited by their scarce computational resources, storage and energy [1]. For these reasons, predicting data becomes a key enabler in improving WSN performance as it makes it possible to reduce communications and to prolong network lifetime. Moreover, in the context of Ambient Intelligence, predictions could be also exploited by reasoners to control actuators in order to satisfy user needs.

The main contribution of this chapter lies in the implementation of *Short-Term Sensory Data Prediction* (STSDP), i.e. a software module able to predict the physical phenomena monitored by a Wireless Sensor Network, with the combined effect of actuators.

Figure 1 shows our chosen reference AmI architecture and the placement of the STSDP module: we assume that the WSN is arranged as a set of clusters and that each cluster-head runs STSDP; the computational burden of the sensor nodes is kept as low as possible as they are only required to communicate the sensed readings to the cluster-head. The cluster heads relay their predictions to the gateway that is responsible for aggregating and communicating them to the upper layer. The intelligent modules (reasoners) are responsible for performing complicated tasks such as controlling the actuators and computing long-term data predictions. The context (sensors, actuators and physical phenomena) is modeled using *Ontology Web Language* (OWL), which allows STSDP to build an updated representation of the environment.

The remainder of this chapter is organised as follows: Sect. 2 depicts the current state of the art methods for predicting data in wireless sensor networks; Sect. 3 provides mathematical details of STSDP as well as the description of the OWL ontology that models the environmental context. Section 4 describes the experiments we carried out to validate our software module, and finally Sect. 5 presents our conclusions.

2 Related Works

Predicting data is a widely studied topic in wireless sensor networks as it makes it possible to prolong network lifetime by aggregating and compressing data, lowering network communications and reducing the amount of storage required.

Any prediction algorithm for wireless sensor networks is specified from three different perspectives: *scope*, *topology* and *methodology*. The temporal and spatial scope of a prediction algorithm are respectively short (minutes or hours/meters) or long (days or months/kilometers). Broadly speaking, short-term prediction algorithms fit the requirements of a WSN rather well, as they require a little knowledge of the past and few computational resources. Moreover their degree of precision within the temporal and spatial scopes designed is very high. Long-term prediction algorithms are very complex if compared to the short-term ones and are usually performed by devices with more computational resources (as an example the reasoners of the reference AmI architecture chosen here may be good candidates for performing such complex computations). The precision of such algorithms is quite constant over time, although it is less than that of short-term prediction algorithms designed only for limited scopes.

The topology perspective defines "who" is responsible for carrying out predictions. *Centralized* approaches assume that a central base station gathers readings from the surrounding nodes and then builds a global model of the monitoring field [3, 17], whereas *distributed* approaches [16] focus on local data processing and are very precise as compared to the former methods. Their main drawback is their elavated computational complexity, which makes them unsuitable for resource constrained devices like wireless sensor nodes.

The methodological perspective defines how to predict data; the *stochastic* approach models each physical phenomenon as a random process based on a set of observable and unobservable parameters. Such parameters are associated with previously learned prior probability distribution functions (PDF) and the predictions are drawn from the posterior PDF conditioned on the observed variables [11, 12, 15]. The *deterministic* approach assumes some kind of mathematical law that links past readings to future ones. Common implementations are "Time Series" [13, 16] and "Regression models" [10]. Stochastic approaches are suitable for all those applications that require long-term predictions, but their prediction error is very high as compared to deterministic approaches. On the other hand, deterministic approaches require a limited amount of storage and computation and produce smaller errors when used as short-term predictors.

We chose to implement STSDP using a centralized approach to minimize the computational burden on sensor nodes and to prolong their lifetime as much as possible. The decision to use a deterministic implementation was straightforward, as our module was specifically designed to perform short-term data prediction.

3 Proposed Approach

This Section discusses our implementation of the STSDP software module together with the OWL ontology adopted to describe the environmental context.

Let us assume that the WSN is arranged as a set of clusters and that the nodes behave as *cluster head* or *leaf*. Leaf nodes are characterized by their scarce computational resources and are responsible for gathering measurements and relaying them to the cluster head. Cluster heads are not limited by energy or computational resources and act as the so called *micro-servers* (e.g. Stargate nodes). They are responsible for building the spatio-temporal representation and prediction of the monitored phenomena.

Each cluster is associated with an *area of interest*—a spatial portion of the monitoring field—bounded by the convex-hull defined by its own sensor nodes. Moreover, we assume that the areas of interest do not overlap with each other.

Figure 2 shows the steps followed by STSDP to predict physical phenomena. The *Context Generation* submodule reads the Ontology and creates the description of the context: sensor nodes (e.g. sampling rate, position, status), actuators (e.g. affected phenomenon, position) and the phenomena (e.g. light, temperature or humidity). The *Prior estimates* submodule creates a rough representation of the phenomenon using the readings sensed by each node during the previous 24 hours; the current readings add fresh information and allow to build the more precise *Posterior estimators*; the *Fusion* step mixes the Posterior estimators to generate a spatio-temporal mesh of the area of interest from which predictions are collected. In order to make the approach suitable for ambient intelligence scenarios we added the *Actuators correction* module that integrates the effect of the actuators on the monitored phenomena. Finally, the *Context Update* module updates the ontology to keep track of possible changes in the context (e.g. dead nodes, active/inactive actuators).

The mathematical details of STSDP as well as the Ontology implemented will be described by referring to a single cluster of nodes. The generalization to more clusters is straightforward.

3.1 Context Generation and Update Modules

Ontologies are useful tools which enable us to describe a domain of interest (concepts and relationships between them) using a formal language (in our case we chose to adopt the Ontology Web Language, OWL).

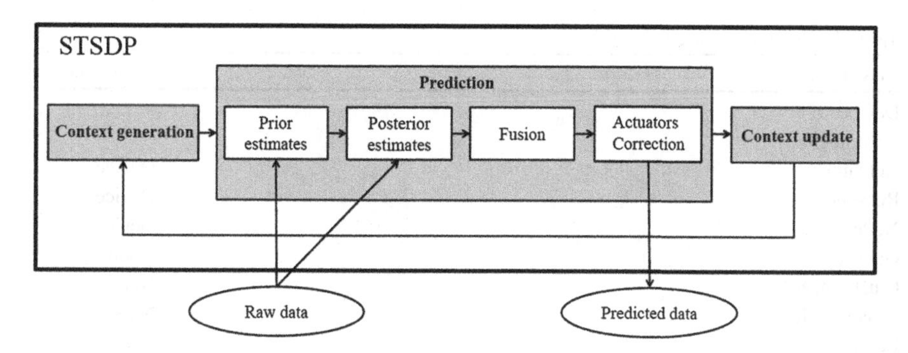

Fig. 2 Sequence diagram of the STSDP module

Classes represent categories of concepts classified by using "isa" hierarchies, whilst *properties* represent relationships among classes and are identified by *domain* and *range* class.

The Simple Web Rule Language (SWRL) implements logical inference and updates the OWL ontology as a function of changes in the environment. Each rule is written in the form $antecedent \rightarrow consequent$ where both *antecedent* and *consequent* are conjunctions of one or more atoms, where each atom is a property or a class.

The context generation module reads the content of the ontology and gathers all the information needed by the prediction module to carry out predictions (e.g. the number of nodes of the cluster, their sampling rate, the active actuators and so on). The context update module modifies the ontology as a function of any environmental changes. For example a node with a battery level under a given threshold should be excluded by the cluster head as its readings could be incorrect.

Table 1 represents in details the "isa" hierarchies we devised for the classes of interest. The root classes are the *Device* and the *Phenomenon*. Any device could be an *Actuator*, a *Reasoner* or a *Node*. Each *Node* in its turn is a *Gateway*, a *Cluster Head* or a *Sensor Node*. Finally, each *Phenomenon* could be *Light*, *Humidity* or *Temperature*.

Table 2 represents the ontology properties. Whenever the range class is a raw type (e.g. String, Float, Integer, Boolean and so on) it is conventionally named as a *data* property, i.e. an internal attribute of the class. Finally, a property is said to be *functional* when the mapping between domain and range is injective. For instance the "hasID" property is functional because each device has just one identifier; the "manages" property is not functional as a device could manage many devices (e.g. each cluster head manages many sensor nodes).

The context update submodule implements three SWRL rules that infer: (i) the observability of a phenomena, (ii) the working status of sensor nodes, and (iii) the manager device of each node (not codified at design time).

The formal representation of the three rules described is as follows:

Table 1 Ontology classes

Class name	Parent class
Device	–
Phenomenon	–
Actuator	Device
Reasoner	Device
Node	Device
Gateway	Node
Cluster head	Node
Sensor node	Node
Light	Phenomenon
Humidity	Phenomenon
Temperature	Phenomenon

Table 2 Ontology properties

Property Name	Domain	Range	Data	Funct.	Description
hasID	Device	Integer	✓	✓	The unique identifier
isAtX	Device		✓	✓	The X-position
isAtY	Device	Float	✓	✓	The Y-position
manages	Device	Device	✗	✗	Controller and controlled device
isManagedBy	Device	Device	✗	✓	Inverse of the manages property
hasSamplingPeriod	Sensor node	Float	✓	✓	Sampling period
hasBatteryLevel	Sensor node	Float	✓	✓	Current battery level
hasMinBatteryLevel	Sensor node	Float	✓	✓	Minimum battery level for considering a node as "active"
hasStatus	Device	Boolean	✓	✓	Working status (active or inactive)
senses	Sensor node	Phenomenon	✗	✗	The phenomena sensed by a sensor node
affects	Actuator	Phenomenon	✗	✗	The phenomenon affected by the given actuator
isObservable	Phenomenon	Boolean	✓	✓	Observability of a phenomenon

Rule 1:
Phenomenon(?x) $\wedge$ senses(?y, ?x) $\wedge$ hasStatus(?y,True) $\rightarrow$ isObservable(?x,True)

Rule 2:
Node(?y) $\wedge$ hasBatteryLevel(?y,?t) $\wedge$ hasMinBatteryLevel(?y,?z) $\wedge$ lessOrEqual (?t,?z) $\rightarrow$

hasWorkingStatus(?y,False)

Rule 3:
manages(?x,?y) → isManagedBy(?y,?x)

3.2 Prediction Submodule

Let us assume that each sensor node i gathers readings with sampling rate Δt and that $r_i(t)$ is the reading at time t. Moreover the node location is (x_i, y_i).

The prior estimator $f_{x_i,y_i}^{prior}(t)$ roughly assumes that the current representation of the monitored phenomena is identical to that of the previous day; the past 24 h of readings are fitted using a Gaussian Mixture as follows:

$$f_{x_i,y_i}^{prior}(t) = \sum_{k=1}^{K} w_k N(t|\mu_k, \sigma_k^2) \tag{1}$$

where the parameters μ_k, σ_k^2 and w_k represent the mean, the variance and the importance weight of the k-th Gaussian. Such parameters minimize the square error between the fitted curve and the sensed readings and are recomputed with 24-h time steps using the Nelder-Mead optimization algorithm [14] as follows:

$$(\mu_1, \sigma_1, w_1, ..., \mu_K, \sigma_K, w_k) = \underset{\substack{\mu_k, \sigma_k, w_k \\ k \in [1..K]}}{argmin} \sum_t [r_i(t) - f_{x_i,y_i}^{prior}(t)]^2. \tag{2}$$

The prior estimator enables STSDP to learn the *shape* of the monitored phenomenon and to constrain the trend of the posterior estimator. Its intrinsic limitations are related to poor performance as a short-term predictor and to the covered space that is the point location x_i, y_i.

The posterior estimator integrates the trend of the current readings to the information provided by the above step. Let us consider the time window pinpointed by the last W sensed readings $r(t)$, $t \in [t_0 - (W-1)\Delta t, t_0]$; then the posterior estimator is computed as a geometrical transformation of the prior one as follows:

$$f_{x_i,y_i}^{post}(t) = \beta[f_{x_i,y_i}^{prior}(t) - \gamma] + \gamma, \tag{3}$$

where $\beta \in [\beta_l, \beta_u]$ and $\gamma \in [0, \gamma_u]$ represent respectively the scaling and translation parameters; the upper and lower bounds should be set by a field expert and limit the range of possible transformation of the prior estimator.

The geometrical parameters are computed for non-overlapping time windows of size W using the same approach as Eq. 2:

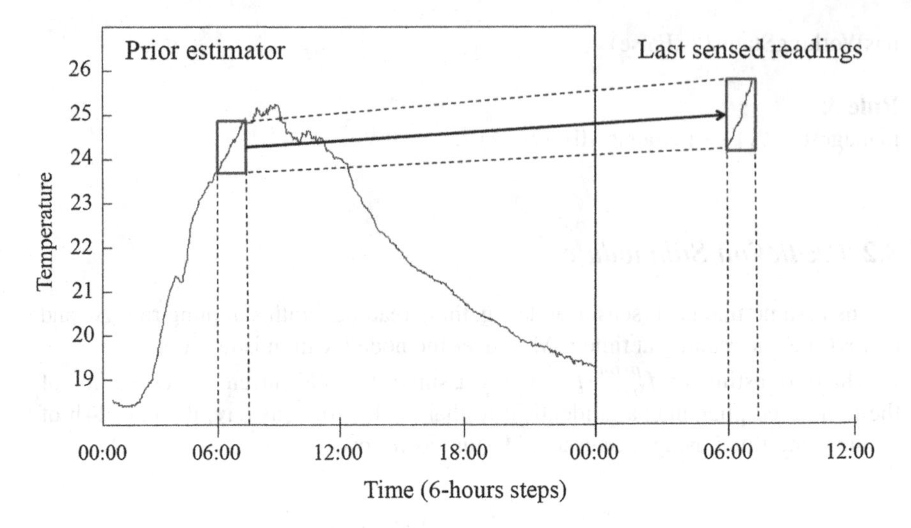

Fig. 3 Geometrical transformation of a prior estimator into a posterior one

$$(\beta, \gamma) = \underset{\beta, \gamma}{argmin} \sum_{t=t_0-(W-1)\Delta t}^{t_0} \{\beta[f_{x_i, y_i}^{prior}(t) - \gamma] + \gamma - r_i(t)\}^2 \quad (4)$$

Figure 3 shows how the computation works: the last W sensed readings identify the current time window (on the right); the same time instants also identify the portion of the prior estimator that is used to fit the current readings (on the left). Then, the geometrical transformation tries to compute the best match between prior estimator and current readings.

The posterior estimators perform better than prior ones as they integrate the knowledge from the previous day's readings and the current behavior of the phenomenon; however they are still limited by punctual spatial coverage.

The fusion step extends the spatial coverage of the posterior predictors to the entire area of interest. The continuous function $f^{fuse}(x, y, t)$ is computed as a normalized linear interpolation of the posterior estimators as follows:

$$f^{fuse}(x, y, t) = \frac{\sum_{i=1}^{I} w_i(x, y) f_{x_i, y_i}^{post}(t)}{\sum_{i=1}^{I} w_i(x, y)}, \quad (5)$$

where $w_i(x, y) = e^{-[(x-x_i)^2+(y-y_i)^2]}$ and I is the number of nodes within the cluster. The function covers the convex-hull of the cluster.

3.3 Modeling the Effect of Actuators

Environmental actuators are an important component of AmI applications that manipulate physical phenomena according to user needs. A crucial aspect of any predictor is its capability of integrating the effect of the actuators on the environment to draw more precise predictions.

The current implementation of the actuators correction submodule considers the effect of rolling shutters and neon lights on indoor environments. We are working to include the support for actuators affecting temperature and humidity such as air conditioners and radiators.

We assessed experimentally that the rolling shutter gives a multiplicative contribution to f^{fuse} and was modeled as a compression function $R : [0, 1] \rightarrow [0, 1]$ that accepts as input the value of $h \in [0, 1]$, i.e. the portion of the window that is not covered by the rolling shutter(0 means totally closed and 1 means totally open), and gives as output $c \in [0, 1]$ the compression factor. We did not include dependence on the space position in $R(h)$ as experimental results (Sect. 4) have shown that such knowledge only makes a negligible contribution to reduce prediction error.

The effect of neon lights is additive, location-dependent and was modeled as a function $N(x, y) : R^2 \rightarrow [0, \infty]$ that accepts as input the 2-D point of the area of interest and gives as output the increment in light exposure. The mathematical expression is as follows:

$$N(x, y) = \frac{\sum_{i=1}^{I}\sum_{l=1}^{L} w_i(x, y) s_l N_l(x_i, y_i)}{\sum_{i=1}^{I}\sum_{l=1}^{L} s_l w_i(x, y)}, \tag{6}$$

where l is the subscript that identifies the neon light, L is the number of neon lights, $w_i(x, y) = e^{-[(x-x_i)^2+(y-y_i)^2]}$ is the importance weight, s_l indicates whether the light is turned off/on, and $N_l(x_i, y_i)$ is the punctual light increment at the node location (x_i, y_i) due to the l-th neon light.

The output of the actuator correction submodule $f^{light}(x, y, t)$ is therefore computed accordingly to the previous considerations:

$$f^{light}(x, y, t) = R(h) \times f^{fuse}(x, y, t) + N(x, y). \tag{7}$$

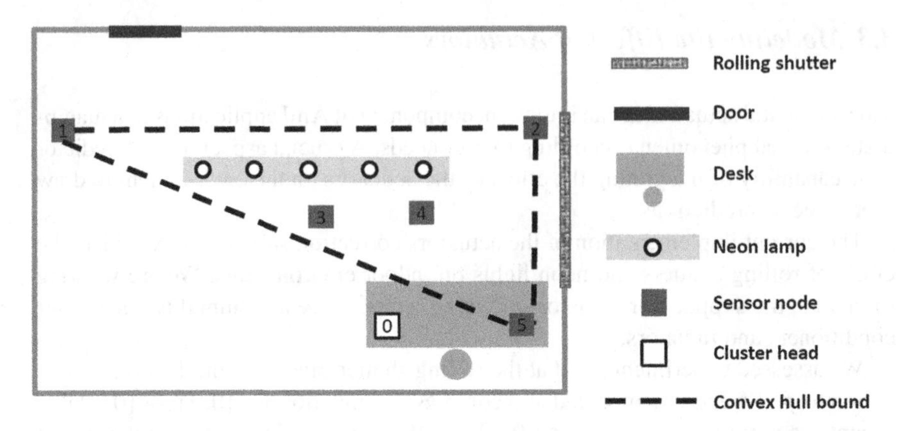

Fig. 4 The Ambient Intelligence scenario we used for evaluating the performance of STSDP

4 Experimental Results

The aim of this Section is to evaluate the performance of the STSDP module with respect to the spatial and temporal precision of the predicted data.

Figure 4 shows the Ambient Intelligence scenario we adopted to test the performance of STSDP. We deployed a single cluster within a 6 m × 5 m office at our department: the WSN was made up of five Mica2Dot sensor nodes equipped with light, temperature and humidity sensors while the cluster head was a FitPC2i (mini-computer). The monitoring field was bounded by the convex hull of the sensor nodes belonging to the cluster and is marked by a dashed line. Sensor nodes gathered measurements from 07-28-2013 to 08-02-2013 with a sampling rate of 30 s. The first day of readings was used to learn the prior models of the sensor nodes, while from 07-29-2013 to 07-31-2013 we assessed the performances of the predictor without the effect of the actuators. The days 08-01-2013 and 08-02-2013 were used to assess the performance of STSDP under the effect of the light exposure actuators. We set the number of mixing components of the prior estimators to $K = 5$.

4.1 Prediction Performance

In order to evaluate the ability of STSDP to predict data over space, we computed the function f^{fuse} by relying only on the measurements gathered from nodes 1, 2 and 5. We then compared the readings from nodes number 3 and 4 with the output provided by f^{fuse} and computed the mean and standard deviation of the prediction error.

Light, temperature and humidity were measured respectively in Lux, Celsius degrees and Percentage. Both the mean and standard deviation were normalized

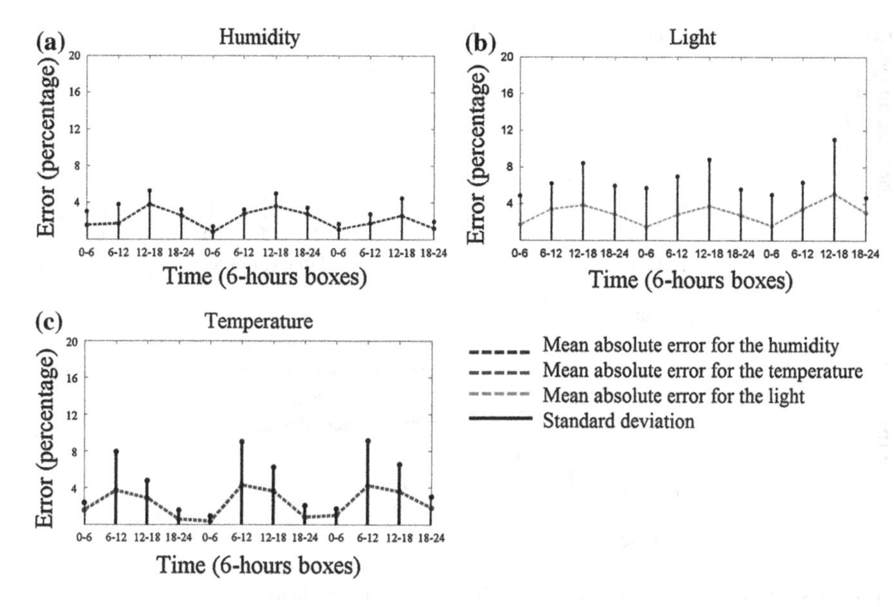

Fig. 5 Assessment of the performance of STSDP as spatial predictor

with respect to the minimum and maximum value of the observed phenomenon: the light ranged from 0 to 1600 Lux, the temperature from 17 to 35 °C, and humidity from 0 to 100 %. Figure 5 shows the results obtained: the x-axis ranges over 3 days and each step aggregates the errors of 6 h (military time). The y-axis contains the percentage mean absolute error and the standard deviation of the error. Experimental results show that the mean and standard deviation are very low and have peaks of about 4 % for all of the phenomena observed.

The temporal prediction performance were assessed by comparing the sensed readings and the value of the posterior predictor f^{post} for each sensor node with different sampling rates (from 0.5 to 60 mins).

Figure 6 shows the mean and standard deviation of the error: both the indicators are above 4 % for all the observed phenomena when the sampling rates range from 0.5 to 5 mins; the performance values are still acceptable for a sampling rate of 60 mins. In particular temperature/humidity and light have a mean error of about 8 and 12 % respectively, meaning that, in general, light exposure is less predictable and has a greater variance than humidity or temperature, a conclusion we had already reached in previous works [5, 6].

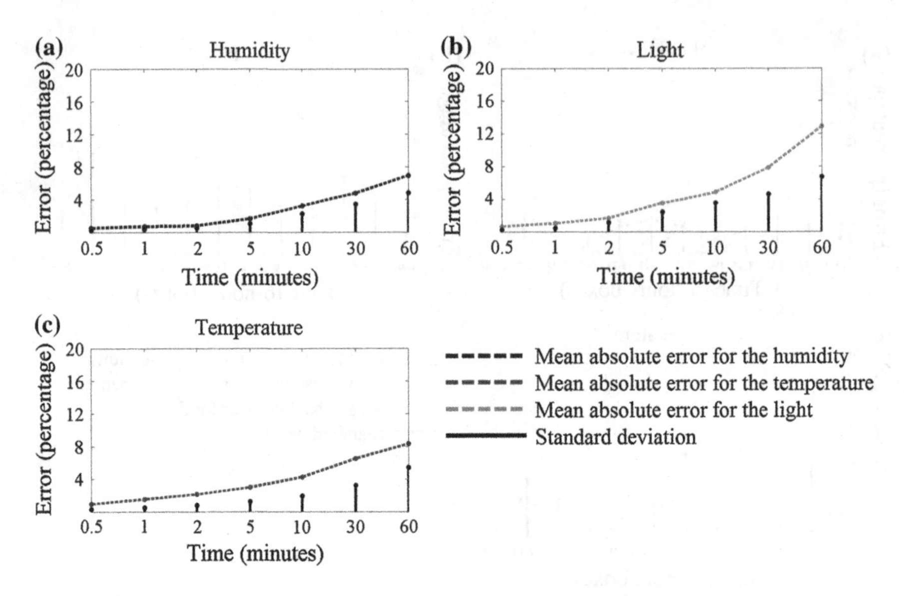

Fig. 6 Assessment of the performances of STSDP as temporal predictor

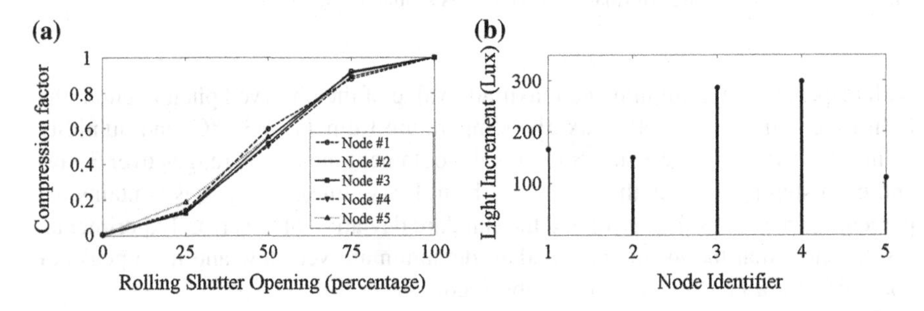

Fig. 7 Effects of the light exposure actuators on the sensor nodes

4.2 Effect of Light Exposure Actuators

The performance of STSDP under the effect of light exposure actuators (neon and rolling shutter) was evaluated during the day 08-02-2013. The functions $R(h)$ and $N(x, y)$ were learned using the readings gathered the previous day. At fixed steps of 1 h, we opened the rolling shutter and positioned it at five different locations (from 0 to 100 % opened) and let the nodes record the differences in the sensed light. We also carried out the same procedure for the neon light which was turned on and off. The results obtained were averaged for each node i over the 24 recorded values and the resulting curves $R_i(h)$ and $N(x, y)$ are reported in Fig. 7.

The compression functions are very similar to each other, so we excluded spatial dependence and computed $R(h)$ as the mean of the learned curves. The additive

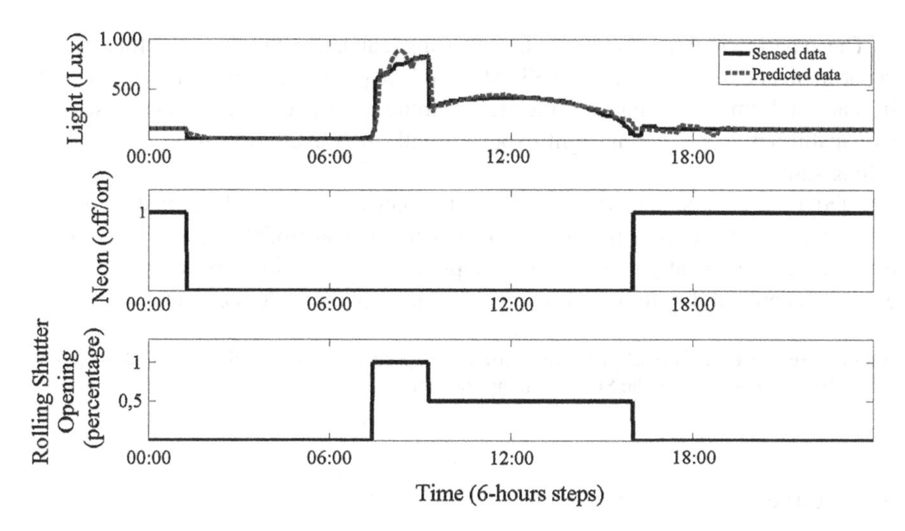

Fig. 8 Prediction of light exposure under the effect of a rolling shutter and a neon light

terms $N(x_i, y_i)$ are location dependent and based on the distance between the sensor node and the neon light; sensor 3 and 4 present two peaks as they are closer to the neon than the other nodes.

Figure 8 shows the performance of the actuators correction module for sensor node 5.

The rolling shutter was kept opened from 07:00 a.m. to 16:30 a.m. while the neon light was turned on for the remaining hours. The sensed and the predicted data are represented by the solid and dashed lines respectively.

The performance of the light exposure predictor appeared to be very encouraging and showed small errors even in correspondence of the transitions caused by actuators. At 08:20 am the predictions became unreliable, but recovered after about 20 mins. The problem was caused by a sequence of suboptimal solutions provided by the optimization algorithm that computes the geometrical transformation parameters (see Eq. 4).

5 Conclusions

This work proposes the implementation of Short-Term Sensory Data Prediction (STSDP), a software module for Ambient Intelligence scenarios. The module was able to predict common physical phenomena like temperature, humidity and light exposure even with the effect of environmental actuators.

The OWL ontology made it possible to describe the environmental context and the relationships among the components of the AmI reference architecture whilst keeping information about the state of the sensor network and actuators updated.

The experimental results were achieved using real data gathered in an office at our department and demonstrated that STSDP is able to provide reliable predictions both in space and time with ranges of meters and minutes respectively. We also assessed its capabilities in predicting light exposure with the effects of a neon light and a rolling shutter.

STSDP was implemented as a set of interconnected sub-modules that could be independently improved using more refined mathematical models. As a further development we are currently integrating the support for air conditioners and radiators to extend its applicability to more complex Ambient Intelligence scenarios.

Acknowledgments This work has been partially supported by the PO FESR 2007/2013 grant G73F11000130004 funding the SmartBuildings project.

References

1. Akyildiz, I., Su, W., Sankarasubramaniam, Y., Cayirci, E.: A survey on sensor networks. IEEE Commun. Mag. **40**(8), 102–114 (2002)
2. Augusto, J.C., Nugent, C.D.: The use of temporal reasoning and management of complex events in smart homes. In: ECAI, vol. 16, p. 778 (2004)
3. Chong, S.K., Gaber, M.M., Krishnaswamy, S., Loke, S.W.: Energy-aware data processing techniques for wireless sensor networks: a review. In: Transactions on Large-Scale Data-and Knowledge-Centered Systems III, pp. 117–137. Springer (2011)
4. De Paola, A., Gaglio, S.: Lo Re, G., Ortolani, M.: Sensor9k: A testbed for designing and experimenting with WSN-based ambient intelligence applications. Pervasive Mob. Comput. **8**(3), 448–466 (2012)
5. De Paola, A., Lo Re, G., Milazzo, F., Ortolani, M.: Adaptable data models for scalable ambient intelligence scenarios. In: International Conference on Information Networking (ICOIN), pp. 80–85 (2011)
6. De Paola, A., Lo Re, G., Milazzo, F., Ortolani, M.: Predictive models for energy saving in wireless sensor networks. In: IEEE International Symposium on a World of Wireless, Mobile and Multimedia Networks (WoWMoM), pp. 1–6 (2011)
7. De Paola, A., Lo Re, G., Morana, M., Ortolani, M.: An intelligent system for energy efficiency in a complex of buildings. In: Sustainable Internet and ICT for Sustainability (SustainIT), pp. 1–5 (2012)
8. Ducatel, K., Bogdanowicz, M., Scapolo, F., Leijten, J., Burgelman, J.C.: Scenarios for ambient intelligence in 2010. Office for official publications of the European Communities (2001)
9. Gaggioli, A.: Optimal experience in ambient intelligence. Ambient intelligence pp. 35–43 (2005)
10. Guestrin, C., Bodik, P., Thibaux, R., Paskin, M., Madden, S.: Distributed regression: an efficient framework for modeling sensor network data. In: IEEE Third International Symposium on Information Processing in Sensor Networks (IPSN), pp. 1–10 (2004)
11. Jain, A., Chang, E.Y., Wang, Y.F.: Adaptive stream resource management using kalman filters. In: Proceedings of the 2004 ACM SIGMOD International Conference on Management of Data, pp. 11–22 (2004)
12. Kanagal, B., Deshpande, A.: Online filtering, smoothing and probabilistic modeling of streaming data. In: IEEE 24th International Conference on Data Engineering, pp. 1160–1169 (2008)
13. Le Borgne, Y.A., Santini, S., Bontempi, G.: Adaptive model selection for time series prediction in wireless sensor networks. Signal Process. **87**(12), 3010–3020 (2007)

14. Nelder, J.A., Mead, R.: A simplex method for function minimization. Comput. J. **7**(4), 308–313 (1965)
15. Oldewurtel, F., Parisio, A., Jones, C.N., Gyalistras, D., Gwerder, M., Stauch, V., Lehmann, B., Morari, M.: Use of model predictive control and weather forecasts for energy efficient building climate control. Energy Buildings **45**, 15–27 (2012)
16. Tulone, D., Madden, S.: Paq: Time series forecasting for approximate query answering in sensor networks. In: Wireless Sensor Networks, pp. 21–37. Springer (2006)
17. Ye, F., Zhong, G., Cheng, J., Lu, S., Zhang, L.: Peas: A robust energy conserving protocol for long-lived sensor networks. In: IEEE Proceedings of the 23rd International Conference on Distributed Computing Systems, pp. 28–37 (2003)

A Structural Approach to Infer Recurrent Relations in Data

Pietro Cottone, Salvatore Gaglio and Marco Ortolani

Abstract Extracting knowledge from a great amount of collected data has been a key problem in Artificial Intelligence during the last decades. In this context, the word "knowledge" refers to the non trivial new relations not easily deducible from the observation of the data. Several approaches have been used to accomplish this task, ranging from statistical to structural methods, often heavily dependent on the particular problem of interest. In this work we propose a system for knowledge extraction that exploits the power of an ontology approach. Ontology is used to describe, organise and discover new knowledge. To show the effectiveness of our system in extracting and generalising the knowledge embedded in data, we have built a system able to pick up some strategies in the solution of complex puzzle game.

1 Introduction

During the last decades the ever-decreasing cost of wireless sensors and actuators has allowed an increasing diffusion of pervasive networks to monitor and control every kind of environment. In this context a new paradigm was conceived, namely the *Internet of Things* (*IoT*). The aim of this paradigm is to allow to a large set of different appliances in the environment to interact with each other and cooperate to get common goals [1], through an Internet-like structure and a unique addressing scheme. The availability of such technologies has pushed for the creation of better Decision Support Systems (DSS), able to take advantage of the richness present in

P. Cottone · S. Gaglio · M. Ortolani (✉)
DICGIM, Viale delle Scienze, University of Palermo, Edificio 6, 90128 Palermo, Italy
e-mail: marco.ortolani@unipa.it

S. Gaglio
e-mail: salvatore.gaglio@unipa.it

P. Cottone
e-mail: pietro.cottone@unipa.it

S. Gaglio and G. Lo Re (eds.), *Advances onto the Internet of Things*,
Advances in Intelligent Systems and Computing 260, DOI: 10.1007/978-3-319-03992-3_8,

data. *Ambient Intelligence* (*AmI*) is an emerging framework in this scenario, whose aim is to make the environment aware of the user presence and thus supporting them to perform every-day activities. Meeting the goals of *AmI* means understanding what is happening in the monitored area, in order to plan a set of actions on the actuators so as to modify the environment conditions according to user's desires. An *AmI* system, therefore, must deal with high level concepts, expressed through simple sensor readings. It analyses a great amount of rough data (i.e, sensor measures), coming from the environment, and summarises them in a high-level representation, through some concepts and their relations. In other words, the *AmI* system must be able to extract knowledge from a great amount of sensory data, giving an explanation of data itself. According to some proposals, an *AmI* system acts like an agent [2, 3], but an agent needs a model of the environment to operate; therefore, the designer of an *AmI* system has to embed some a-priori knowledge into the system, in order to code this model. Obviously, this task can take advantage of the analysis of such a great amount of sensory data, but comprehension of the data and the following translation into usable knowledge is not an easy task. In fact, "measuring" does not directly translate into "understanding", so sensory data provided from pervasive networks can not be easily turned in a corresponding new knowledge. Moreover, extracted knowledge would be easily generalizable: similar problems have similar solutions, so knowledge can be summarily defined as the common structure shared among similar problem solution, to construct a general model of the environment. Thus, constructing a new model for every instance of similar problems may be redundant. Nowadays, this kind of knowledge is only saved in the experience of the designer and there is no automatic system to extract it or to aid the designer in this task. However, new challenges coming from *IoT* or *AmI* call for the creation of system able to learn from experience, that is capable of capturing the hidden structure of the data, in terms of relations between its key components.

All this problems are related to knowledge *representation*, *management* and *reuse*, i.e. they are an *ontology* problem [4]. The ontology notion comes from philosophy, where it refers to the metaphysical study of the nature of being and existence. In computer science, and more specifically in the field of knowledge engineering, ontologies are used for modelling concepts and relationships on some expertise domain. Thus, building an ontology of the most relevant entities of a scenario in a semi-automatic fashion is a key problem in emerging technologies and a cutting-edge challenge of Artificial Intelligence. It involves different research areas (e.g., data mining, planning, etc), but the most interesting formulation, according to our vision, is the one arising from machine learning. In fact, this problem can be formulated as the creation of a system able to construct and recognise likely explanation of a great amount of data, unveiling their hidden structure. All the approaches used nowadays (statistical, syntactic, logic, etc) are showing their limits and inappropriateness. In fact, it is very difficult (and maybe impossible) to use only one of these approaches to manage very high level concept, to model the living world and make sense of it. Moreover, the development of domain ontology has been a task entirely based on human intervention. But new applications in *IoT* require the management of such a large number of concepts that is impossible to be performed by a human alone [5].

So, the availability of semi-automated (or, less likely, full-automated) ontology systems for the management and discovery of new knowledge is a key point in the development of actually useful DSS. There are several approaches to create semi-automated ontology learning systems, but none of them has been applied to the field of sensory data. Most of them have been used on semantic web data or huge text corpora [5, 6].

We claim that the expressive power of structural approaches is the key to handle the complexity of acquiring knowledge from unstructured data and related to every-day situation. The idea of the proposed work is to describe a general framework to deal with this problem, using as example application the problem of finding strategies. Given a problem description, whose solution is obviously unknown, and a set of solution examples (our rough data), we aim to abstract general guidelines about problem solution. This implies highlighting the common characteristic of solution and obtaining a general description of the solution itself, in terms of its key components. In particular, we address the problem of finding a good heuristic for the well-known slide tile puzzle, using structural information.

The rest of the paper is organised as follows. Section 2 summarizes some of the approaches presented in literature, with regards to knowledge extraction tools. Section 3 describes the problem of Ontology Learning and our proposal to deal with unstructured data corpora, such as sensory data. Section 4 proposes a testbed to evaluate our approach. Finally, in Sect. 5, our conclusions are reported.

2 Related Work

The need for coupling semantics with a sequence of sensor readings is well-known in literature. In fact, inferring knowledge from data is an open issue in Computer Science, and in particular in the area of data mining [7]. In this context, defining what can be deemed as *interesting* knowledge is a hard problem, because it implies to find what can be interpreted as an important information. Historically, a first debate on the most profitable way to extract useful information (i.e., knowledge) from a data collection was opened by John Tukey [8]. In the seventies, he proposed the *Exploratory Data Analisys* (*EDA*), as opposite to the *Confirmatory Data Analysis* (*CDA*) or *Statistical Hypothesis Testing*, that was the standard approach in those years. In the *EDA* approach, data are analysed with different techniques to summarize their characteristics. Unlike *CDA*, Tukey suggests to let hypotheses emerge from data themselves, rather than using data only to test a-priori hypotheses. The *Exploratory Data Analisys* is just an approach, not a set of techniques, i.e. a suggestion about how data analysis should be carried out and what its goals should be. Most of the techniques inspired to *EDA* use a graphical approach, because it represents a very powerful instrument to reveal the structure of the data to the analyst, offering new and often unexpected insights. In other words, it empowers the analyst's natural pattern-recognition capabilities. Therefore, it was the seminal work of modern approaches to data mining and pattern recognition.

One of the contemporary and independent developed research carried out on the track of *EDA* is the so-called *GUHA* (*General Unary Hypotheses Automaton*) principle [9]. The aim is to describe all assertions which may be hypotheses, verify each of such assertions and found the "interesting" ones, based on collected data. These systems generate systematically all interesting hypotheses with respect to the given data (hypotheses describing relations among properties of objects) via a standard computer system, and therefore represent a first attempts to formalise an automatic inductive approach. Logic is used to formulate hypotheses, coded as association of properties. Each object is represented by a row in a rectangular matrix, whose column are properties of the object. Analysing this data structure is possible to discover dependencies between different properties. The whole process is composed by three steps: *preprocessing*, *kernel* and *post-processing*. In the first step, matrix is arranged in a form suitable for a quick hypothesis generation. In the *kernel* phase, hypotheses are generated and evaluated, while in the last step hypotheses are analysed in order to interpret them.

It is crucial to note that the problem of letting structure and explanation emerge from data itself and not from a-priori hypotheses was central since the beginning of data analysis history, and has gained more relevance over the years, due to the ever-increasing size and heterogeneity that have characterised the data to analyse. Nowadays, the collected sensory data make it impossible to promote a-priori hypotheses to describe events of interest. The discussion between *EDA* and *CDA* approaches has renewed in the machine learning. In fact, two different approaches have grown in importance: *inductive* and *deductive* learning. This distinction reflects the differences and goals already underlined by Tukey, with a special focus of attention to the learning matter. The inductive approaches state the learning problem as finding a hypothesis that agrees with the examples, preferring the most simple one. It includes a variety of algorithms, such as instance-based learning, Support Vector Machines, Naïve Bayes, Artificial Neural Network, etc. Each of these approaches stresses different aspects of learning problem, but they relieve the analyst and designer from formulating an a-priori hypothesis about data. On the other hand, their responses are not useful to increase the knowledge regarding a particular problem because they can be considered as a black-box that can be applied on unseen data, but the model of the data they use is not human interpretable.

The deductive learning approaches constitute the other class of machine learning algorithms. For example, a method to infer general concepts from examples is known as *Explanation-Based Generalisation* (*EBG*) [10]. This deductive approach explains why a training example is a member of the concept being learned. This approach relies on four main components: a goal concept, training example, domain theory and operational criterion. Explanations are represented by Horn-clause inference rules arranged in proof trees. The goal concept is described through high-level properties that are not directly found in the example. Training example is a representation of a specific example in terms of lower level features. The domain theory is made up of a set of inference rules and axioms about the domain of interest. Domain theory is used to demonstrate the validity of the example. The operational criterion indicates how a concept must be expressed to be recognised. The aim of the system is just to generalise concepts from examples. A slightly different approach is that proposed by [11]. In this case the system is not only able to generalise a concept, but to check where a generalisation fails for a particular example, so that the system can refine it. Therefore it is possible not only to infer a general concept, but also to check whether an example is coherent with that generalisation, or why it is not; in other words, the system is able to learn. This approach is called *Explanation-Based Learning* (*EBL*). An evolution of the *EBL* is proposed in [12]. This approach tries to merge the old *EBL* engine, based on symbolic knowledge representation, with the statistical approach. The proposed system aims to take advantage of the robustness of statistical approach respect to real word problems, but at the same time it exploits the expressive power of symbolic knowledge representation.

An alternative approach to generalisation uses formal languages, and is known as *syntactic pattern recognition* [13]. In these systems, concepts are decomposed into simpler parts and their description relies on a grammar. A grammar is formally defined by the quadruple (Σ, N, P, S), where:

- Σ, the *alphabet*, is the set of the so-called terminal symbols, i.e. the basic elements of the grammar;
- N is the set (disjoint from Σ) of the nonterminal symbols; each of these symbols represents one or more strings of terminal and nonterminals symbols.
- P is the set of the *production rules*, composed by a head, represented by a nonterminal, and a body made up of a sequence of terminals and/or non terminals.
- $S \in N$ is a special symbol, known as the start symbol.

The set P represents possible and interesting structures, i.e. frequent patterns. The problem of inferring knowledge is stated as the problem of design a learning machine for pattern recognition, where a pattern is a particular structure included into the grammar. The system infers a grammar from training examples and applies it on the new data, in order to verify if the string of terminal symbols belongs to the learned grammar. This kind of approach requires preliminary work by the designer in ontology domain definition, in order to identify the key elements of the representation. The major drawback with this methods is the high computational cost needed to infer grammars. Historically, these approaches has been considered as alternatives to statistical learning systems, but during last decades many efforts have been made to unify statistical and syntactic pattern recognition (see [14]).

Other authors consider traditional approaches inadequate to cope with the complexity of managing knowledge and its evolution in complex phenomena. However, they believe that these scenarios cannot be modeled only by mathematical or statistical means. For example, Evolving Transformation System (*ETS*) is a formalism that tries to unify the syntactic and statistical pattern recognition, in order to create a new kind of class representation. The definition of *class*, according to the author, rests on the generative side: objects belonging to the same class share similar generative histories. In this context, a generative system is a nondeterministic system operating on actual entities and assembling them into larger entities (and eventually into class objects), guided by some hierarchical description of the class [15]. This kind of representation is focused on the problem of giving a structural representation to the data. Each object in this formalism is thought of as a temporal structural process and the representation of each element of a class evolve with the description of the class itself. *ETS* is a work in progress framework, limited by the lack of new mathematical instruments to deal with the complexity of a structural description.

In [16], Chazelle proposes a new vision to deal with phenomena arising from life sciences, stating that means used in physical science are not adequate. According to his work, algorithms are more suitable for these purpose, due to their rich and expressive language. Moreover, the author claims that some problems can take an enormous advantage from the novelties introduce by a new perspective, taking into account the peculiarities of complex non physical systems. In the case of sensory data used to investigate and predict human habits and behaviour, the complexity is very high, because of the high number of variables to include in the model. Chazelle introduces the *natural algorithms* to model these systems. This approach relies on the so-called *influence systems*, i.e. networks of agents that perpetually rewire themselves. These networks are specified by two functions: f and G; the function f calculates the position of an agent, taking as input the location of its neighbour agent, given by function G. The output of G is function of the state of the whole system, that is the position of all agents. In this approach is possible to note how the information travels through the system, in a way that separates its syntactic or structural component and its semantic. In other words,this method models complex systems exploiting equally qualitative and structural information.

Our proposal differs from those presented in literature, because it aims to extract knowledge from a large set of unstructured data (such as sensory data), translating it in a machine-understandable form. Many approaches have attempted to deal with the complexity of such kind of data. In particular, many systems have been proposed in the area of Ambient Intelligence, which typically deals with sensor readings, and their interpretation. For example, in [17], the authors suggest a three-tier paradigm for knowledge extraction. In particular, this paradigm cuts irrelevant details off from raw sensor readings, in order to obtain more refined data that can be analyzed by the reasoning module, at the top of this processing hierarchy. Statistical methodologies (e.g., correlation analysis, clustering) are used in [18] to cope with the complexity of large sensor reading dataset. The proposed system models and learns user habits through his interactions with the actuators deployed in the environment. According to the authors, user habits are coded into sensor readings, thus they can be inferred analyzing sensory data and discovering relations between environmental conditions and user.

Our system uses a similar approach, but, at the same time, aims to use as little a-priori knowledge as possible, because this is hard to obtain in the great part of real problem belonging to sensory data mining. In fact, such data is hardly understandable according to simple user description of the phenomenon that has generated it. So, it is very hard to formulate a-priori hypotheses, as in deductive approaches; this may force to use a too specific and detailed model, with a high risk of overfitting. On the other hand, an expert of the application domain possesses some knowledge, which can represent a very important resource. The actual problem is translating it to be usable by the system. With a pure inductive approach, this task would include a very hard empirical work, in order to tune up all the parameters used in the chosen method. In this case, the risk of overfitting is very high, because the relations between sensory data and model parameters are not very clear, so it is impossible to distinguish when the over fitting starts. A structural approach is less prone to this problem, because it can be more simple build measure of complexity over the model it uses, due to its representation.

3 Ontology Learning

The most common ontology definition can be found in [4]; the author defines an ontology as the specification of a conceptualization, that is a model of the concepts and relations describing a particular domain. According to [6], this definition can be formally translated in a structure $\mathscr{O}$:

$$\mathscr{O} := (C, \leq_C, R, \sigma_R, \mathscr{A}, \sigma_A, \mathscr{T})$$

- four disjoint sets: concept identifiers (C), relation identifiers (R), attribute identifiers ($\mathscr{A}$), data types ($\mathscr{T}$);
- a semi-upper lattice $\leq_C$ on C;
- a *relation signature* function $\sigma_R : R \rightarrow C^+$;
- a *relation hierarchy* partial order $\leq_R$ on R;
- a function $\sigma_{\mathscr{A}} : \mathscr{A} \rightarrow C \times \mathscr{T}$.

Thus, learning an ontology means specifying all these elements, by inferring them from data. This problem can be decomposed into the following subtasks:

1. acquisition of the relevant terminology;
2. identification of synonym terms;
3. formation of concepts;
4. hierarchical organization of the concepts (concept hierarchy);
5. learning relations, properties or attributes, together with the appropriate domain and range;
6. hierarchical organization of the relations (relation hierarchy);
7. instantiation of axiom schemata;
8. definition of arbitrary axioms.

This schema has been formulated mainly in the context of ontology learning in text corpora. Obviously, in the context of sensory data these steps need some changes, in order to adapt them to the different characteristics of the new scenario. However, the general structure of the process remains the same. Thus, in this work we propose an adaptation of this schema, modifying it according to the new kind of data and the differences between text and sensory data.

Undoubtedly, the goal we aim at is very challenging, and many issues are to be addressed; some of those are related to theoretical open issues in computer science, so it is impossible to known if they are practically solvable. We do believe, though, that knowledge extraction can take advantage of an ontology learning approach, because this might exploit the information potential embedded in sensory data. The basic idea is that data collected from sensors share an underlying language, i.e. it can be considered as generated by a particular language describing some phenomena. Similarly to what described in [19], we assume that data are drawn from a process that can be modeled by a Turing Machine (TM). This means, according to Chomsky, that there is a language that can describe such data. Likely, this language is very complex and, moreover, data is corrupted by noise, so reconstructing the original language from data is a very tricky task.

If we compare the problem of grammar induction (i.e., learning) and the problem of ontology learning, we discover that they have much in common. This can be explained if we consider the ontology problem as the semantic side of grammatical induction. The literature has shown that this two aspects are strictly coupled and it is impossible to solve one of them discarding the other.

Given these caveats, it is simple to describe and motivate the changes we made to the general ontology learning schema. From a sensory data point of view, we are not very interested in the relation discovery and our focuses is principally on concepts and their hierarchy.

The proposed approach is composed by:

1. individuating a set of basic properties (axioms) and features to discovery significant patterns;
2. discovery of relevant elementary patterns as terminology;
3. abstraction of patterns as concepts;
4. inferring hierarchical concept organisation;

Setting axioms In the first step, the key elements of the ontology are defined; it is the only phase of the system that requires human intervention. The analyst has to specify a description of the goal concepts in terms of general properties, like time, space or other very general features. The main difference from others approaches is how these features are described. For example, consider the user activity recognition task. A user activity may be defined as *a recurrent sequence of actions, that can be recursive decomposed in simpler subtasks*. Two approaches can translate this definition in features on data: the deductive, and the inductive one. According to the deductive approach, the definition is transformed into an abstract and general model, that hypothesises sensor reading interactions that identify executions of the same activity. In the inductive approach, the analyst translates the definition in terms of properties the sensors can measure, such as time duration; so, an activity is treated as a recurrent pattern in data, whose instances have similar structures and time durations. No attempts at generating a general activity model are made, but the model will emerge from data. The feature selection depends on the experience of the analyst, but it is a simpler task than the other approaches. Moreover, it is less prone to error and axioms chosen can be simpler checked.

Discovery terminology The second step faces the problem of finding data with the properties defined at the previous step. There exist several techniques to accomplish this task, but a data mining approach is the best choice. In fact, a statistical analysis of data can aid the most significant pattern to emerge, so the basic level of the ontology can be identified with the smallest, but more recurrent patterns in sensory data. Moreover, these techniques guarantee robustness, reliability, and thoroughness, since they have been extensively used in many different systems in the last decades. The a-priori based algorithms are an example of these techniques.

This step can be thought as a data fusion, i.e. the system associates data coming from different source and of different types. This is a very common approach in this kind of systems (e.g., [20, 21]) and it is useful to process data in a multi-sensor context, in order to explode relations between different sensor triggers.

Pattern abstraction Pattern abstraction allows to obtain a generalization from the instances of patterns present in data. The system addresses the problem of *synonyms*, grouping similar instances of the same patterns. In particular, the best choice for this step is the use of a technique coming from the statistical learning. In fact, these approaches allow a statistical description of data, summarising it according to the most emergent properties. Unsupervised clustering algorithm and subsequent statistical classifier can be used in order to discover and then recognise most common patterns. This kind of systems are well-known in literature and have obtained a successful application in many research area. So, at the end of this step the system will be able to associate a statistical model to each pattern and a trained classifier to distinguish between them.

Inferring concept hierarchy In this step, the system finds recurrent terminology structures. A grammar induction algorithm is used: the system considers the data as a language and looks for the suitable grammar to generate it. According to [22], a grammar can be defined as a device that enumerates the sentences of a language. In other words, it is a function, F, that generates every sentence of a language, $\mathcal{L}$. The expressive power of these devices depends on the restrictions imposed on F. If no restriction is imposed, then the function belongs to the set of General Turing Machine. On the contrary, if a strict restriction on F is imposed, such as the constraint on each grammar to be a finite Markovian source, then it is possible to prove that some language (e.g., natural language) can not be generated by these grammars. So, it is important to choose the adequate restriction, based on the aim set for the language. In this context, we assume that a grammar represented by a Finite Automaton may have the sufficient power to model sensor language.

Representing each base concept, that is each term, with a terminal symbol of a grammar, it is possible to reconstruct complex concepts as sentences of the same language. In fact, once the system infers a grammar from data, the syntactic representation of a complex concept is represented by the derivation tree of the sentence.

In [23], the task of grammatical inference is widely debated. Many learning system have been proposed and each of them is specialized for a particular target, with respect to the kind of data and the application domain. Despite most systems focus on learning grammar from text, there are others that have been applied to bioinformatics or computer vision. This means that grammatical inference is possible even if data are less structured than text data and results obtained in this area encourage the use of these techniques in the scenario of sensory data.
Moreover, a representation of a sensory data through a grammar can be seen as a compression of the data. The Occam's Razor gives us a simple guideline to choose between the possibly different grammars that can represent the data equally well. In fact, the simplest explanation is always the best, so the shortest grammar is the best to represent data. Measures of complexity in the sense of Occam's Razor are Minimum Description Length (MDL) principle [24] or Kolmogorov complexity [25]. Even if it is impossible to calculate the MDL or similar measures, they represents a good guideline for the choice among different representations.

4 A Proof of Concept: The Slide Puzzle

Extracting knowledge from sensory data is a very complex and difficult challenge and may prove not too useful to test the effectiveness our approach for the present discussion. In fact, a too complex problem does not allow to study the details of every single part of the system and their effects on whole results. A more manageable problem, whose difficulties can be tuned according to very specific tasks, appears a more reasonable choice. Thus, a very similar problem was chosen to test the proposed system, even though it is still representative of the original target scenario, due to their very similar features. Extracting knowledge from sensory data shares in fact many similarities with the heuristic learning problem, because the success of both of them relies on the ability to discover hidden and counter-intuitive relations in data.

In fact, heuristic learning is a very interesting research area in structural analysis. In some cases, it is difficult to unveil the structure of a problem, in order to improve solution search algorithms. This is the case, for instance, of the n-puzzle slide game.

The n-puzzle problem is a generalisation of the more common 15-puzzle. In its original form, the puzzle consists of 16 squares numbered from 1 to 15 and arranged in a 4×4 box, with one position of the box left empty. A legal move consists of sliding an adjacent block into the empty space. The goal is to reposition the squares from a given arbitrary starting arrangement by sliding them one at a time into the configuration shown in Fig. 1.

Ratner and Warmuth [26] showed that finding an optimal solution to n-puzzle, i.e one involving as few moves as possible, is an NP-complete problem. Moreover, getting the goal position, that is solving the puzzle, is not possible for all initial arrangements. Only instances corresponding to even permutations of the goal configuration are solvable, that is only half of all possible configuration can be transformed into the goal configuration (setting as initial configuration the one with the empty tile on the

Fig. 1 Goal position of 15-puzzle

1	2	3	4
5	6	7	8
9	10	11	12
13	14	15	

bottom right corner). The 8 instance of the problem has been solved with a breadth-first search, and the 15 one has been solved with IDA* search algorithm [27]. The heuristic used by this search algorithms are based on Manhattan distance, but there is no suggestion as regards a general solution algorithm, i.e. one that is independent of the dimensionality of the problem. For higher dimension instance of the problem, such as 24-puzzle, many ad-hoc solutions were used, based on the particular properties of the problem. The main difficulty with high dimensional instances of n-puzzle concerns the very wide search space, that make an efficient search impossible.

Cutting-edge research has been devoted to finding a suitable solution to the 24- and higher instance of this game. The first proposed solution used some particular features of the problem to speed-up search and obtaining an optimal solution [27]. In particular, this approach encode more knowledge of the problem in the form of improved heuristics, in an automatic fashion. The drawback with this system is that its heuristic does not guarantee a good execution time on every instance of the problem, but some random instances take very long time to be solved or worse they have not been solved at all. Later tries to solve the problem have been turned on the problem of finding better heuristics, created with the aid of machine learning. Discarding the strong limitation of finding admissible heuristics, Ernandes and Gori proposed a machine learning approach using an Artificial Neural Network. Given a rich set of training example, *ANN* learns how far a configuration, i.e. a node, is from the goal state. This approximation is used to improve the heuristic, with a remarkable speed up for the search in the 8 and 15 instances of the game. In the case of 24-puzzle, the lack of a good training set make an effective use of the *ANN* impossible. The authors proposed but did not implement a bootstrapping technique to overcome this problem. That is, they assert that is possible to specialise an initially weak heuristic function in an iterative manner, using example generated at each step by each function. The authors of [28] continued the work by Ernandes and Gori, implementing the bootstrapping approach. In particular, they used incremental bootstrapping process augmented by a random walk method for generating successively more difficult problem instances, obtaining good results in the solution of random generated instances of 24-puzzle. The great drawback with this approaches is the long time the bootstrapping phase takes. The author proposed a solution only in the case of a single instance problem: in that case the bootstrapping are interleaved with a classic heuristic search to lower total execution time. A similar approach can be found in [29]. The authors propose a system that learns a heuristic to solve a single instance, and the bootstrapping is done over successive failed attempts to solve the instance. They show the effectiveness of their approach in the 15-puzzle game.

In [30] a system is proposed that mixed multiple heuristics to improve them. In this work, the key idea is to merge the knowledge coming from different heuristic, merging them into a better one that summarise all the others. This knowledge mesh-up is produced by an *ANN*, that is by a statistical learning. A very interesting work is that presented by Korf and Felner in [31]. The authors deal with the problem of finding an accurate admissible heuristic. They propose a memory-based approach, that utilises partial puzzle resolution to obtain an estimate of the cost. The scheme is dived into many parts, and for each a number of move to goal state is saved in a database. With a carefully choose of the subparts, is possible to obtain a good estimate and therefore a tight (and better) heuristic. The authors obtain relevance results on the 24-puzzle, proving the goodness of their idea. Also, this means that the recurring relations and interactions between subproblems are the key of this problem, and thus capturing them leads to a better solution.

4.1 Hidden Relations: Knowledge to Enhance Heuristic

The work produced in this area shows as the solution of n-puzzle game, with $n > 15$ requires a heuristic enriched with some knowledge inferred by example of simpler instance solutions. In our opinion, this means that the proposed systems are trying to summarise some unveiled properties and relations of the application domain to empowered search algorithm, as demonstrated by the use of machine learning approach and in particular of statistical learning (e.g., *ANN*).

We agree with this vision and we proceed a step further: the system has to learn the structure of known solutions and has to find relevant relations between them in order to go beyond combinatorial explosion nature of the problem. Obviously, we do not assert that we are able to find an optimal solution for the problem, but we only want to drive our system to “comprehend” the essence of the game and could suggest a possible heuristic for a higher dimensionality game, focusing computational efforts especially on more promising areas of the search space. This implies the identification of recurrent structure that constitute the sub problem to solve and a model to describe how they interact.

The basic idea is to describe a solution as the evolution of a set of properties evaluated on each state it goes through. Each of these properties expresses a particular feature related to the state. It can be a static feature, such as the Manhattan distance to the goal solution, or a dynamic one, i.e. the row of the last moved tile. We suppose that a good set of properties can capture some general structures shared by solutions. This choice is similar to the multiple heuristic approach presented in [30], and is motivated by the same reasons: we are trying to mix different properties in order to unveil not evident and counter-intuitive features of the solution.

Given a set of solutions to the 8-puzzle, 15-puzzle and some simple instances of the 24-puzzle, we can translate them in their feature evolution representation.

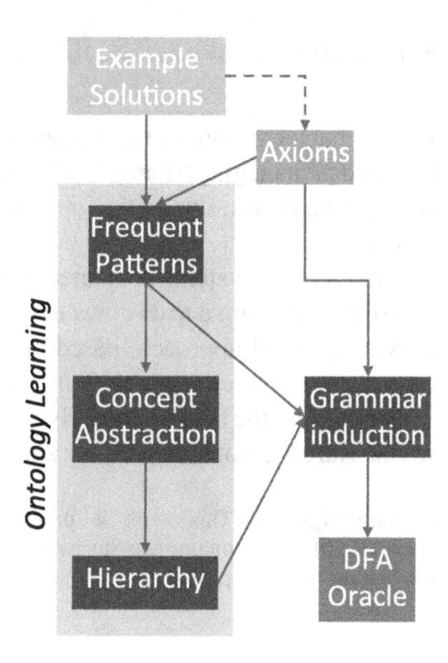

Fig. 2 Sketch of the proposed system

Assigning a unique symbol to each possible feature combination, we can describe a solution as a string made up of these symbols. In this way, we can describe each solution as a string generated by the language of solutions. In this language, we the terminal symbol are represented by the unique symbol chosen for each state.

In the next step, we try to obtain the *Discrete Finite Automaton* (DFA) that recognize that language. In this manner, we have found a machine able to suggest us the most probable path in the tree to find the solution. At each node, we can evaluate the solution we are building and choose next step as that most probable according to our DFA. The DFA acts like an oracle, able to guide in the path toward the solution.

In Fig. 2, the proposed system is shown. The two main blocks of the system are the *Ontology Learning* and the *Grammar Induction*. According to the previous description of the Ontology Learning approach, proposed in Sect. 3, the main steps of this process are presented in the figure. The first step, that is stating axioms, is the only non fully-automated step of the process and, therefore, it is placed outside of the *Ontology Learning*: domain expert's analysis of example data is required in order to obtain axioms. The whole process of *Ontology Learning* produces the input data for the Grammar induction module. In particular, terminal symbols (axioms), an initial approximations of nonterminals (concepts) and production rules (hierarchy), are elaborated by the *Grammar induction* module to produce the *DFA Oracle*.

At the end of the processing chain, the system builds up the oracle, and an insight into the structure of input data that has generated it. This ontology can be used to evaluated the quality of the solution and to suggests to the designer new tuning to the axioms definition, in order to capture the actual structure of the problem.

5 Conclusion

In this chapter we presented a proposal for a system to cope with the complexity of knowledge managing, sharing and reuse in the sensory data scenario. Nowadays, this is a very hard task, due to the great amount of raw data and to their lack of evident structure.

Unlike other systems presented in literature, we try to infer an ontology directly from data and using it to discover new intelligible knowledge. The system we propose uses a structural approach, based on the use of grammatical inference, in order to emerge the most relevant relations present in data.

We propose the heuristic search as a controlled testbed for our system. In particular, we measure the potential of sour approach in the n-puzzle game.

Acknowledgments This work is partially supported by the PO FESR 2007/2013 grant G73F11000130004 funding the SmartBuildings project and by the PON R&C grant MI01_00091 funding the SeNSori project

References

1. Atzori, L., Iera, A., Morabito, G.: The internet of things: a survey. Comput. Netw. **54**(15), 2787–2805 (2010)
2. Cook, D., Youngblood, M., Heierman E.O., Gopalratnam, K., Rao, S., Litvin, A., Khawaja, F.: Mavhome: an agent-based smart home. In: Proceedings of the First IEEE International Conference on Pervasive Computing and Communications (PerCom 2003), pp. 521–524, (2003)
3. Doctor, F., Hagras, H., Callaghan, V.: A fuzzy embedded agent-based approach for realizing ambient intelligence in intelligent inhabited environments. IEEE Trans. Syst. Man Cybern. (Part A: Systems and Humans) **35**(1), 55–65 (2005)
4. Gruber, T.R.: A translation approach to portable ontology specifications. Knowl. Acquis. **5**(2), 199–220 (1993)
5. Velardi, P., Faralli, S., Navigli, R.: Ontolearn reloaded: A graph-based algorithm for taxonomy induction. Comput. Linguist. **39**(3), 665–707 (2012)
6. Cimiano, P.: Ontology Learning and Population from Text: Algorithms, Evaluation and Applications. Springer, New York (2006)
7. Yang, Q., Wu, X.: 10 challenging problems in data mining research. International Journal of Information Technology & Decision Making (IJITDM) **5**(04), 597–604 (2006)
8. Tukey, J.W.: Exploratory Data Analysis. Addison-Wesley, Menlo Park (1977)
9. Hájek, P., Havránek, T.: Mechanizing hypothesis formation : Mathematical Foundations for a General Theory (Universitext). Springer, New York (1978)
10. Mitchell, T., Keller, R., Kedar-Cabelli, S.: Explanation-based generalization: A unifying view. Mach. Learn. **1**(1), 47–80 (1986)
11. Dejong, G., Mooney, R.: Explanation-based learning: an alternative view. Mach. Learn. **1**(2), 145–176 (1986)
12. DeJong, G.: Toward robust real-world inference: a new perspective on explanation-based learning. In: Proceedings of the 17th European conference on Machine Learning (ECML'06), pp. 102–113. Springer, Heidelberg (2006)
13. Fu, K.S.: Syntactic Methods in Pattern Recognition, Mathematics in science and engineering, vol. 112, Academic press, New York (1974)

14. Tsai, W.H., Fu, K.S.: Attributed grammar-a tool for combining syntactic and statistical approaches to pattern recognition. IEEE Trans. Syst. Man Cybern. B Cybern. **10**(12), 873–885 (1980)
15. Goldfarb, L.: Representation before computation. Nat. Comput. **9**(2), 365–379 (2010)
16. Chazelle, B.: Natural algorithms and influence systems. Commun. ACM **55**(12), 101–110 (2012)
17. De Paola, A., Gaglio, S., Lo Re, G., Ortolani, M.: An ambient intelligence architecture for extracting knowledge from distributed sensors. In: Proceedings of the 2nd International Conference on Interaction Sciences: Information Technology, Culture and Human, ACM, pp. 104–109, (2009)
18. Augello, A., Ortolani, M., Lo Re, G., Gaglio, S.: Sensor mining for user behavior profiling in intelligent environments. In: Advances in Distributed Agent-Based Retrieval Tools, pp. 143–158. Springer, Heidelberg (2011)
19. Gaglio, S., Lo Re, G., Ortolani, M.: Cognitive meta-learning of syntactically inferred concepts. In: Samsonovich, A.V., Johannsdottir, K.R. (eds.) BICA, Frontiers in Artificial Intelligence and Applications, vol. 233, pp. 118–123. IOS Press, Amsterdam (2011)
20. De Paola, A., Cascia, M., Lo Re, G., Morana, M., Ortolani, M.: User detection through multi-sensor fusion in an ami scenario. In: 15th International Conference on Information Fusion (FUSION 2012), pp. 2502–2509, (2012)
21. De Paola, A., Gaglio, S., Lo Re, G., Ortolani, M.: Multi-sensor fusion through adaptive Bayesian networks. In: AI* IA 2011: Artificial Intelligence Around Man and Beyond, pp. 360–371. Springer, Berlin (2011)
22. Chomsky, N.: On certain formal properties of grammars. Inf. Control **2**(2), 137–167 (1959)
23. de la Higuera, C.: Grammatical Inference: Learning Automata and Grammars. Cambridge University Press, New York (2010)
24. Rissanen, J.: Minimum description length principle. In: Encyclopedia of Machine Learning. Springer, New York (2010)
25. Li, M., Vitányi, P.M.: An Introduction to Kolmogorov Complexity and Its Applications, 3rd edn. Springer, Heidelberg (2008)
26. Ratner, D., Warmuth, M.K.: Finding a shortest solution for the n $\times$ n extension of the 15-puzzle is intractable. In: Proceedings of the 5th National Conference on Artificial Intelligence (AAAI), pp. 168–172, (1986)
27. Korf, R.E., Taylor, L.A.: Finding optimal solutions to the twenty-four puzzle. In: Proceedings of the 13th National Conference on Artificial intelligence (AAAI'96), Vol. 2, pp. 1202–1207. AAAI Press, Menlo Park (1996)
28. Arfaee, S.J., Zilles, S., Holte, R.C.: Learning heuristic functions for large state spaces. Artif. Intell. **175**(16–17), 2075–2098 (2011)
29. Humphrey, T., Bramanti-Gregor, A., Davis, H.: Learning while -solving problems in single agent search: Preliminary results. In: Gori, M., Soda, G. (eds.) Topics in Artificial Intelligence, Lecture Notes in Computer Science, vol. 992, pp. 56–66. Springer, Heidelberg (1995)
30. Samadi, M., Felner, A., Schaeffer, J.: Learning from multiple heuristics. In: Proceedings of the 23rd National Conference on Artificial intelligence (AAAI'08), vol. 1, pp. 357–362. AAAI Press, Menlo Park (2008)
31. Korf, R.E., Felner, A.: Disjoint pattern database heuristics. Artif. Intell. **134**(1–2), 9–22 (2002)

Hardware and Software Platforms for Distributed Computing on Resource Constrained Devices

Gloria Martorella, Daniele Peri and Elena Toscano

Abstract The basic idea of distributed computing is that it is possible to solve a large problem by using the resources of various computing devices connected in a network. Each device interacts with each other in order to process a part of a problem, contributing to the achievement of a global solution. Wireless sensor networks (WSNs) are an example of distributed computing on low resources devices. WSNs encountered a considerable success in many application areas. Due to the constraints related to the small sensor nodes capabilities, distributed computing in WSNs allows to perform complex tasks in a collaborative way, reducing power consumption and increasing battery life. Many hardware platforms compose the ecosystem of WSNs and some lightweight operating systems have also been designed to ease application deployment, to ensure efficient resources management, and to decrease energy consumption. In this chapter we focus on distributed computing from several points of view emphasizing important aspects, ranging from hardware platforms to applications on resource constrained devices.

1 Introduction

Distributed computing in Wireless Sensor Networks (WSNs) represents an emerging scenario obtaining the attention of researchers. Performing individually a smaller task, sensor nodes of a WSN collaborate exchanging information, data or partial

---

G. Martorella · D. Peri (✉)
Dipartimento di Ingegneria Chimica Gestionale Informatica Meccanica, Università degli Studi di Palermo, Viale delle Scienze, Edificio 6, 90128 Palermo, Italy
e-mail: daniele.peri@unipa.it

E. Toscano
Dipartimento di Matematica e Informatica, Università degli Studi di Palermo, via Archirafi 34, 90123 Palermo, Italy
e-mail: elena.toscano@unipa.it

S. Gaglio and G. Lo Re (eds.), *Advances onto the Internet of Things*,
Advances in Intelligent Systems and Computing 260, DOI: 10.1007/978-3-319-03992-3_9,

results to achieve a global goal. A WSN is composed of a variable number of autonomous sensor nodes deployed in the environment. Basically, a sensor node includes a processor unit, a radio transceiver for data transmission, small memory for data storage and sensor modules for data gathering. Moreover, a battery with a limited energy budget is often the only power source. Therefore, sensor nodes present several constraints in terms of storage, processing and energy. They are often deployed in a hostile environment that individuals cannot easily reach and consequently changing or charging batteries is unsuitable.

Due to the sensor's small memory and to the low processing capability, solving complex problems individually is often impossible and thus, some form of cooperation is required. Therefore, distributed computing in WSN allows to achieve complex targets by using cheap sensor nodes with a considerable cost saving and without a single point of failure. The advantages include higher performance, reliability, collaboration and scalability. In spite of this, distributed computing defines a scenario still full of challenges since resources management techniques and power saving strategies are required. Research is thus naturally focused on the creation of new platforms for specific application fields, and on the development of distributed computing techniques and algorithms improving the performances of WSNs.

Sensor nodes have earned the attention of the market with considerable success because of their low price and their potential in different fields. In addition to the traditional well known platforms, there are several low-cost hardware platforms and starter kits on the market, often including lightweight operating systems (OS) that allow anyone to create their own sensor network. In this chapter we provide an overview about distributed computing on low resources constrained devices such as sensor nodes of a WSN. Many chapters in literature discuss distributed computing but they focus just on one significant aspect such as algorithms, applications, challenges or issues. Unlike these papers, we cover several significant aspects of distributed computing ranging from low to high level, from hardware platforms to distributed applications, also describing the underlying algorithms and issues. We also discuss a typical distributed application showing how even a trivial problem may become complicated on resource-constrained hardware.

The remainder of this chapter is organized as follows. Section 2 presents some recent hardware platforms for WSN, while Sect. 3 describes some of the existing operating systems for sensor node's resources management. A common representation of sensor nodes is provided by sensor ontologies, briefly discussed in Sect. 4. Distributed computing algorithms and applications in WSN are surveyed in Sect. 5, while Sect. 6 provides a brief characterization of distributed synchronization as a real example of the issues and challenges related to distributed computing on low resources devices. Finally Sect. 7 concludes this chapter.

2 Hardware Platforms

Distributed computing cannot be discussed separately from the specific underlying hardware implementation. In recent years many hardware platforms for WSN have been designed, tested and brought to market. These platforms present some common features since basically they include a microcontroller, a transceiver, low memory, power supply and expansion pins. These platforms differ in the on-board hardware components and in their actual use in several applications. Many available platforms have been used in several research projects such as the well-known MicaZ, Iris, Mica2 and Telos platforms. More recently, other hardware platforms are earning the market thanks to their low cost and high expandability through a large set of sensor boards.

In [22], the authors presents the commercially available **Atlas** sensor platform for the creation of pervasive smart spaces. It is composed of three basic layers: the Processing Layer, the Communication Layer, and the Device Connection Layer. Other add-ons provide additional capabilities to the node. The Processing Layer is based on Atmega 128L microcontroller including 128 KB Flash memory, 4 KB SRAM, 4 KB EEPROM. In addition, 64 KB of expanded SRAM is added. The Communication Layer is responsible of data transfer over the network. This layer can be based on wired 10BaseT Ethernet, Bluetooth, 802.11b WiFi and USB. The Connection Layer allows to connect up to 32 sensors and actuators to the platform. Atlas node is configured through an interface allowing the selection of the currently connected sensors and actuators. The authors of [6] describe Atlas and cases study based on it.

SHIMMER [7] is a platform designed for healthcare monitoring. It is based on Texas Instruments MSP430 microcontroller including 10 KB RAM, 48KB Flash, 8 ADC channels, USART and SPI connections. SHIMMER uses a CHIPCON CC2420 radio transceiver and Roving Networks RN-41 Class 2 Bluetooth module for communications. A MICRO SD Flash for additional storage is also on-board. Internal and external connectors allows the integration of other boards adding kynematics, ECG and GSR sensing capabilities, while a three axis accelerometer is already built-in.

Waspmote [41] is a general-purpose board to develop applications for WSNs. It is based on an Atmega1281 microcontroller running at 14 MHz, 8 KB SRAM, 4 KB of EEPROM, 128 KB of FLASH and a 2 GB microSD. A temperature sensor and an accelerometer, a radio socket, a SPI-UART socket, a solar socket and a battery socket are included. A 802.15.4/Zigbee module is used for communication with other nodes in the network and for node programming. Nodes can be programmed with Over The Air Programming (OTA) through 802.15.4 modules or 3G/GPRS/WiFi modules via FTP. Secure communication can also be implemented on Waspmotes by using AES 128, RSA 1024 and AES 256 encryption library provided implementations. Waspmote is well expandable with several modules for communication like RFID, Bluetooth for nodes discovery, radio expansion boards (for using two radios at the same time) and GPS. Sensor boards provide additional sensing capabilities to the node. Case studies affect smart agriculture projects, smart water and smart cities projects. Waspmote has also been used in the deployment of long distance WSNs [44].

Arduino is an open source general-purpose platform for developing a wide range of applications involving sensors, actuators and communication with other devices. The philosophy of Arduino is based on prototyping, tinkering and patching concepts. Arduino boards can be purchased pre-assembled or can be assembled manually. There are 14 commercially available Arduino versions. Typically an Arduino board includes an Atmel AVR8 microcontroller and it is expandable using other shields. Only Arduino Due includes an Atmel SAM3X8E ARM Cortex-M3 CPU. For example the Arduino Uno board [2] is based on Atmega328 running at 16 MHz including 32 KB Flash memory, 2 KB SRAM and 1 KB EEPROM. The board can be powered through either the USB connector or an external power supply. On board, digital IO pins, analog input and output pins are available. An integrated development environment (IDE) is freely available to developers. Thanks to this IDE, programs written in the Wiring programming language are easily cross-compiled on a host OS such as Windows, Macintosh and Linux, and deployed on Arduino boards through USB programming. Developers can add several Arduino kits including all the necessary for developing applications. In addition several shields, actuators and sensors are available for this platform. Several researches concerning WSN are based on this platform, while projects for beginners can be found in [4].

Zolertia **Z1** [45] is a low-power general-purpose platform based on Texas Instruments MSP430F2617 microcontroller running at 16 MHz including 92 KB flash and 8 KB RAM. Z1 is equipped with a 2.4 GHz CC22420 transceiver, a TMP102 temperature sensor and a ADXL345 3 axis digital accelerometer, 52-pin expansion connector, 2 ports for connecting Phidget sensors, a micro-USB connector and a ceramic antenna. It can be powered via USB or through batteries. Z1 is expandable with several on-board sensors like a temperature-humidity sensor, a light sensor, a barometric sensor, and it can also be connected to a wide range of sensor modules producted by Phidget Inc. Z1 has be used in some researches and applications such as environmental and agriculture monitoring, healthcare and power consumption monitoring.

3 Operating Systems for Sensor Node Devices

In this section, we briefly discuss operating systems for sensor nodes. Since sensor devices present several constraints in terms of memory, processing and energy, operating systems for WSNs must be designed to manage resource efficiently, taking care of specific issues arising from use cases. For instance, in most applications batteries are often the only source of energy and changing or charging batteries is not suitable. Consequently, an OS for sensor devices must not only limit power waste, ensuring optimized energy consumption, but also be able to recover from power faults. Over the years, several lightweight operating systems for WSNs have been created. Although they all share the same goals, architectural and implementation choices differ. Architecture, programming model, programming language and energy saving support are the main varying factors. Several chapters in literature as [9, 16], survey operating systems for WSNs while [9, 34], provide a comparative study of them.

Some of these operating systems such as **TinyOs** [27] and **Contiki** [14] present an event-driven programming model where processes use the same stack in order to reduce memory requirements. Instead, a multithreaded model requires additional resources as each thread needs its own stack. A package for multithreading support in TinyOS is presented in [23]. For multithreading, Contiki uses a lightweight thread model called protothread. Protothreads share a stack and their implementation requires only 2 bytes. A comparison of TinyOs and Contiki can be found in [34]. Multithreading-based operating systems are **MantisOs** [5], **LiteOs** [8]. Scheduling management deals with resource allocation. For example, TinyOs provides a FIFO scheduling. Tasks are enqueued and executed until their completion without preemption by other tasks. A task gets preempted only if an event or interrupt occurs. MantisOs provides a priority-based round robin scheduling within each of five available priority levels. The first scheduled thread gets the highest priority and it is executed until it completes or its time slice expires. If any thread is ready for execution, the system goes into sleep mode. In Contiki, preemptive multithreading is implemented as a library that can be linked with applications. LiteOs implements both priority-based scheduling and round robin scheduling. Operating systems for WSNs can also differ in kernel architecture. TinyOs presents a component-based architecture where every component is specified in terms of used and offered interfaces, while Contiki's kernel is composed of a lightweight scheduler and a poll event handler supporting synchronous and asynchronous events. A Contiki system is divided into two parts: the core system and the loaded programs. The core system includes the kernel, libraries and the program loader. MantisOs system presents a layered architecture where each level provides services to the higher levels. The lower level is the hardware level. The second level includes the kernel and the scheduler, the communication level and the device drivers. Over this level, Mantis System API provides services to network stack, command services and user's threads. The architecture of LiteOs is composed of three main components: LiteShell, LiteFS, and the kernel on the node. LiteShell subsystem runs on a PC station and allows the user to interact directly to a sensor node through a Unix shell providing commands for file processing and device management. LiteFS subsystem mounts a sensor node to the root filesystem of the base station. **MansOs** [37] derives from LiteOs but it is designed to be easy portable to new platforms. For this purpose, it presents three levels providing different functionalities. As described in [37] the Hardware Presentation Level (HPL) is the hardware dependent level and thus it contains the specific code for the target hardware platform. The Hardware Abstraction Layer (HAL) contains the representation of the platforms in terms of platform constants, pin assignments, device assignments and function binding. The Hardware Independent Layer (HIL) provides user hardware-independent functions abstracting the lower levels. Operating systems for WSNs must also provide support to communication. For example, Contiki presents two communication stacks: the *UIP* stack for communication over Internet and *Rime* aimed at low power radio applications with addressing, broadcast and unicast communication. MantisOS handles the communication through a layered network stack implemented with user level threads. MantisOS *COMM* layer

Table 1 Summary table of some operating systems for sensor nodes. They differ mainly in architecture, programming model and scheduling

	Architecture	Programming model	Scheduling	Language
TinyOs	Monolithic (component-based)	Event-driven	2 levels scheduling	NesC
Contiki	Modular	Event-driven and thread-based	2 levels scheduling hierarchy and preemptive multithreading	C
MantisOs	Layered architecture	Thread-based	Priority-based thread scheduling	C
LiteOs	Partioned into three parts: LiteShell, LiteFs and Kernel	Thread-based	Priotity-based and Round-robin	C

supports the MAC protocol. This layer provides a common interface for different communication device drivers and manages packet buffering.

These operating systems support distributed computing since they provide primitives needed by distributed applications. The choice of the operating system may in fact depend on the application requirements. Table 1 shows the main differences between some operating systems for sensor nodes.

4 Sensor Ontologies

Ontologies are used to represent formally the concepts of a domain through their properties and the relationships existing between them. Several ontologies for sensor nodes can be found in literature with the aim of providing a common representation of sensor nodes. These ontologies aim to describe characteristics, platforms, properties, sensing capabilities of a sensor node in a WSN. SensorML is a standard XML language for sensors definition. It is generic and therefore it supports a wide range of sensors. It provides archiving of sensor parameters, plug-n-play, support for tasking, observation, auto-configuration, sensor accuracy definition. CSIRO Sensor Ontology [11] is written in the Web Ontology Language (OWL) with the aim of representing sensors in a generic domain. A sensor is described in terms of its physical properties (platform, serial ID, weight, dimension) and its concrete implementation. The ontology also models the sensor's functionalities and its sensing capabilities (data flow, sensor's response, input, output). In WIreless Sensor Networks Ontology (WISNO) [3] a sensor node is described through its physical components (radio, memory, cpu and power supply) modeling both its properties and the existing relationship with other domain's concepts. OntoSensor is based on SensorML and it represents components like sensing units, radio and data acquisition board. In the

Semantic Sensor Network (SSN) ontology [10] a sensor is considered as a device able to detect changes or events from the environment to process them producing a result. The ontology can also represents the concept of single observation, platform, deployment, data and energy consumption. Therefore, distributed applications in WSNs can reach a certain degree of interoperability using ontologies. Ontologies allow to represent concept of a specific domain formally and thus heterogeneous applications can share information having the same representation of concepts and consequently a common base of knowledge.

5 Distributed Algorithms and Applications in WSNs

Since sensor nodes present many constraints both in storage and processing, distributed computing can overcome these limitations increasing the performance of the entire WSN. In a distributed application, sensor nodes achieve complex tasks by a cooperative effort. Cooperation implies that sensor nodes need to communicate with each other to perform a given task. Radio communication, however, involves a considerable energy expenditure. Furthermore, managing routing tables for a large set of nodes is challenging. Therefore, researchers have designed several distributed routing protocols in order to increase sensor's battery lifetime by reducing the number of transmissions. Distributed applications are often based on clustering and grouping, on neighbors' discovery and localization, but also on synchronization and on data collection algorithms. Data-centric routing protocols aims to reduce transmissions between nodes removing the need of a node addressing mechanism using a query-based approach involving just a neighbor-to-neighbor communication. Flooding [20], gossiping [20] and SPIN [25] are few examples of data-centric routing protocols. Unlike data-centric algorithms, hierarchical routing protocols are characterized by clusters or groups of nodes and each cluster has a leader. Multi-hop communication is allowed within the same cluster or group, and communication between different clusters or groups is managed by the leader. Data fusion and aggregation are also performed to reduce the number of communications to the base station. Other hierarchical routing protocols are described in [36]. Location-based protocols focus on the idea that knowing the position of the nodes, a query can be addressed just to a limited region eliminating unnecessary transmissions. One significant advantage is that these protocols consider node mobility. Other protocols are based on network flow or Quality of Service (QoS) modeling. A survey on routing protocols for WSNs can be found in [1]. Multicast routing protocols could reduce the number of the messages exchanged, by transmitting simultaneously just to a subset of nodes. A multicast distributed routing algorithm is proposed in [18]. Data aggregation is also a fundamental task in WSNs because aggregation eliminates redundant transmissions saving the WSN's global energy. When sensory data are aggregated and then transmitted, collisions may occur during aggregation. A survey on data aggregation algorithms can be found in [33]. The authors of [33] also discuss the influence of the network topology and the routing protocol on data aggregation. In [43], the authors present a distributed scheduling algorithm to improve the energy management decreasing the time latency. The same chapter also proposes an adaptive version of the algorithm to deal with network topology changes. Detecting faults in WSNs makes it possible to identify occurred error situations in order to handle them properly. A distributed bayesian approach for fault detection is suggested in [30], where cooperation allows a node to infer the nodes source of incorrect data. Reducing the considerable energy consumption caused by channel listening improves node's lifetime. For this reason, several algorithms involving duty cycling techniques have been proposed. In this way, a node alternates sleep and wakeup times. Some of these protocols are based on sleep and wakeup time synchronization. Other protocols do not require synchronization being thus more flexible. The authors of [39] propose the PW-MAC algorithm based on prediction of receiver's wakeup. Distributed data storage overcomes problems related to node's low storage capability. Many approaches rely on the idea of a distributed database. In this context, the authors of [32] propose a cooperative middleware providing a distributed data storage for WSNs. For example, in Hood [42] the node shares its data only with some of its neighbors. In TinyPeds [19] the cluster head receives aggregated data from its cluster's members and stores them in a neighbor cluster head. In [21], the choose of the backup node is probabilistic in order to make the network more robust and efficient. WSNs often handle sensitive or critical data that must be protected from malicious attacks. Since many security algorithms rely on random number generator, the authors of [29] present a distributed algorithm for selecting a node as a number random generator using a leader election approach. The authors of [12] present a sample Ambient Intelligence (AmI) system for minimizing energy consumption in indoor environments according to users' preferences and needs. The system's middleware is distributed on sensory devices for low-level data gathering management, while higher level functionalities are implemented in a centralized manner. In [31] a WSN system including collaboration in the healthcare field is presented. The proposed system is composed of wearable and ambient sensors cooperating to monitor user's vital signs. Ambient sensors provide information about context and the information from wearable and ambient sensors are then fused. In [24] an intrusion detection system involving collaboration among sensors has been proposed. The sensor tracking is an important research aspect exploited by many application, like animal or person tracking. Other collaborative applications can involve data filtering techniques like Kalman filter to estimate the node position [35]. In [26] a system for environmental monitoring is presented. The authors used local storage and Constrained Application Protocol (CoAP) as communication protocol, concluding that a completely distributed algorithm decreases consumptions and increases reliability. In [40] a surveillance system relies on a peer to peer distributed sensor network infrastructure implementing collaborative on-line learning and target localization. The state of art about collaborative WSNs can be found in [28] while a survey about collaborative tracking is provided in [13].

6 An Example of Distributed Application: Synchronization

In this section, we aim to make the reader aware of distributed computing on resources constrained devices, such as sensor nodes of a WSN, showing how a trivial application, such as synchronization, becomes a complex goal on these platforms. Synchronization is a typical distributed application where sensors must align their own clock to share the same notion of time. Since sensors cooperate, having the same time axis is fundamental. Different microcontrollers must perform an action simultaneously or periodically together with the other nodes of the network. Even if the nodes are turned on at the same time, their clocks will drift increasingly because of the tolerances in clock generators. Therefore, synchronization poses some important questions such as how to estimate the frequency drift or how frequently sensor nodes must synchronize their clocks. In addition, to achieve common time knowledge, nodes cooperate through messages and consequently the transmission delay must also be considered. The drift and the propagation delay are thus important parameters to compute in order to perform synchronization with relative time measure. In this case, sensor nodes estimate their relative drift and the propagation delay adjusting their clocks accordingly. Since distributed synchronization relies on communication, a synchronization application must handle channel noise, unreliable packet delivery, packet losses, asymmetrical links between sender and receiver, and must implement some kind of fault tolerance. In literature, several synchronization protocols have been proposed. Some of them are based on hierarchical network topology or they use a probabilistic approach rather then a deterministic synchronization. Figure 1 shows the synchronization phase of the Timing-sync Protocol for Sensor Networks [17] occurring once the hierarchical topology in the network has been established. Synchronization is performed between pairs of nodes. In [15], the authors consider some evaluation metrics such as energy utilization, precision, lifetime, scope and availability, cost and size. A high energy conservation is provided by post-facto synchronization updating clocks only when it is strictly necessary. A synchronization application can be evaluated on the basis of the time the network remains synchronized before performing synchronization again, while cost and size indicate hardware costs. The authors of [38] evaluate the synchronization protocols on the basis of quantitative criteria such as precision, computational complexity, message complexity, convergence time, number of nodes and sleep mode support. In this context, they provide a comparison between some protocols based on these evaluation metrics. They also consider qualitative parameters for synchronization evaluation such as accuracy, scalability and overall complexity including storage requirements, communication overhead and fault tolerance. This brief discussion about synchronization shows that distributed computing applications on low resources devices are usually challenging. These applications must cope with the issues related to cooperation, coordination, scheduling but also with the typical constraints of WSNs such as limited hardware and computing capabilities, reduced energy and bandwidth availability, and latency.

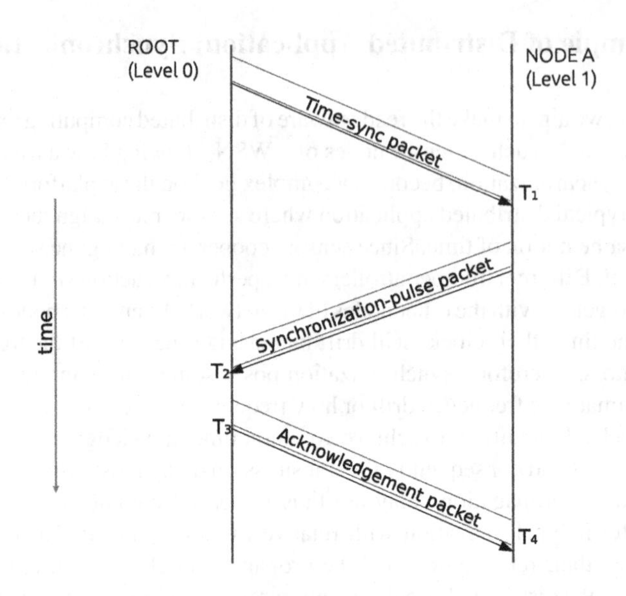

Fig. 1 Synchronization phase of the Timing-sync Protocol for Sensor Networks [17]. Once the root node has sent a time-sync packet to all of its neighbors, each neighbor of level 1 waits for a random time interval before sending a synchronization-pulse packet containing its level and its local time to the root. The root replies with its level, the time received from node A, its local time of arrival and its time of sending. Node A thus estimates the drift and the propagation delay and adjusts its clock accordingly. This process propagates through the network from nodes of level 1 to nodes of level 2 and so on

7 Conclusion

In this chapter, we provided an overview about a new emerging scenario concerning distributed computing on limited resource devices. We discussed it from several points of view, from recently appeared hardware platforms and operating systems to distributed algorithms and applications. We also provided a summary description of some existing ontologies for sensor nodes. We also briefly described a typical distributed application to show how even simple applications may be challenging in this scenario.

Acknowledgments This work has been partially supported by the PON R&C grant MI01_00091 funding the SeNSori project.

References

1. Akkaya, K., Younis, M.: A survey on routing protocols for wireless sensor networks. Ad hoc Netw. **3**(3), 325–349 (2005)
2. Arduino uno http://arduino.cc/en/Main/arduinoBoardUno. (2013). Accessed 7 Sept 2013
3. Avancha, S., Patel, C., Joshi, A.: Ontology-driven adaptive sensor networks. In: MobiQuitous, Boston, pp. 194–202, 2004
4. Banzi, M.: Getting Started with Arduino. O'Reilly Media, Inc., Newton (2009)
5. Bhatti, S., Carlson, J., Dai, H., Deng, J., Rose, J., Sheth, A., Shucker, B., Gruenwald, C., Torgerson, A., Han, R.: Mantis os: An embedded multithreaded operating system for wireless micro sensor platforms. Mob. Netw. Appl. **10**(4), 563–579 (2005)
6. Bose, R., King, J., El-Zabadani, H., Pickles, S., Helal, A.: Building plug-and-play smart homes using the atlas platform. In: Proceedings of the 4th International Conference on Smart Homes and Health Telematic (ICOST), Belfast, June 2006, citeseer(2006)
7. Burns, A., Greene, B.R., McGrath, M.J., O'Shea, T.J., Kuris, B., Ayer, S.M., Stroiescu, F., Cionca, V.: Shimmer-a wireless sensor platform for noninvasive biomedical research. IEEE J. Sens. **10**(9), 1527–1534 (2010)
8. Cao, Q., Abdelzaher, T., Stankovic, J., He, T.: The liteos operating system: Towards unix-like abstractions for wireless sensor networks. In: Proceedings of 7th International Conference on Information Processing in Sensor Networks (IPSN '08), pp. 233–244, April 2008
9. Chien, T.V., Chan, H.N., Huu, T.N.: A comparative study on operating system for wireless sensor networks. In: IEEE International Conference on Advanced Computer Science and Information System (ICACSIS'11), pp. 73–78, (2011)
10. Compton, M., Barnaghi, P., Bermudez, L., García-Castro, R., Corcho, O., Cox, S., Graybeal, J., Hauswirth, M., Henson, C., Herzog, A., et al.: The ssn ontology of the w3c semantic sensor network incubator group. Web Semant. Sci. Serv. Agents on the World Wide Web **17**, 25–32 (2012)
11. Compton, M., Neuhaus, H., Taylor, K., Tran, K.N.: Reasoning about sensors and compositions. In: Proceedings of Semantic Sensor Network, pp. 33–48, (2009)
12. De Paola, A., Gaglio, S., Lo Re, G., Ortolani, M.: Sensor9k: a testbed for designing and experimenting with WSN-based ambient intelligence applications. Pervasive and Mob. Comput. **8**(3), 448–466 (2012)
13. Demigha, O., Hidouci, W.K., Ahmed, T.: On energy efficiency in collaborative target tracking in wireless sensor network: a review. IEEE Commun. Surv. Tutorials **15**(3), 1210–1222 (2013)
14. Dunkels, A., Gronvall, B., Voigt, T.: Contiki - a lightweight and flexible operating system for tiny networked sensors. In: Proceedings of the 29th Annual IEEE International Conference on Local Computer Networks, pp. 455–462, (2004)
15. Elson, J., Römer, K.: Wireless sensor networks: a new regime for time synchronization. ACM SIGCOMM Comput. Commun. Rev. **33**(1), 149–154 (2003)
16. Farooq, M.O., Kunz, T.: Operating systems for wireless sensor networks: a survey. Sensors **11**(6), 5900–5930 (2011)
17. Ganeriwal, S., Kumar, R., Srivastava, M.B.: Timing-sync protocol for sensor networks. In: Proceedings of the 1st international conference on Embedded networked sensor systems (SenSys'03), pp. 138–149. ACM, New York (2003)
18. Gatani, L., Lo Re, G., Gaglio, S.: An efficient distributed algorithm for generating multicast distribution trees. In: International Conference on Parallel Processing Workshops (ICPP 2005 Workshops), pp. 477–484, (2005)
19. Girao, J., Westhoff, D., Mykletun, E., Araki, T.: TinyPEDS: tiny persistent encrypted data storage in asynchronous wireless sensor networks. Ad Hoc Netw. **5**(7), 1073–1089 (2007)
20. Hedetniemi, S.M., Hedetniemi, S.T., Liestman, A.L.: A survey of gossiping and broadcasting in communication networks. Networks **18**(4), 319–349 (1988)
21. Jardak, C., Osipov, E., Mahonen, P.: Distributed information storage and collection for wsns. In: IEEE Internatonal Conference on Mobile Adhoc and Sensor Systems (MASS'07), pp. 1–10, (2007)

22. King, J., Bose, R., Yang, H.I., Pickles, S., Helal, A.: Atlas: A service-oriented sensor platform: Hardware and middleware to enable programmable pervasive spaces. In: Proceedings of 31st IEEE Conference on Local Computer Networks, pp. 630–638, (2006)
23. Klues, K., Liang, C.J.M., Paek, J., Musaloiu-Elefteri, R., Levis, P., Terzis, A., Govindan, R.: Tosthreads: thread-safe and non-invasive preemption in tinyos. In: SenSys, vol. 9, pp. 127–140, (2009)
24. Krontiris, I., Benenson, Z., Giannetsos, T., Freiling, F.C., Dimitriou, T.: Cooperative intrusion detection in wireless sensor networks. In: Wireless Sensor Networks, pp. 263–278. Springer, Berlin (2009)
25. Kulik, J., Heinzelman, W., Balakrishnan, H.: Negotiation-based protocols for disseminating information in wireless sensor networks. Wireless Netw. **8**(2/3), 169–185 (2002)
26. Larios, D., Mora-Merchan, J., Personal, E., Barbancho, J., León, C.: Implementing a distributed wsn based on ipv6 for ambient monitoring. Int. J. Distrib. Sens. Netw. (2013)
27. Levis, P., Madden, S., Polastre, J., Szewczyk, R., Whitehouse, K., Woo, A., Gay, D., Hill, J., Welsh, M., Brewer, E., et al.: Tinyos: an operating system for sensor networks. In: Ambient intelligence, pp. 115–148. Springer, Heidelberg (2005)
28. Li, W., Bao, J., Shen, W.: Collaborative wireless sensor networks: a survey. In: IEEE International Conference on Systems, Man, and Cybernetics (SMC'11), pp. 2614–2619, (2011)
29. Lo Re, G., Milazzo, F., Ortolani, M.: Secure random number generation in wireless sensor networks. In: Proceedings of the 4th international conference on Security of Information and Networks, pp. 175–182, (2011)
30. Lo Re, G., Milazzo, F., Ortolani, M.: A distributed bayesian approach to fault detection in sensor networks. In: Proceedings of the IEEE Global Communications Conference (GLOBECOM'12), pp. 634–639, (2012)
31. Nakamura, M., Nakamura, J., Lopez, G., Shuzo, M., Yamada, I.: Collaborative processing of wearable and ambient sensor system for blood pressure monitoring. Sensors **11**(7), 6760–6770 (2011)
32. Piotrowski, K., Langendoerfer, P., Peter, S.: tinydsm: a highly reliable cooperative data storage for wireless sensor networks. In: International Symposium on Collaborative Technologies and Systems (CTS'09), pp. 225–232, (2009)
33. Rajagopalan, R., Varshney, P.K.: Data aggregation techniques in sensor networks: a survey. IEEE Comm. Surv. Tutorials **8**, 48–63 (2006)
34. Reusing, T.: Comparison of operating systems tinyos and contiki. Sens. Nodes-Operation, Netw. Appli. (SN) **7** (2012)
35. Ribeiro, A., Giannakis, G.B., Roumeliotis, S.I.: Soi-kf: distributed kalman filtering with low-cost communications using the sign of innovations. IEEE Trans. Sig. Process. **54**(12), 4782–4795 (2006)
36. Singh, S.K., Singh, M., Singh, D.: A survey of energy-efficient hierarchical cluster-based routing in wireless sensor networks. Int. J. Adv. Networking and Appl. (IJANA) **2**(02), 570–580 (2010)
37. Strazdins, G., Elsts, A., Selavo, L.: Mansos: easy to use, portable and resource efficient operating system for networked embedded devices. In: Proceedings of the 8th ACM Conference on Embedded Networked Sensor Systems (SenSys '10), pp. 427–428, ACM (2010)
38. Sundararaman, B., Buy, U., Kshemkalyani, A.D.: Clock synchronization for wireless sensor networks: a survey. Ad Hoc Netw. **3**(3), 281–323 (2005)
39. Tang, L., Sun, Y., Gurewitz, O., Johnson, D.B.: Pw-mac: an energy-efficient predictive-wakeup mac protocol for wireless sensor networks. In: INFOCOM, 2011 Proceedings IEEE, pp. 1305–1313 (2011)
40. Wang, X., Wang, S., Bi, D.: Distributed visual-target-surveillance system in wireless sensor networks. IEEE Trans. Syst. Man Cybern. B Cybern. **39**(5), 1134–1146 (2009)
41. Waspmote datasheet: Available online at http://www.libelium.com/downloads/documentation/waspmote_datasheet.pdf(2013). Accessed 7 Sept 2013
42. Whitehouse, K., Sharp, C., Brewer, E., Culler, D.: Hood: a neighborhood abstraction for sensor networks. In: Proceedings of the 2nd international conference on Mobile systems, applications, and services, pp. 99–110. ACM (2004)

43. Yu, B., Li, J., Li, Y.: Distributed data aggregation scheduling in wireless sensor networks. In: IEEE INFOCOM 2009, pp. 2159–2167 (2009)
44. Zennaro, M., Bagula, A., Gascon, D., Noveleta, A.B.: Long distance wireless sensor networks: simulation vs reality. In: Proceedings of the 4th ACM Workshop on Networked Systems for Developing Regions, pp. 12:1–12:2. ACM (2010)
45. Zolertia z1 datasheet: Available online at http://zolertia.sourceforge.net/wiki/images/e/e8/Z1_RevC_Datasheet.pdf (2013). Accessed 7 Sept 2013

From IEEE 802.15.4 to IEEE 802.15.4e: A Step Towards the Internet of Things

Domenico De Guglielmo, Giuseppe Anastasi and Alessio Seghetti

Abstract Wireless Sensor and Actuator Networks (WSANs) are expected to have a key role in the realization of the future Internet of Things that will connect to the Internet any kind of devices, living beings, and things. A number of standards have been released over the last years to support their development and encourage interoperability. In addition IETF has defined a set of protocols to allow the integration of sensor and actuator devices into the Internet. In this chapter we focus on the 802.15.4e, released by IEEE in 2012 to enhance and add functionality to the previous 802.15.4 standard, so as to address the emerging needs of embedded industrial applications. We describe how the limitations of the 802.15.4 standard have been overcome by the new standard, and we also show some simulation results to better highlight this point.

1 Introduction

In the future Internet of Things (IoT) a very large number of real-life objects will be connected to the Internet, generating and consuming information. IoT elements will no longer be only computers and personal communication devices, as in the current Internet, but all kinds of devices (e.g., cars, robots, machine tools), living beings (persons, animals, and plants) and things (e.g., garments, food, drugs, etc.). A key role in the realization of the IoT paradigm will be played by wireless sensor/actuator

D. De Guglielmo (✉) · G. Anastasi · A. Seghetti
Department of Information Engineering, University of Pisa, Pisa, Italy
e-mail: domenico.deguglielmo@iet.unipi.it

G. Anastasi
e-mail: giuseppe.anastasi@iet.unipi.it

A. Seghetti
e-mail: seghetti@iet.unipi.it

S. Gaglio and G. Lo Re (eds.), *Advances onto the Internet of Things*,
Advances in Intelligent Systems and Computing 260, DOI: 10.1007/978-3-319-03992-3_10,

networks (WSANs) that will behave as a sort of digital skin, providing a virtual layer through which any computational system can interact with the physical world [1, 2].

A WSAN consists of a number of sensor and actuator devices deployed over a geographical area and interconnected through wireless links. Sensor devices gather information from the physical environment or a monitored system (e.g., temperature, pressure, vibrations), optionally perform a preliminary local processing of acquired information, and send (raw or processed) data to a controller. Based on the received information, the controller performs appropriate actions, through actuator devices, to change the behavior of the physical environment or the monitored system.

WSANs are already used in many application domains, ranging from traditional environmental monitoring and location/tracking applications to more constrained applications such as those in the industrial [3] and healthcare domain [4]. In the industrial field WSAN applications include factory automation [5], distributed and process control [6–8], real-time monitoring of machinery health, detection of liquid/gas leakage, radiation check [9] and so on. In the healthcare domain WSANs have been considered for the monitoring of physiological data in chronicle patients and transparent interaction with the healthcare system.

In many application domains *energy efficiency* is usually the main concern in the design of a WSAN. This is because sensor/actuator devices are typically powered by batteries with a limited energy budget and their replacement can be expensive or, even, impossible [10]. However, in some relevant application domains additional requirements need to be considered, such as *timeliness, reliability, robustness, scalability,* and *flexibility* [3, 11]. *Reliability* and *timeliness* are very critical issues for industrial and healthcare applications. If data packets are not delivered to the final destination, correctly and within a pre-defined deadline, the correct behavior of the system (e.g., the timely detection of a critical event) may be compromised. The maximum allowed latency depends on the specific application. Typical values ranges from tens of milliseconds (e.g., for discrete manufacturing and factory automation), to seconds (e.g., for process control), and even minutes (e.g., for asset monitoring) [11].

In recent years many standards have been issued by international bodies to support the development of WSANs in different application domains. They include IEEE 802.15.4 [12], ZigBee [13], Bluetooth [14], WirelessHART [15] and ISA-100.11a [16]. At the same time, the Internet Engineering Task Force (IETF) has defined a number of protocols to facilitate the integration of smart objects (i.e., sensor and actuator devices) into the Internet. The most important of them are the *IPv6 over Low power WPAN* (6LoWPAN) [17] adaptation layer protocol that allows the integration of smart objects into the Internet, the *Routing Protocol for Low power and Lossy networks* (RPL) [18], and the *Constrained Application Protocol* (CoAP) [19] that enables web applications on smart objects.

In this chapter we focus on the IEEE 802.15.4 standard [12] that defines the physical and Medium Access Control (MAC) layers of the OSI reference model and is complemented by the ZigBee specifications [13] covering the networking and application layers. The 802.15.4 standard was originally conceived for applications without special requirements in terms of latency, reliability and scalability. In order to overcome these limitations, in 2008 the IEEE set up a Working Group (named

802.15e WG) with the aim of enhancing and adding functionality to the 802.15.4 MAC, so as to address the emerging needs of embedded industrial applications [20]. The final result was the release of the 802.15.4e standard in 2012. In the following sections, after emphasizing the limitations and deficiencies of the 802.15.4 standard, we will show how they have been overcome in the new standard. Specifically, we will describe the new access modes defined by 802.15.4e, with special emphasis on the *Time Slotted Channel Hopping* (TSCH) mode. We will also present some simulation results to better highlight the performance limitations of 802.15.4 and show that they are overcome by 802.15.4e.

The remainder of this chapter is organized as follows. Section 2 describes the 802.15.4 standard. Section 3 highlights its main limitations and deficiencies. Section 4 describes the new functionalities provided by the 802.15.4e standard. Section 5 compares the performance of 802.15.4 and 802.15.4e in a simple scenario through simulation. Finally, Sect. 6 concludes the chapter.

2 IEEE 802.15.4 Standard

IEEE 802.15.4 [12] is a standard for low-rate, low-power, and low-cost Personal Area Networks (PANs). A PAN is formed by one PAN coordinator which is in charge of managing the whole network, and, optionally, by one or more coordinators that are responsible for a subset of nodes in the network. Regular nodes must associate with a (PAN) coordinator in order to communicate. The supported network topologies are *star* (single-hop), *cluster-tree* and *mesh* (multi-hop).

The standard defines two different channel access methods: a *beacon enabled* mode and a *non-beacon enabled* mode. The beacon enabled mode provides a power management mechanism based on a duty cycle. It uses a superframe structure (see Fig. 1) which is bounded by *beacons,* i.e., special synchronization frames generated periodically by the coordinator node(s). The time between two consecutive beacons is called *Beacon Interval* (BI), and is defined through the *Beacon Order* (BO) parameter ($BI = 15.36 \cdot 2^{BO}$ ms, with $0 \leq BO \leq 14$).[1] Each superframe consists of an active period and an inactive period. In the active period nodes communicate with the coordinator they are associated with, while during the inactive period they enter a low power state to save energy. The active period is denoted as *Superframe Duration* (SD) and its size is defined by the *Superframe Order* (SO) parameter ($SD = 15.36 \cdot 2^{SO}$ ms, with $0 \leq SO \leq BO \leq 14$). It can be further divided into a *Contention Access Period* (CAP) and a *Contention Free Period* (CFP). During the CAP a slotted CSMA-CA algorithm is used for channel access, while in the CFP communication occurs in a Time Division Multiple Access (TDMA) style by using a number of *Guaranteed Time Slots* (GTSs), pre-assigned to individual nodes. In the non-beacon enabled mode there is no superframe, nodes are always active (energy conservation is delegated to

[1] Throughout the chapter we assume that the sensor network operates in the 2.4 GHz frequency band.

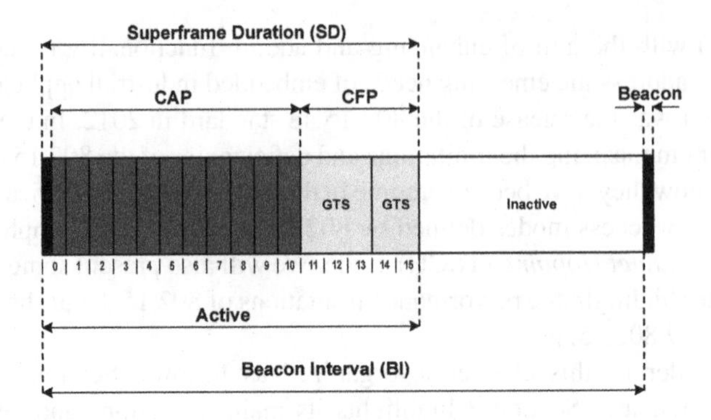

Fig. 1 IEEE 802.15.4 Superframe Structure

the layers above the MAC protocol) and use an unslotted CSMA-CA algorithm for channel access.

2.1 CSMA-CA Algorithm

The CSMA-CA algorithm is used in both the *beacon enabled* mode (during the CAP portion of the active period) and the *non-beacon enabled* mode. In the beacon-enabled mode a slotted scheme is used—i.e., all operations are aligned to backoff period slots (whose duration is 320 μ s)—while in the non-beacon enabled mode there is no such alignment.

Upon receiving a data frame to be transmitted, the CSMA-CA algorithm performs the following steps.

1. A set of state variables is initialized, i.e., the contention window size ($CW = 2$, only for the slotted variant), the number of backoff stages carried out for the on-going transmission ($NB = 0$), and the backoff exponent ($BE = macMinBE$).
2. A random backoff time, uniformly distributed in the range $[0, 2^{BE} - 1] \cdot 320\,\mu$s, is generated and used to initialize a backoff timer. In the beacon-enabled mode, the starting time of the backoff timer is aligned with the beginning of the next backoff slot. In addition, if the backoff time is larger than the residual CAP duration, the backoff timer is stopped at the end of the CAP and resumed at the beginning of the next superframe. When the backoff timer expires, the algorithm proceeds to step 3.
3. A Clear Channel Assessment (CCA) is performed to check the state of the wireless medium.
    (a) If the medium is busy, the state variables are updated as follows: $NB = NB + 1$, $BE = min(BE + 1, macMaxBE)$ *and* $CW = 2$ (only for the slotted variant). If the number of backoff stages has exceeded the maximum admis-

sible value (i.e. $NB > macMaxCSMABackoffs$), the frame is dropped. Otherwise, the algorithm falls back to step 2.

(b) If the medium is free and the access mode is unslotted, the frame is immediately transmitted.

(c) If the medium is free and the access mode is slotted, then $CW = CW - 1$. If $CW = 0$ then the frame is transmitted.[2] Otherwise the algorithm falls back to step 3 to perform a second CCA.

It should be noted that, unlike the algorithm used in 802.11 WLANs, the 802.15.4 slotted CSMA-CA does not guarantee a transmission at the end of the backoff time after the channel is found clear. Instead, transmission occurs only if the wireless medium is found free for two consecutive CCAs. The complete CSMA-CA algorithm, both in the slotted and unslotted version, is depicted in Fig. 2.

The 802.15.4 CSMA-CA algorithm also includes an optional retransmission mechanisms for improving reliability. When retransmissions are enabled, the destination node must send an acknowledgement whenever it correctly receives a data frame (the acknowledgement is not sent in case of collision and corrupted frame reception). On the sender side, if the acknowledgment is not (correctly) received within the pre-defined timeout, a retransmission is scheduled. The frame can be re-transmitted up to a maximum number of times, specified by the MAC parameter *macMaxFrameRetries*. Upon exceeding these value, the data frame is rejected and a failure notification is sent by the MAC sublayer to the upper layers.

3 Limitations of IEEE 802.15.4 MAC

The performance of the 802.15.4 MAC protocol, both in BE mode and NBE mode, have been thoroughly investigated in the past. As a result of this extended study, a number of limitations and deficiencies have been identified, the main of which are discussed below.

- *Unbounded Delay*. Since the 802.15.4 MAC protocol, both in BE mode and NBE mode, is based on the CSMA-CA algorithm it cannot guarantee any bound on the maximum delay experienced by data to reach the final destination. This makes 802.15.4 unsuitable for time-critical application scenarios where a low and deterministic delay is required (e.g., industrial and medical applications).
- *Limited communication reliability*. The 802.15.4 MAC in BE mode provides a very low delivery ratio, even when the number of nodes is not so high which make it unsuitable for critical application scenarios. This is mainly due to the random-access method (i.e., CSMA-CA algorithm) and the synchronization introduced by the periodic Beacon. A similar behavior also occurs in the NBE mode when

[2] In the beacon-enabled mode, before starting the frame transmission, the algorithm calculates whether it is able to complete the operation within the current CAP. If there is not enough time, the transmission is deferred to the next superframe.

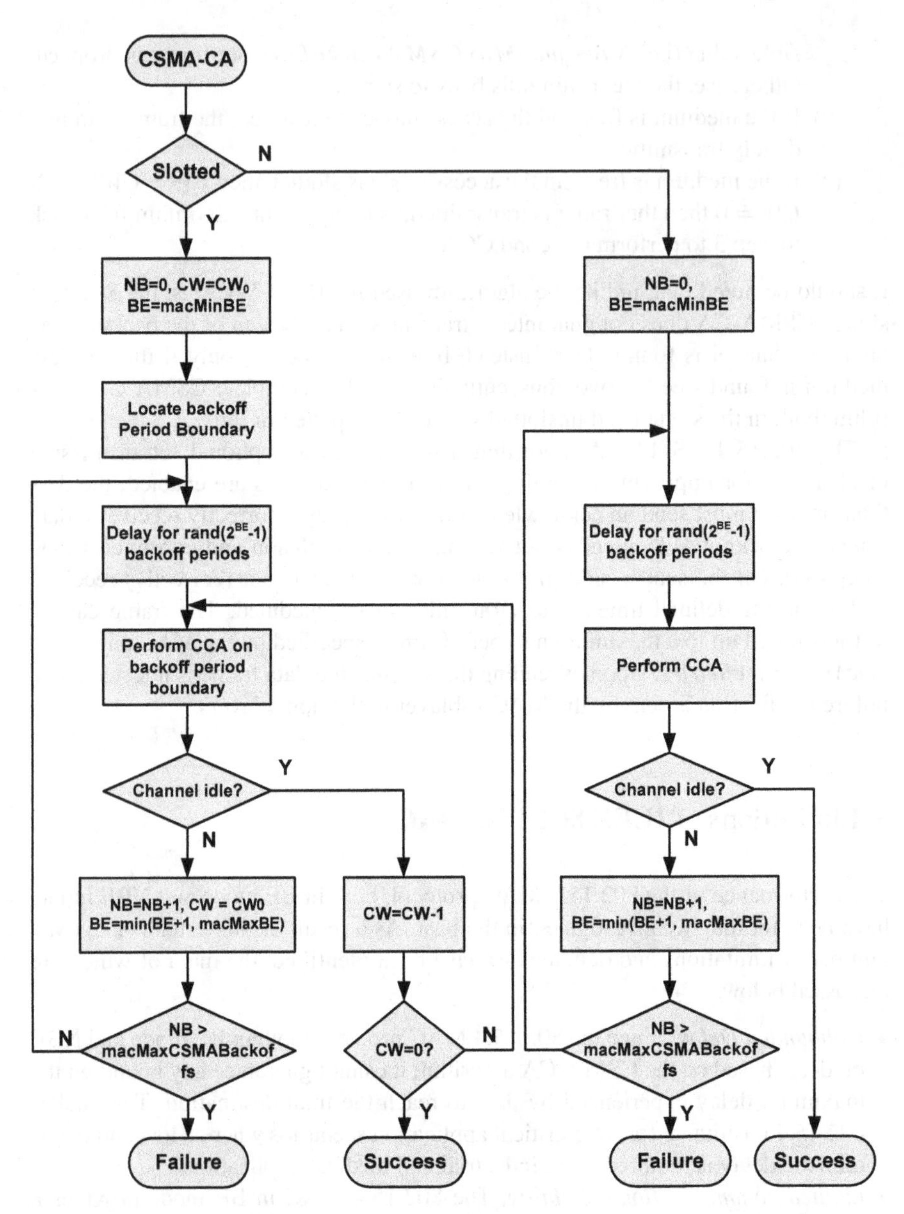

Fig. 2 CSMA-CA algorithm

a large number of nodes start transmitting simultaneously (e.g., in event-driven applications).

- *No protection against interferences/fading.* Interferences and multi-path fading are very common phenomena, especially in application scenarios where sensor/actuator networks are expected to be used. Unlike other wireless network

technologies (e.g., Bluetooth [14], ISA 100.11a [16] and WirelessHART [15]), the 802.15.4 MAC takes a single-channel approach and has no built-in frequency hopping mechanism to protect against interferences and multi-path fading. Hence, the network is subject to frequent instabilities and may also collapse. This make 802.15.4 unsuitable to be used in critical application scenarios (e.g., industrial or healthcare applications).
- *Powered relay nodes.* The 802.15.4 support both single-hop (star) and multi-hop (peer-to-peer) topologies. In principle, the BE mode could be used to form multi-hop PAN with a tree topologies where intermediate node do not need to stay active all the time. In practice, intermediate relay nodes in 802.15.4 networks (both with tree and mesh topologies) need to keep their radio on all the time, which leads to a large energy consumption.

4 IEEE 802.15.4e Standard

To overcome the limitations of the 802.15.4 standard, emphasized in the previous section, the 802.15.4e Working Group was created by IEEE in 2008 to redesign the existing 802.15.4 MAC protocol. The goal was to define a low-power multi-hop MAC protocol, capable of addressing the emerging needs of embedded industrial applications. The final result was the IEEE 802.15.4e MAC Enhancement Standard document [20], approved in 2012. Specifically, the 802.15.4e standard extends the previous 802.15.4 standard by introducing two different categories of MAC enhancements, namely *MAC behaviors* to support specific application domains and *general functional improvements* that are not tied to any specific application domain. In practice, 802.1.5.4e borrows many ideas from existing standards for industrial applications (i.e., WirelessHART [15] and ISA 100.11.a [16]), including slotted access, shared and dedicated slots, multi-channel communication, and frequency hopping.

The MAC behavior modes defined by the 802.1.5.4e standard are listed below. They will be described in the next section.

- *Radio Frequency Identification Blink (BLINK).* intended for applications such as item and people identification, location, and tracking;
- *Asynchronous multi-channel adaptation (AMCA).* targeted to application domains where large deployments are required (e.g., process automation/control, infrastructure monitoring, etc.);
- *Deterministic and Synchronous Multi-channel Extension (DSME).* aimed to support industrial and commercial applications with stringent timeliness and reliability requirements;
- *Low Latency Deterministic Network (LLDN).* intended for applications requiring very low latency requirement (e.g., factory automation, robot control)
- *Time Slotted Channel Hopping (TSCH).* targeted to application domains such as process automation.

The general functional enhancements, not specifically tied to a particular application domain, are as follows.

- *Low Energy (LE).* This mechanism is intended for applications that can trade latency for energy efficiency. It allows a device to operate with a very low duty cycle (e.g., 1 % or below), while appearing to be *always on* to the upper layers. This mechanism is extremely important for enabling the Internet of Things paradigms as Internet protocols have been designed assuming that hosts are always on. However, it may be useful also in other applications scenarios (e.g., event-driven and/or infrequent communications, networks with mobile nodes).
- *Information Elements (IE).* The concept of IEs was already present in the 802.15.4 standard. It is an extensible mechanism to exchange information at the MAC sublayer.
- *Enhanced Beacons (EB).* Extended Beacons are an extension of the 802.15.4 beacon frames and provide a greater flexibility. They allow to create application-specific beacons, by including relevant IEs, and are used in the DSME and TSCH modes.
- *Multipurpose Frame.* This mechanism provides a flexible frame format that can address a number of MAC operations. It is based on IEs.
- *MAC Performance Metrics* are a mechanism to provide appropriate feedback on the channel quality to the networking and upper layers, so that appropriate decision can be taken. For instance the IP protocol running on top of 802.15.4e MAC may implement dynamic fragmentation of datagrams depending on the channel conditions.
- *Fast Association (FastA).* The 802.15.4 association procedure introduces a significant delay in order to save energy. For time-critical application latency has priority over energy efficiency. The FastA mechanism allows a device to associate in a reduced amount of time.

4.1 802.15.4e MAC Behavior Modes

In this section we describe the MAC behavior modes that have been introduced in the previous section. The description is necessarily brief for the sake of space. The reader can refer to [20] for details.

The *Radio Frequency Identification Blink (BLINK)* mode is intended for application domains such as item/people identification, location, and tracking and is, thus, very relevant in the perspective of Internet of Things. Specifically, it allows a device to communicate its ID (e.g., a 64-bit source address) to other devices. The device can also transmit its alternate address and, optionally, additional data in the payload. No prior association is required and no acknowledgement is provided to the sending device. The BLINK mode is based on a minimal frame consisting only of the header fields that are necessary for its operations. The BLINK frame can be used by "transmit only" devices to co-exist within a network, utilizing an Aloha protocol.

The *Asynchronous multi-channel adaptation (AMCA)* mode is targeted to application domains where large deployments are required, such as smart utility networks, infrastructure monitoring networks, and process control networks. In such networks using a single, common, channel for communication may not allow to connect all the devices in the same PAN. In addition, the variance of channel quality is typically large, and link asymmetry may occur between two neighboring devices (i.e., a device may be able to transmit to a neighbor but unable to receive from it). The AMCA mode relies on asynchronous multi-channel adaptation and can be used only in non Beacon-Enabled PANs.

The *Deterministic and Synchronous Multi-channel Extension (DSME)* mode is intended for the support of industrial applications (e.g., process automation, factory automation, smart metering), commercial applications (such as home automation, smart building, entertainment) and healthcare applications (e.g., patient monitoring, telemedicine). This kind of applications requires low and deterministic latency, high reliability, energy efficiency, scalability, flexibility, and robustness [20]. As mentioned in Sect. 2, the 802.15.4 standard provides *Guaranteed Time Slots (GTSs)*. However, the GTS mode has a number of limitations. It only includes up to seven slots and, thus, it is not able to support large networks. In addition, it relies on a single frequency channel. DSME enhances GTS by grouping multiple superframes to form a multi-superframe and using multi-channel operation. Like GTS, DSME runs on Beacon-enabled PANs. All the devices in the PAN synchronize to multi-superframes via beacon frames. A multi-superframe is a cycle of superframes, where each superframe includes the beacon frame, the Contention Access Period, and Contention Free Period (i.e., GTS slot). A pair of nodes wakes up at a reserved GTS slot to exchange a data frame and an ACK frame. In order to save energy, DSME uses CAP reduction, i.e., the Contention Access Period (CAP) is only in the first superframe of the multi-superframe, while it is suppressed in subsequent superframes.

The *Low Latency Deterministic Network (LLDN)* mode is mainly targeted to industrial and commercial applications requiring low and deterministic latency. Typical application domains addressed by LLDN include factory automation (e.g., automotive manufacturing), robots, overhead cranes, portable machine tools, milling machines, computer-operated lathes, automated dispensers, cargo, airport logistics, automated packaging, conveyors. In this kind of applications typically there are a large number of sensors/actuators observing and controlling a system, e.g., a production line or a conveyor belt. In addition, applications have very low requirements in terms of latency (transmission of sensor data in 5–50 ms, and low round-trip time) [20]. To guarantee stringent latency requirements of target applications LLDN only supports the star (i.e., single hop) topology, and uses a *superframe*, based on timeslots, with small packets. Keeping the size of packets (and, hence, timeslots) short leads to superframes with short duration (e.g., 10 ms). Obviously, the number of timeslots in a superframe determines the number of devices that can access the channel. Since the number of devices may very large (there may be more than 100 devices per PAN coordinator) LLDN allows the PAN coordinator to use multiple transceivers on different channels. In the LLDN mode each superframe consists of a *beacon timeslot*, *management timeslots* (if present), and a number of *base timeslots* of equal size. Base timeslots include uplink timeslots and bidirectional timeslots. There are two categories of base timeslot, namely *dedicated* and *shared group* timeslots. Dedicated timeslots are assigned to a specific node (owner) that has the exclusive access on them, while shared group timeslots are assigned to more than one device. The devices use the slotted CSMA-CA algorithm described in Sect. 2 to contend for shared group timeslots. In addition, they use a simple addressing scheme with 8-bit addresses in. The LLDN mode includes a *Group ACK* (GACK) function to reduce the bandwidth overhead. GACK is sent by the PAN coordinator in a superframe to stimulate the retransmission of failed transmission in uplink timeslots.

The *Time Slotted Channel Hopping (TSCH)* mode is mainly intended for the support of process automation applications with a particular focus on equipment and process monitoring. Typical segments of the TSCH application domain include oil and gas industry, food and beverage products, chemical products, pharmaceutical products, water/waste water treatments, green energy production, climate control [20]. TSCH combines *time slotted access*, already defined in the IEEE 802.15.4 MAC protocol, with *multi-channel* and *channel hopping* capabilities. Time slotted access increases the potential throughput that can be achieved, by eliminating collision among competing nodes, and provides deterministic latency to applications. Multi-channel allows more nodes to exchange their frames at the same time (i.e., in the same time slot), by using different channel offsets. Hence, it increases the network capacity. In addition, channel hopping mitigates the effects of interference and multipath fading, thus improving the communication reliability. Hence, TSCH provides increased network capacity, high reliability and predictable latency, while maintaining very low duty cycles (i.e., energy efficiency) thanks to the time slotted access mode. TSCH is also topology independent as it can be used to form any network topology (e.g., star, tree, partial mesh or full mesh). It is particularly well-suited for multi-hop networks where frequency hopping allows for efficient use of the available resources.

4.2 Time Slotted Channel Hopping (TSCH) Mode

Among the various access modes defined by the 802.15.4e standard, *TSCH* is certainly the most complex and interesting one. Hence, in the following we will provide a more detailed description of it.

In the *TSCH* mode nodes synchronize on a periodic slotframe consisting of a number of timeslots. Figure 3 shows a slotframe with 4 timeslots. Each timeslot allows a node to send a maximum-size data frame and receive the related acknowledgement (Fig. 4). If the acknowledgement is not received within a predefined timeout, the retransmission of the data frame is deferred to the next time slot assigned to the same (sender-destination) couple of nodes.

One of the main characteristics of TSCH is the multi-channel support, based on channel hopping. In principle 16 different channels are available for communication. Each channel is identified by a *channelOffset* i.e., an integer value in the range [0:15].

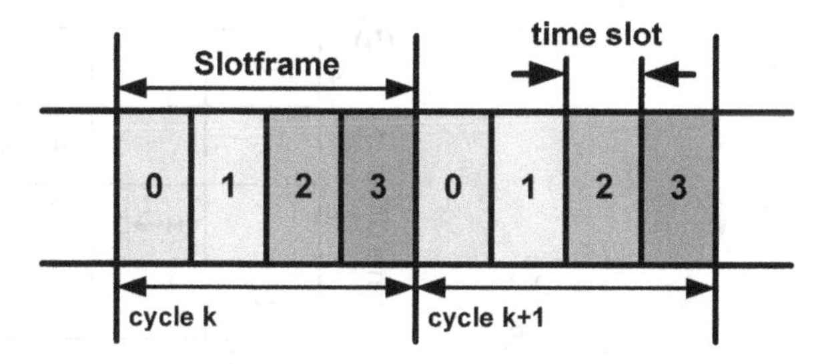

Fig. 3 Slotframe

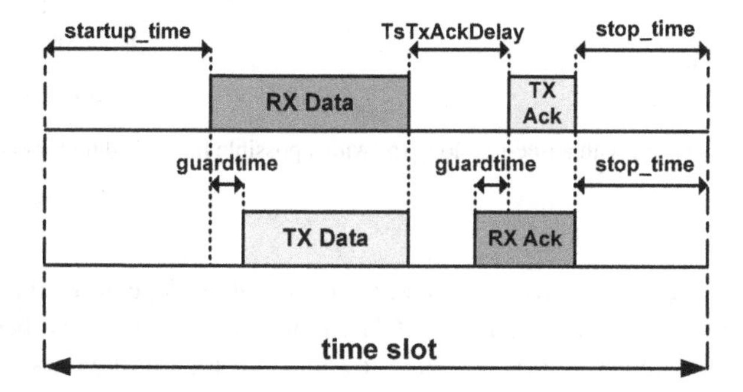

Fig. 4 Timeslot

However, some of these frequencies could be blacklisted (because of low quality channel) and, hence, the total number of channels $N_{channels}$ available for channel hopping may be lower than 16. In TSCH a link is defined as the pairwise assignment of a directed communication between devices in a given timeslot on a given channel offset [20]. Hence, a link between communicating devices can be represented by a couple specifying the timeslot in the slotframe and the channel offset used by the devices in that timeslot. Let denote a link between two devices. Then, the frequency f to be used for communication in timeslot of the slotframe is derived as follows.

$$f = F[(ASN + channeloffset) \% N_{channels}] \quad (1)$$

where is the *Absolute Slot Number*, defined as the total number of timeslots elapsed since the start of the network (or an arbitrary start time determined by the PAN coordinator). It increments globally in the network, at every timeslots, and is thus used by devices as timeslot counter. Function F can be implemented as a lookup table. Thanks to the multi-channel mechanism several simultaneous communications can take place in the same timeslot, provided that different communications use different channel offsets. Also, Eq. 1 implements the channel hopping mechanism by returning a different frequency for the same link at different timeslots.

Figure 5 shows a possible link schedule for data collection in a simple sensor network with a tree topology. We have assumed that the slotframe consists of four

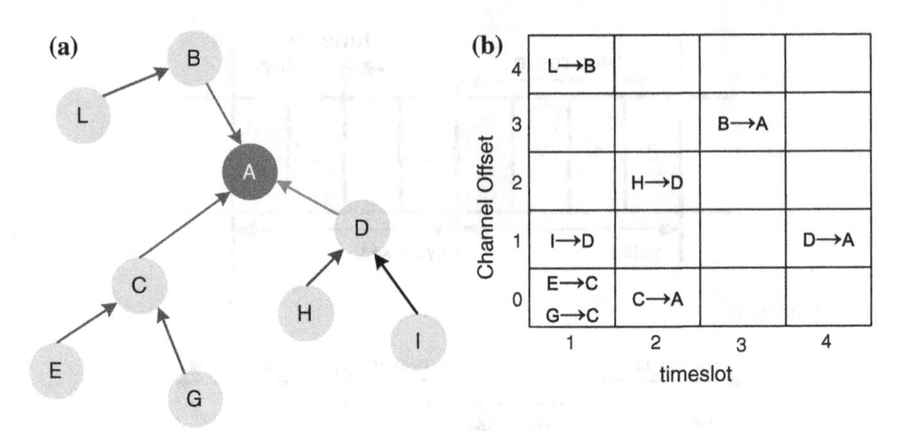

Fig. 5 A sensor network with a tree-topology (**a**) with a possible link schedule for data-collection (**b**)

timeslots and there are only five channel offsets available. We can see that, thanks to the multi-channel approach used by TSCH, eight transmissions have been accommodated in a time interval corresponding to four timeslots. In the allocation shown in Fig. 5 all links but one are *dedicated* links, i.e., allocated to a single device for communication. The 802.15.4e standard also allows *shared* links, i.e., links intentionally allocated to more than one device for transmission. This is the case of the link [1,0] allocated to both nodes E and G.

Since shared links can be accessed by more than one transmitter, collisions may occur that result in a transmission failure. To reduce the probability of repeated collisions, the standard defines a retransmission backoff algorithm. The latter is invoked by a sending device whenever a data frame is transmitted on a shared link and the related acknowledgment is not received. The data frame will be retransmitted in the next link assigned to the sending device and with the same destination, which may be either a shared link or a dedicated link. The retransmission algorithm relies on a backoff delay and works as follows. The retransmission backoff only applies to the transmission on shared links, whereas dedicated links are accessed without any delay. The retransmission backoff is calculated using an exponential algorithm analogous to that described in Sect. 2 for CSMA-CA (it is still based on *macMaxBE* and *macMinBE*). However, in TSCH the backoff delay is expressed in terms of number of shared links that must be skipped. The backoff window increases for each consecutive failed transmission in a shared link, while it remains unchanged when a transmission failure occurs in a dedicated link. A successful transmission in a shared link resets the backoff window to the minimum value. The backoff window does not change when a transmission is successful in a dedicated link but there are still other frames to transmit (the transmission queue is not empty). The backoff window is reset to the minimum value if the transmission in a dedicated link is successful and the transmit queue is then empty.

A key element in TSCH is the link schedule, i.e., the assignment of links to nodes for data transmissions. Of course, neighboring nodes may interfere and, hence, they should not be allowed to transmit in the same timeslot and with the same channel offset. The multi-channel mechanism makes the link scheduling problem easier with respect to the traditional scenario where a single channel is used. However, finding out an optimal schedule may not be a trivial task, especially in large networks with multi-hop topology. The problem is even more challenging in dynamic networks where the topology changes over time (e.g., due to mobile nodes). It may be worthwhile emphasizing here that the derivation of an appropriate link schedule is out of the scope of the 802.15.4e standard. The latter just defines mechanisms to execute a link schedule, however, it does not specify how to derive such a schedule. This is left to upper layers.

A number of link scheduling algorithms have been specifically proposed for TSCH [21–23]. Also previous solutions for slotted multi-channel systems can be easily adapted to TSCH. Link scheduling algorithms can be broadly classified into two different categories, namely *centralized* and *distributed* algorithms. In centralized solutions [22] there is a specific node in the network (typically, the PAN coordinator) that is in charge of creating and updating the link schedule, based on information received by network nodes (about neighbors and generated traffic). Since the PAN coordinator has a global knowledge of the network status, in terms of network topology and traffic matrix, it can create very efficient link schedules. However, the link schedule has to be re-computed each time the network conditions change. Hence, the centralized approach is not very appealing for dynamic networks (e.g., networks with mobile nodes), where a distributed approach is typically more suitable. In a distributed link scheduling algorithm [21, 23] each node decide autonomously which link to activate with its neighbors, based on local and, hence, partial, information.

5 Performance Comparison

To measure the potential performance improvements that can be achieved when using IEEE 802.15.4e, instead of IEEE 802.15.4, we performed a set of simulation experiments using the ns2 simulation tool [24]. Specifically, we considered the 802.15.4 MAC in Beacon Enabled (BE) mode and Non Beacon Enabled (NBE) mode, and compared its performance to that of the 802.15.4e MAC in TSCH mode. To make the comparison fair and, also, to better emphasize the performance improvements that can be achieved with 802.15.4e, in TSCH we did not consider the multi-channel and frequency hopping mechanisms, i.e., we assumed a single channel frequency. Under this assumption TSCH reduces to a simple TDMA scheme.

In our analysis we considered a sensor network with star topology, where the sink node acts as the PAN coordinator and sensor nodes are placed in a circle centered at the PAN coordinator, 10 m far from it. The transmission range was set to 15 m, while the carrier sensing range was set to 30 m (according to the model in [25]). We considered a periodic reporting application where data acquired by sensors have to be reported periodically to the PAN coordinator. Time is divided into communication periods of duration T and each sensor node generates one data packets every T seconds.

To evaluate the performance of the different access modes, we derived the following performance indices.

- *Latency*, defined as the average time from when the packet transmission is started at the source node to when the same packet is correctly received by the PAN coordinator. It characterizes the *timeliness* of the system.
- *Delivery ratio*, defined as the ratio between the number of data packets correctly received by the PAN coordinator and the total number of data packets generated by *all* sensor nodes. It measures the network *reliability* in the data collection process.
- *Energy per packet*, defined as the total energy consumed by each sensor node divided by the number of data packets correctly delivered to the PAN coordinator. It measures the *energy efficiency* of the system.

The energy consumed by a sensor node was calculated using the model presented in [26], based on the Chipcon CC2420 radio transceiver [27]. This model supports the following radio states: *transmit*, *receive*, *idle* (the transceiver is on, but it is not transmitting nor receiving, i.e., it is monitoring the channel) and *sleep* (the transceiver is off and can be switched back on quickly).

The operating parameter values used in our experiments are shown in Table 1. The acknowledgement mechanism was always enabled in all the considered modes. When using the 802.15.4 BE mode the communication period corresponds to the Beacon period. We set BO = 6, which corresponds to a Beacon period of approximately 1 s (0.983 s to be precise). To make the comparison fair we used the same T value also for NBE and TSCH. In our experiments, for each simulated scenario, we performed 10 independent replications, where each replication consists of 1000 communication periods. For each replication we discarded the initial transient interval (10 % of the overall duration) during which nodes associate to the PAN coordinator node and start generating data packets. The results shown below are averaged over all the different replications. We also derived confidence intervals through the independent replication method. However, they are so small that they cannot be appreciated in the figures below.

Figures 6, 7 and 8 show the performance of the different MAC modes, for an increasing number of sensor nodes, in terms of delivery ratio, average latency, and energy efficiency, respectively. As expected, TSCH outperforms both BE and NBE for all the considered indices. Specifically, it performs a 100 % delivery ratio, with low (and fixed) latency and minimal energy consumption. In addition, its performance do not depend on the number of sensor nodes, at least until this number is less than or equal to the number of timeslots in the slotframe. Conversely, the 802.15.4 BE mode exhibits very poor performance, even when the number of sensor nodes is relatively high (e.g., with 20 nodes). This is because in BE mode nodes synchronizes to the periodic beacon emitted by the PAN coordinator. Hence, *all* sensor nodes having data to transmit compete for channel access at the beginning of the beacon

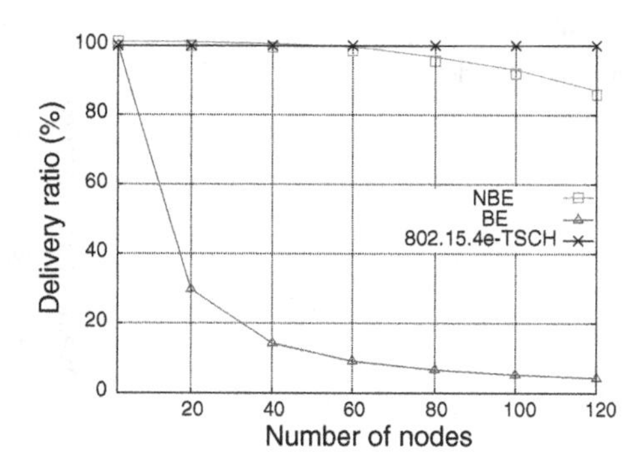

Fig. 6 Delivery ratio versus number of nodes

Table 1 Operating parameters

Parameter	Value
Communication Period (T)	0.983 s
Data frame size	127 bytes
ACK frame size	11 bytes
macMaxFrameRetries	3
macMaxCSMABackoffs	4
macMaxBE	5
macMinBE	3
P_{rx}	35.46 mW
P_{tx}	31.32 mW
P_{idle}	0.77 mW
P_s	36 μW

period. This maximizes the competition among nodes and results in high latencies and energy consumption. Also, a large percentage of frames is discarded due to exceeded number of backoff trials [28]. The NBE mode performs better than BE because, unlike BE, there is no synchronization and sensor nodes access the channel asynchronously, when they have a data packet ready for transmission. This reduces the competition among nodes even if conflicts can still occur. Hence, NBE performs similarly to TSCH when the number of nodes is low and there are no conflicts, while the performance gap between NBE and TSCH increases very quickly as the number of nodes grows up. It must be emphasized that, while TSCH provides a deterministic latency, thanks to its *slotted* access scheme, NBE is not able to guarantee a bounded latency, even when the number of nodes is low, since it implements a *contention-based* access scheme. For the same reasons, it is not able to guarantee a 100 % delivery ratio when the number of nodes is large or under high traffic conditions. Hence, NBE is not suitable for application scenarios where low and deterministic latency and/or high reliability are required. On the other side, being based on contention-based access, NBE does not require any preliminary link schedule to work and is, thus, more flexible and easy to manage, especially in network with dynamic topology.

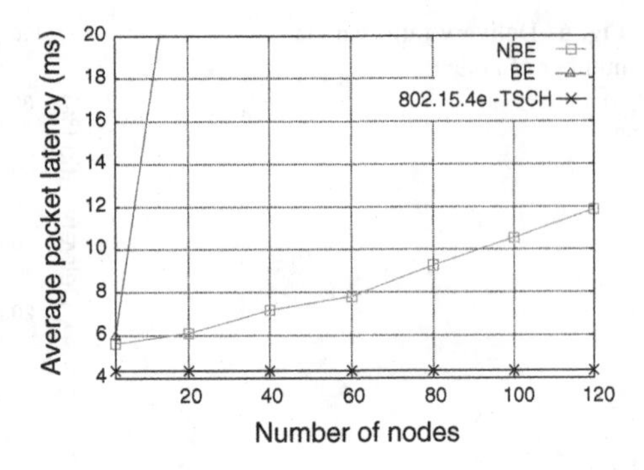

Fig. 7 Average latency versus number of nodes

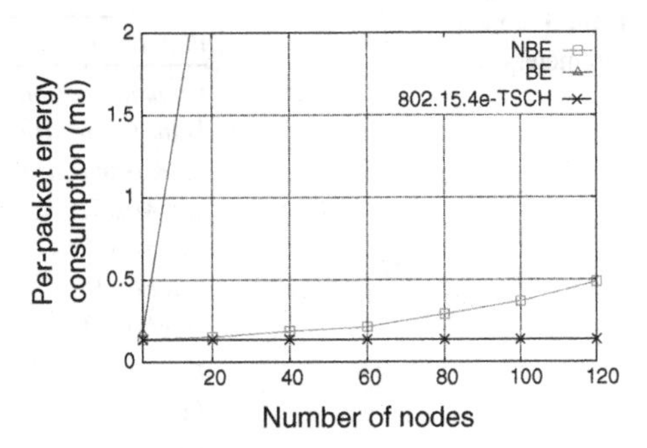

Fig. 8 Energy per packet versus number of nodes

Therefore, it can be preferred to TSCH in all application scenarios where latency and/or reliability requirements are not so stringent.

6 Conclusions

In this chapter we have focused on the 802.15.4e standard, recently released by IEEE to enhance and add functionality to the 802.15.4 standard so as to address the emerging needs of embedded industrial applications. The 802.15.4 standard was conceived for applications without special requirements in terms of timeliness, reliability, robustness, and scalability. Therefore, it is unsuitable for application domains such as applications in the industrial and healthcare fields. We have highlighted the main limitations and deficiencies of the 802.15.4 standard and shown how these limitations have been overcome in the new standard. We have also presented some simulation results to better highlight the performance improvements allowed by the new standard.

References

1. Alcaraz, C., Najera, P., Lopez, J., Roman, R.: Wireless sensor networks and the internet of things: Do we need a complete integration? In: 1st International Workshop on the Security of the Internet of Things. (SecIoT), Tokyo, Japan (2010)
2. Akyildiz, I.F., Kasimoglu, I.H.: Wireless sensor and actor networks: research challenges. Adhoc Netw. **2**(4), 351–367 (2004)
3. Willig, A.: Recent and emerging topics in wireless industrial communications: A selection. IEEE Trans. Ind. Inform. **4**(2), 102–124 (2008)
4. Milenković, A., Otto, C., Jovanov, E.: Wireless sensor networks for personal health monitoring: issues and an implementation. Comput. Commun. **29**(13), 2521–2533 (2006)
5. Miorandi, D., Uhlemann, E., Vitturi, S., Willig, A.: Guest Editorial: Special section on wireless technologies in factory and industrial automation, part I. IEEE Trans. Ind. Inf. **3**(2), 95–98 (2007)
6. Lemmon, M., Ling, Q., Sun, Y.: Overload management in sensor-actuator networks used for spatially-distributed control systems. In: Proceedings of the 1st International Conference on Embedded Networked Sensor Systems, pp. 162–170, ACM (2003)
7. Sinopoli, B., Sharp, C., Schenato, L., Schaffert, S., Sastry, S.S.: Distributed control applications within sensor networks. Proc. IEEE **91**(8), 1235–1246 (2003)
8. Platt, G., Blyde, M., Curtin, S., Ward, J.: Distributed Wireless Sensor Networks and Industrial Control Systems—A New Partnership. In: The Second IEEE Workshop on Embedded Networked Sensors. EmNetS-II, pp. 157–158, IEEE (2005)
9. Low, K. S., Win, W. N. N., Er, M. J.: Wireless sensor networks for industrial environments. In: International Conference on Computational Intelligence for Modelling, Control and Automation, 2005 and International Conference on Intelligent Agents, Web Technologies and Internet Commerce, Vol. 2, pp. 271–276, IEEE (2005)
10. Anastasi, G., Conti, M., Di Francesco, M., Passarella, A.: Energy conservation in wireless sensor networks: a survey. AdHoc Netw **7**(3), 537–568 (2009)
11. Zurawski, R.: Networked Embedded Systems. CRC press, Boca Raton (2009)
12. IEEE Standard for Information technology, Part 15.4; Wireless Medium Access Control (MAC) and Physical Layer (PHY) Specifications for Low-Rate Wireless Personal Area Networks (LR-WPANs), IEEE Computer Society (2006)
13. ZigBee Alliance, The ZigBee Specification version 1.0 (2007)
14. Wireless Medium Access Control (MAC) and Physical Layer (PHY) Specifications for Personal Area Networks (WPANs), IEEE Standard 802.15.1 (2005)
15. HART Field Communication Protocol Specification. HART Communication Foundation Std., version 7.4, revised in 2012. http://www.hartcomm.org/, (2007)
16. Wireless Systems for Industrial Automation: Process Control and Related Applications, International Society of Automation (ISA) Standard ISA-100.11a (2009)
17. RFC 4944: Transmission of IPv6 Packets over IEEE 802.15.4 Networks. http://tools.ietf.org/html/rfc4944
18. Accettura, N., Grieco, L. A., Boggia, G., Camarda, P.: Performance analysis of the RPL routing protocol. In: IEEE International Conference on Mechatronics (ICM), pp. 767–772, IEEE (2011)
19. Shelby Z., Hartke K., Bormann C., Frank B.: Constrained Application Protocol (CoAP), draft-ietf -core-coap-13. https://datatracker.ietf.org/doc/draft-ietf-core-coap/
20. IEEE std. 802.15.4e, Part. 15.4: Low-rate wireless personal area networks (LR-WPANs) amendament 1: MAC sublayer. IEEE Comput. Soci. (2012)
21. Tinka, A., Watteyne, T., Pister, K.: A decentralized scheduling algorithm for time synchronized channel hopping. Ad Hoc Networks, pp. 201–216. Springer, Berlin (2010)
22. Palattella, M. R., Accettura, N., Dohler, M., Grieco, L. A., Boggia, G.: Traffic Aware Scheduling Algorithm for reliable low-power multi-hop IEEE 802.15. 4e networks. In: IEEE 23rd International Symposium on Personal Indoor and Mobile Radio Communications (PIMRC), pp. 327–332, IEEE (2012)

23. Accettura, N., Palattella, M. R., Boggia, G., Grieco, L. A., Dohler, M.: Decentralized Traffic Aware Scheduling for multi-hop Low power Lossy Networks in the Internet of Things. In: IEEE 14th International Symposium and Workshops on a World of Wireless, Mobile and Multimedia Networks (WoWMoM), pp. 1–6, IEEE (2013)
24. Network Simulator Ns2 http://www.isu.edu/nsnam/ns
25. Anastasi, G., Borgia, E., Conti, M., Gregori, E., Passarella, A.: Understanding the real behavior of Mote and 802.11 ad hoc networks: an experimental approach. Pervasive Mob. Comput. **1**(2), 237–256 (2005)
26. Bougard, B., Catthoor, F., Daly, D. C., Chandrakasan, A., & Dehaene, W.: Energy efficiency of the IEEE 802.15. 4 standard in dense wireless microsensor networks: modeling and improvement perspectives. In Design Automation and Test in Europe, pp. 221–234. Springer The Netherlands (2008)
27. Chipcon CC2420 Website http://focus.ti.com/docs/prod/folders/print/cc2420.html
28. Anastasi, G., Conti, M., Di Francesco, M.: A comprehensive analysis of the MAC unreliability problem in IEEE 802.15. 4 wireless sensor networks. IEEE Trans. Ind. Inform. 7(1), 52–65(2011)

Extracting Structured Knowledge From Sensor Data for Hybrid Simulation

Marco Ortolani

Abstract Obtaining continuous and detailed monitoring of indoor environments has today become viable, also thanks to the widespread availability of effective and flexible sensing technology; this paves the way for the design of practical Ambient Intelligence systems, and for their actual deployment in real-life contexts, which require advanced functionalities, such as for instance the automatic discovery of the activities carried on by users. Novel issues arise in this context; on one hand, it is important to reliably model the phenomena under observation even though, to this end, it is often necessary to craft a carefully designed prototype in order to test and fine-tune the theoretical models.

The work described here proposes to use sensor nodes to capture environmental data related to users' activities; the representation of the environment will rely on an ontology expressed in a well-established ad-hoc formalism for sensor devices. An activity model will be produced by analyzing the effect of users' actions on the collected measurements, in order to infer the underlying structure of sensor data via a *linguistic* approach based on formal grammars. It will finally be shown how such model may be profitably used in the context of a hybrid simulator for wireless sensor networks in order to obtain a scalable and reliable tool.

1 Introduction

Recent developments in sensing technology make it possible to provide non-intrusive and easily deployable solutions for detecting the most diverse observations, spanning from simple phenomena to complex events; this has given rise to the new paradigm of the Internet of Things (IoT), whose fundamental idea is that the objects pervading our everyday environment may make themselves identifiable and may be turned into

M. Ortolani (✉)
DICGIM, University of Palermo, Viale delle Scienze, Edificio 6, 90128 Palermo, Italy
e-mail: marco.ortolani@unipa.it

S. Gaglio and G. Lo Re (eds.), *Advances onto the Internet of Things*,
Advances in Intelligent Systems and Computing 260, DOI: 10.1007/978-3-319-03992-3_11,

smart objects thanks to their ability to communicate with each other and to access other objects' aggregate information [4]. Among available enabling technologies, commonly available wireless sensor networks [2] are noteworthy.

Such developments broaden the horizon of research to include novel fields, such as that of Ambient Intelligence (AmI), a branch of Ambient Intelligence that focuses on adapting the environmental conditions to maximize the user's comfort, and aims to do so transparently by applying methods and ideas borrowed from such fields as pervasive and ubiquitous computing [11, 21]. A key functionality for many AmI applications in a home setting is the automatic recognition of human activities (e.g. cooking, sleeping or working at one's desk), as it represents the premise for the design of complex systems; for instance, this kind of contextual information is fundamental in systems designed for energy saving in buildings, where the activation/deactivation of energy-hungry devices heavily depends on whether the occupants are present or not.

A typical approach to creating predefined models for the most common activities is to use supervised classification methods, but the present proposal substantially differs in that it specifically aims to relieve the designer from the task of creating a detailed model for each activity to track. Even though a network of wireless sensor nodes is apt to represent the connection between the system and the real world, an extensive deployment would not be practically viable, especially during the initial design phase. A common approach to assessing the validity of AmI systems is thus to develop a minimal, albeit fully functional, prototype of the intelligent application to be actually deployed into the environment. This is useful in the light of obtaining more detailed understanding of the semantics of sensor data; in particular, the aim is to infer the *underlying structure* of the environmental sensor readings produced as a consequence of users' interactions with their surroundings, i.e. of users' actions. In other words, the problem will be addressed from an algorithmic perspective, rather than a machine learning one, so that no detailed prior knowledge about the specific application scenario is necessary, other than the readings collected from an environmental sensor network, and the description of the nature of such data in the form of an ontology.

Data acquisition in the system described here will be guided by a tool for *hybrid simulation* of wireless sensor networks, able to exploit models constructed from data sensed by a minimal deployment of actual nodes, and augmented with additional virtual nodes to simulate the behavior of a more complex network. Such a tool may be very effective for identifying users activities starting from sensor data, in order to build reliable models; however, several specific issues arise, with different degrees of complexity, and spanning various architectural levels:

physical-level issues: observed sensor data is likely noisy, either deriving by human inaccuracies (e.g. mistakenly operating one of the actuators), or by the sensor system itself, failing to accurately perform data transmission;

data preprocessing issues: the start and end time of a performed activity is often not easily identifiable, due to the intrinsic ambiguity in the observed sensor data with respect to which activity is taking place; users may perform activities in many different ways, thus making it difficult to infer a general description of an activity;

intelligent analysis: the design of a system able to exploit knowledge about user activities is particularly challenging, also because it requires extensive coverage of the sensor infrastructure; this may be overcome by resorting to simulation, but in this case a reliable predictive model for simulating user activities is also required.

The remainder of the chapter is organized as follows: Sect. 2 provides some technical background about documented techniques for creating usable ontologies for sensors, and for detecting users' activities; Sect. 3 describes the current proposal about a system for structural knowledge extraction from raw sensor data, and Sect. 4 describes an application scenario to an AmI context where users' activities are to be identified and modeled; finally, Sect. 5 presents some conclusions and describes further directions for research.

2 Technical Background

Even though sensor networks provide a practical solution for gathering and analyzing measurements due to complex phenomena [7], the lack of integration and communication between different networks, often prevents the extraction of global knowledge about the monitored phenomena. As pointed out in [26], the main difficulty lies in the fact that sensors encoding of observed phenomena is intrinsically opaque, so metadata play an essential role in managing such kind of data: only the addition of *semantic value* to bare data may provide spatial, temporal, and contextual information essential for analyzing sensor readings.

A powerful tool to address this issue is represented by *ontologies*, which, according to [15], may be defined as a set of formally defined concepts and relations that are relevant to a knowledge domain; roughly speaking, the three three types of semantics mentioned above, and normally associated with sensor data are also useful to define ontologies; moreover, ontological models are suitable to perform advanced tasks, such as representing the sensor domain [22, 25]. SensorML [6], for instance, has been designed as a generic data model expressed in UML for capturing classes and associations that are common to all sensors. Its definition falls within a broader attempt by the Open Geospatial Consortium (OGC)[1] to provide access to sensors via distributed applications and services in order to create some kind of "sensor web" [23]. Instances of classes and associations may be then used to define specific sensor profiles in the light of further processing and integration of the collected readings, even from large number of sensors.

As pointed out in [9], however, SensorML does not include formal definitions of the classes or relations it uses, so it provides no logical or axiomatic-grounded

[1] http://www.opengeospatial.org

theory to account for its conceptualizations; strictly speaking, it therefore cannot be considered to be an ontology. On the other hand, SensorML is able to provide a generic data model for expressing knowledge and data about sensors, as well as sensor metadata and sensed attribute values; moreover, it may be considered as the basic tool to define an actual sensor ontology. Instantiations of classes and associations provided by SensorML can be used to create specific sensor profiles in order to implement higher-level services or to interact with more advanced ontologies (e.g. OWL) [23].

Following the same line of thought, the authors of [25] discuss a Semantic Sensor Web (SSW) in which sensor data is annotated with semantic metadata to increase interoperability as well as provide contextual information essential for situational knowledge, coherently with the standardization efforts of the Semantic Web Activity of the World Wide Web Consortium.[2]

Such frameworks lead the way to new interesting developments; as proposed in [28], given an ontology for an environmental domain, it seems reasonable to suggest that sensor data acquisition can be translated into ontological knowledge acquisition, for instance by means of supervised machine learning to map sensor data into ontological concepts. However, extracting meaningful knowledge from sensor data is still an open challenge, despite the availability of low-power, small-size, wireless technology, and of the related communication and programming models [28].

For the purpose of the present discussion, the focus will be on the extraction of user activity models; in fact, automatic discovery of user activities starting from raw data is a well-studied problem in different contexts, ranging from home automation, to elderly care [29]. As pointed out in [10], from a data mining perspective, activity discovery is often seen as the problem of detecting recurring patterns within a sequence of events; however, there are substantial differences between frequent itemsets detection, and discovery of patterns corresponding to activities, especially since in the latter case, ordering and repetitions of elements must typically be considered. In order to take those factors into account, the authors of [19] use a variant of the *apriori* algorithm [1] to discover sequences of events repeating with regular periodicity, besides patterns related to frequent activities. Another approach, proposed in [8], relies on standard apriori and considers the event sequence as a stream of incoming data; after identifying all sequences of predefined size and support, a transform function maps them into models for activities. A hierarchical description is proposed for such models, and activities are divided into tasks and sub-tasks; the bottom of the hierarchy is represented by activities that cannot be further decomposed; activity recognition, as well as description, is carried on in a bottom-up fashion. A similar approach is described in [18], where the authors address the issue of broken or concurrent activities by considering *emerging patterns*, i.e. those patterns able to capture meaningful differences between two classes of data.

The present proposal, however, argues for the the use of a historically alternative approach, namely the *symbolic* one, for identifying sequences of perceptual inputs and recognizing their underlying *syntactic* structure, which may be used to extract

[2] http://www.w3.org/2001/sw/

a more general pattern representing entire sequences. In this approach, high-level concepts may be identified with computational entities, such as for instance Finite State Automata (FSA), in the simplest case. Inferring models for symbolic learning, however, typically requires the availability of large amounts of source data. When wireless sensor networks are used as the data gathering infrastructure, the issue of expensive deployment cannot be disregarded. In such scenario, simulation is a cost-effective choice for prototyping and testing such applications, as the cost, time, and complexity involved in deploying and constantly changing actual large-scale WSNs for experimental purposes may be prohibitively high. Our proposal for an optimized simulator, consisting of a hybrid approach exploiting data provided by a minimal deployment of actual nodes in order to integrate and validate the data computed in simulation, has been described in [13, 16], and its versatility has been proved by using it in different contexts [17]. Building up on those works, the following discussion will show how to deal with the intrinsic heterogeneity in sensors and data to obtain reliable models for high-level concepts, such as users' activities.

3 Simulating User Activities via a Structural Approach

Considering an indoor environment, fully equipped with sensor devices for capturing all relevant measurements, the actions of the occupants will perturb the environment state; moreover they will likely follow some pattern, not known a priori, which will be reflected into the sensor readings. If we are to infer a model for such actions, we need a detailed knowledge about the nature of the sensing devices, their features and how those are related to the yielded measurements. As previously mentioned, all those characteristics may be effectively represented by means of an appropriate ontology, as long as *contextual information* is accounted for. Generally speaking, ontologies, make it possible to assign data a description of its acquisition and communication policies in order to enhance the integration of the sensor data [5]; moreover, ontologies may be easily fit within a more complex reasoning system. The remainder of this section will show how such an ontology may be used in combination with a hybrid simulator for wireless sensor networks able to generate a sufficiently large number of source data, partly sensed from actual devices and partly generated through reliable models.

3.1 Using a Hybrid Simulator to Capture Activity Data

The design of a WSN-based application is a challenging task due to its dependence on the specific scenario. Although a typical choice for testing such kind of networks requires devising ad-hoc testbeds, this is often impractical as it calls for expensive, and hard to maintain deployment of nodes. On the other hand, simulation is a valuable option, as long as the actual functioning conditions are reliably modeled, and carefully replicated. This is especially true in the context of Ambient Intelligence applications,

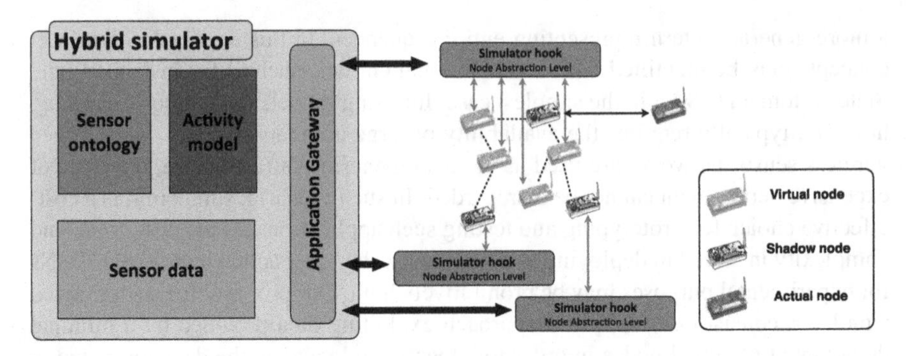

Fig. 1 High-level representation of the structure of the simulation framework

which typically need *scalable* simulations that provide *reliable* sensory data to be used in the preliminary design and test phases.

Although a few general-purpose testbeds have been proposed in the past, WSN applications must be tested on a large scale, and under complex and varying conditions in order to capture a sufficiently wide range of interactions, both among nodes, and with the environment. However, the deployment of a large number of nodes in hostile environments could become prohibitively expensive and practically unfeasible because of maintenance costs. Simulations can address those issues by providing controlled, reproducible environments, and specialized software tools for monitoring and debugging, so that the actual deployment of nodes may be delayed till after algorithms have been thoroughly tested; however they may not deliver fully reliable results, especially due to over-simplistic assumptions about the physical channels, and the node radio models. The issue of implementing an effective simulator for WSNs has been discussed in [16], which describes a framework for supporting the user in early design and testing of a wireless sensor network with an augmented version of a standard WSN simulator, that allows merging actual and virtual nodes seamlessly interacting with each other. One of the key challenges when trying to deliver such kind of realistic simulations is how to ensure fidelity, in terms of ability to reproduce the same behavior both in virtual and real nodes, and accuracy of timing.

Figure 1 shows the architecture of the simulation framework, which will be used also for the present discussion. Once the virtual network is set up by the network administrator, the actual simulation process is handed over to an augmented version of the simulator, customized in order to introduce the notion of *shadow* nodes, i.e. wrappers whose main purpose is to act as interfaces toward their real counterparts, while appearing identical to other virtual nodes from the simulator point of view. The main function of shadow nodes is thus to collect sensed data from the real world, and to re-route communication from virtual nodes to actual ones. In order to ensure correct coordination among virtual, shadow, and actual nodes, simulation timing must be constrained to satisfy at least soft real-time guarantees; this functionality is provided by the application gateway. Introducing actual data into a simulation usually requires constructing complex models, after a considerable amount of data

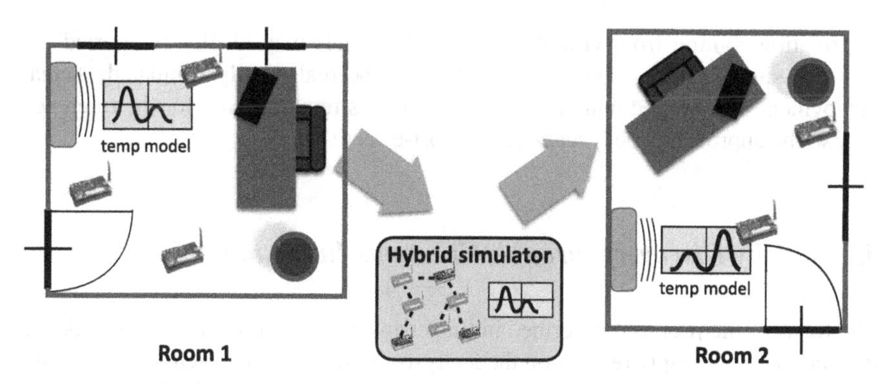

Fig. 2 Scheme of "model transferring"

is preliminarily stored, which is usually very time consuming. The present chapter will instead propose to embed into the hybrid simulator, a description of the nodes characteristics in the form of an ontology, as well as a high-level model for describing information extracted from sensory data, in the form of a linguistic description of user activities.

In order to get basic knowledge about the process or phenomena that generated data, the simulator embeds a "Sensor ontology" submodule, as shown in Fig. 1, with the aim of capturing simple descriptions of the nature of the sensors employed in a specific application scenario, expressed in SensorML formalism. For instance, SensorML can provide a complete and unambiguous description of what is called the *lineage* of an observation, thus delivering a detailed account of the process by which an observation was acquired, possibly also including further postprocessing. This is the first step toward the creation of the basis for intelligent applications grounded into the real world, coherently with the vision of the Internet of Things; in fact, this approach moves the paradigm a step forward in the direction of Semantic Web, i.e. an evolving extension of the World Wide Web in which the semantics of information is formally defined through ontologies, thus allowing for effective automatic interpretation of data content. The most relevant technologies of the Semantic Web include the Resource Description Framework (RDF) data representation model, the ontology representation languages RDF Schema and Web Ontology Language (OWL). A specialization of this paradigm has been proposed for the specific context of the Sensor Web [25]; however, lack of a commonly adopted standard represents an obstacle to a complete realization; to overcome this problem, semantic annotations have further be included so that software applications may be able to "understand" and reason over sensor data consistently, and accurately.

The ultimate purpose for using simulators is the creation of refined models for the sensed data; in particular, this chapter deals with high-level models, which need to be general enough so that they may be exported to different scenarios. The idea is outlined in Fig. 2; initially, reliable models for data sensed in one specific environment taken as reference are extracted, in order to be used as a realistic representation for information coming from virtual nodes. Such models may ideally be "ported" to different sites so that a complex scenario may be realistically simulated. Actual deployment of a limited number of sensor nodes is required only for the reference site, so the approach is both scalable and cost-effective.

3.2 Activity Discovery through Grammar Induction

The focus of the proposal described in this chapter is the recognition of user daily-life activities by simply relying on the analysis of environmental sensory data, with the underlying assumption that a hidden structure may be recognized in such data. A typical approach to creating predefined models for common activities is to use supervised classification methods, i.e. to begin by labelling sample data to train the model, and to go onto look for occurrences of previously found "representatives" in actual data. A few assumptions are implicit with this strategy; namely, all involved users are assumed to carry on the same set of known activities with similar regularity and demeanor; also, a considerable amount of data needs to be collected and consistently labelled, and finally incomplete or discontinuous patterns within original data are not well tolerated. In general, unmediated use of raw sensory data is likely to result in overfitting; in other words, the obtained models are so overly specific for the employed dataset that they are not generalizable to different contexts or users.

The present proposal substantially differs in that it specifically aims to relieve the designer from the task of creating a detailed model for each activity to track; the problem is addressed from an algorithmic perspective, rather than a machine learning one, so that no detailed prior knowledge about the specific application scenario is necessary, other than the readings collected from an environmental sensor network, and the description of the nature of such data in the form of an ontology. The goal is to infer the hidden structure that is inevitably present in data, given that the readings were ultimately caused by the actions of users; this task resembles what human beings do when performing the operations of recognition and classification. In fact, the human brain is very efficient at inferring general patterns from specific examples, a process known as *inductive reasoning*. In the context of algorithmic information theory, an abstract representation for a simplified version of those patterns is provided by formal grammars, and the process of generalization is known as grammatical induction, or grammatical inference; in this case, the specific examples may be regarded as sentences, while the patterns (i.e. sequences with some underlying structure) are captured by the entire grammar.

As pointed out in [27], inferring a language of infinite cardinality from a finite set of examples is not a trivial task, due to the difficulty is finding a good balance between generalization and specialization. A typical technique thus consists in starting with the basic constituting elements of the supposed target language (e.g. symbols directly computed from the raw data) and recursively aggregate them by some kind of merging (e.g. state merging when automata are considered, or non-terminal merging in the case of context-free grammars).

Within computational learning theory, three main formal models have been proposed for inductive inference [24], namely: *identification in the limit* by Gold [14], the *query* learning model by Angluin [3], and finally the *probably approximately correct* (PAC) learning model by Valiant [30].

Identification in the limit regards learning as an infinite process based on the availability of examples from an unknown grammar $\mathscr{G}$ which is to be inferred; positive and negative examples may be provided to the estimation algorithm, which produces a conjecture about the grammar, refining it as long as more examples are provided. If for every set of positive and negative examples from the unknown grammar $\mathscr{G}$, the algorithm guesses a correct grammar which is equivalent to $\mathscr{G}$, and never changes its guess afterwards, then the algorithm is said to identify $\mathscr{G}$ in the limit.

A different approach is the one followed by Angluin, who assumes that an *oracle* is available to the inference algorithm, i.e. some kind of support able to provide the inference algorithm with correct answers to specific queries about the unknown grammar; here, an inference algorithm repeatedly makes queries for the unknown grammar $\mathscr{G}$ to oracles, and eventually halts and outputs a correct grammar in finite time.

Finally, Valiant proposed the probably approximately correct model, which considers a distribution-independent probabilistic model of learning from random examples. In this case, random samples are drawn independently from the input data; the probability distribution is assumed to be arbitrary and unknown. The inference algorithm takes each sample as input and consequently produces a grammar. The quality of identification is measured by an accuracy parameter and a confidence parameter, which may be set by the user.

Not all approaches have the same power, and for instance it has been shown that it is not possible to learn by positive examples alone, with Gold's approach. In the approach proposed here, however, a simplified scenario will be considered and the hybrid simulator will rely on identification in the limit, assuming that the unknown grammar to be inferred may be captured by finite state automata (i.e. it is in fact regular).

4 A Sample Scenario: Predicting User Activities for Energy Saving

In order to highlight the key points of the approach proposed here, a sample scenario will be considered, consisting of an office environment whose occupants' actions are to be identified so that it is possible to reproduce them in analogous contexts; namely, the considered case study will be that of a *smart office* whose overall energy consumption needs to be controlled.

The energy demand for instrumentation in smart offices is nowadays becoming a severe challenge, due to economic and environmental reasons; typical devices with high impact on energy consumption are for instance HVAC systems, computers,

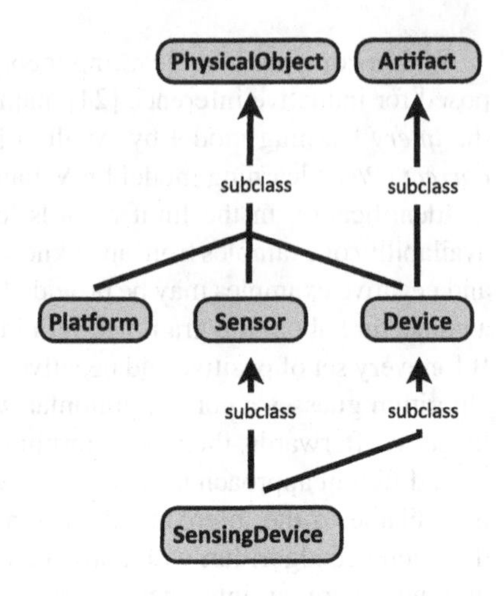

Fig. 3 SensorML representation of a generic sensing device

printers, and so on [12]. Basic usage is heavily related to the occupants' habits, so effective automated approaches must take into account basic information about users, primarily the prediction of their course of actions. In the scenario considered here, it will be assumed that the office premises are made up of smaller areas, which may be considered homogeneous with respect to their intended use and to the environmental conditions therein (e.g. independent rooms, or parts of a larger lab). Users may operate the actuators to modify the environmental conditions of the area they are occupying, and such changes will be reflected into the sensor readings; additional sensors may be deployed to capture the interaction of users with specific devices (e.g. sensor on light switches, weight sensors on chairs, and so on); a complete taxonomy of sensors and their characteristics may be represented by a specialized ontology. The description of available sensors and actuators, and of their main characteristics will be provided to the simulator in the form of a simplified SensorML code, an example of which is reported in Fig. 3.

Analysis of the readings collected by the simulator will be aimed at extracting the underlying pattern identifiable in sensor data and deriving from the action of the users on the environment. The inferred grammar will thus represent a model for the actions themselves. In order to turn raw data into symbols apt to be regarded as the basic elements of a grammar, an approach similar to what described in [10] will be followed.

The main steps of the algorithm for inferring a user activity model are outlined in Fig. 4. Initially, raw sensor triggers are clustered to form basic *events*, which are co-occurrences of triggers, identified via a customized version of an on line algorithm for string matching with wildcards (the alphabet extractor module in the figure). The event sequence needs then to be compressed, and a lossy optimal coding is used for this purpose so that events with low information content will be discarded; in other words, the most relevant patterns will be those that better describe the whole sequence, according to the Minimum Description Length (MDL) principle.

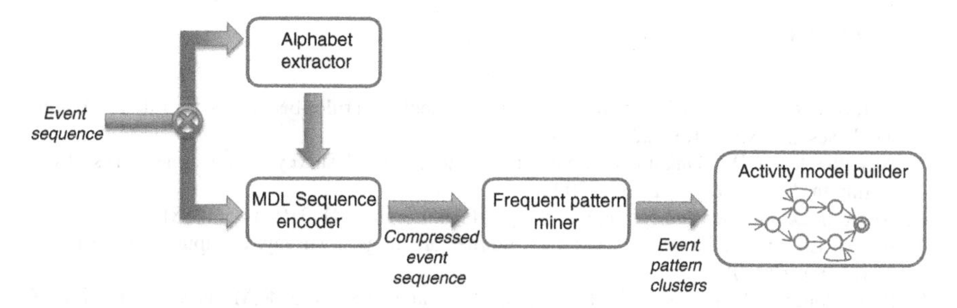

Fig. 4 Outline of the logical blocks for user activity inference

After obtaining a compressed version of the event sequence thanks to the new encoding, the algorithm turns to discovering the most frequent patterns. The proposed approach is similar to Discontinuous Varied-order Sequential Miner (DVSM) [20], which is an apriori-based iterative algorithm. The final frequent pattern set returned by DVSM contains the most relevant patterns, which will be clustered into meaningful classes to obtain the discovered activities, by integrating temporal information with other features of interest, such as composition similarity; this step is accomplished by k-medoids, a variant of the well-known k-means clustering, where representative points are bound to belong to the initial dataset. A dissimilarity measure is computed over all the possible pairs of points, making this algorithm more robust than traditional k-means. The interested reader is referred to [10] for further details.

In the last phase, the features of the obtained clusters are encoded into models representing the discovered activities. The event clusters are regarded as the basic symbols of the unknown target grammar, and positive and negative samples are constructed for the specific domain by using the event patterns, the general SensorML description of the environment, and user knowledge. Following the identification in the limit approach, a grammar representing user actions is finally inferred in the form of a set of FSAs, as depicted in Fig. 4.

5 Conclusion

This chapter described a proposal regarding the use of a tool for hybrid simulation of wireless sensor nodes in the context of Ambient Intelligence scenarios. In particular, considering a *smart office* environment whose occupants' actions are to be identified so that it is possible to model them in order to control the overall energy consumption, the approach proposed here consists in using sensor nodes to capture environmental data related to users' activities. The representation of the environment was assumed to be provided in the form of an ontology expressed in SensorML, a well-established ad-hoc formalism for sensor devices. and it has been shown how an activity model may be produced by analyzing the effect of users' actions on the collected measurements, in order to infer the underlying structure of sensor data via a *linguistic* approach based on formal grammars.

References

1. Agrawal, R., Imieliński, T., Swami, A.: Mining association rules between sets of items in large databases. SIGMOD Rec. **22**, 207–216 (1993)
2. Akyildiz, I., Su, W., Sankarasubramaniam, Y., Cayirci, E.: A survey on sensor networks. IEEE Commun. Mag. **40**(8), 102–114 (2002)
3. Angluin, D.: Queries and concept learning. Mach. Learn. **2**(4), 319–342 (1988)
4. Atzori, L., Iera, A., Morabito, G.: The internet of things: a survey. Comput. Netw. **54**(15), 2787–2805 (2010)
5. Bendadouche, R., Roussey, C., De Sousa, G., Chanet, J.P., Hou, K.M., et al.: Extension of the semantic sensor network ontology for wireless sensor networks: the stimulus-wsnnode-communication pattern. In: Proceedings of the 5th International Workshop on Semantic Sensor Networks (in conj with the 11th ISWC) (2012)
6. Botts, M. et al.: Sensor Model Language (SensorML) for in-situ and remote sensors. Technical report, Open Geospatial Consortium Inc. (2004)
7. Chella, A., Lo Re, G., Macaluso, I., Ortolani, M., Peri, D.: Multi-robot interacting through wireless sensor networks. Lecture Notes in Computer Science (including subseries Lecture Notes in Artificial Intelligence and Lecture Notes in Bioinformatics) **4733 LNAI**, pp. 789–796 (2007)
8. Chikhaoui, B., Wang, S., Pigot, H.: A frequent pattern mining approach for adls recognition in smart environments. In: Proceedings of International Conference on Advanced Information Networking and Applications, pp. 248–255. IEEE Computer Society (2011)
9. Compton, M., Henson, C.A., Neuhaus, H., Lefort, L., Sheth, A.P.: A survey of the semantic specification of sensors. In: Proceedings of Semantic Sensor, Networks, pp. 17–32 (2009)
10. Cottone, P., Gaglio, S., Lo Re, G., Ortolani, M.: User activity recognition for energy saving in smart homes. In: Sustainable Internet and ICT for Sustainability (SustainIT), 2013, pp. 1–9. IEEE (2013)
11. De Paola, A., Farruggia, A., Gaglio, S., Lo Re, G., Ortolani, M.: Exploiting the human factor in a wsn-based system for ambient intelligence. In: Proceedings of International Conference on Complex, Intelligent and Software Intensive Systems., pp. 748–753. IEEE (2009)
12. De Paola, A., Lo Re, G., Morana, M., Ortolani, M.: An intelligent system for energy efficiency in a complex of buildings. In: Sustainable Internet and ICT for Sustainability (SustainIT), 2012, pp. 1–5. IEEE (2012)
13. Gatani, L., Lo Re, G., Ortolani, M., Sorbello, F.: A monitoring framework exploiting the synergy between actual and virtual wireless sensors. In: Proceedings of the International Conference on Parallel Processing Workshops, pp. 361–367 (2006)
14. Gold, E.M.: Language identification in the limit. Inf. Control **10**(5), 447–474 (1967)
15. Gruber, T.R., et al.: A translation approach to portable ontology specifications. Knowl. Acquis. **5**(2), 199–220 (1993)
16. Lalomia, A., Lo Re, G., Ortolani, M.: A hybrid framework for soft real-time wsn simulation. In: 13th IEEE/ACM International Symposium on Distributed Simulation and Real Time Applications (2009)
17. Lo Re, G., Milazzo, F., Ortolani, M.: Secure random number generation in wireless sensor networks. In: Proceedings of the 4th international conference on Security of information and, networks, pp. 175–182 (2011)
18. Pung, H.K.: epsicar: An emerging patterns based approach to sequential, interleaved and concurrent activity recognition. 2009 IEEE International Conference on Pervasive Computing and Communications, pp. 1–9 (2009)
19. Rashidi, P., Cook, D.: Keeping the intelligent environment resident in the loop. In: 4th International Conference on Intelligent Environments (2008 IET), pp. 1–9, (2008)
20. Rashidi, P., Cook, D.J., Holder, L.B., Edgecombe, M.S.: Discovering activities to recognize and track in a smart environment. IEEE Trans. Knowl. Data Eng. **23**, 527–539 (2011)
21. Remagnino, P., Foresti, G.L.: Ambient intelligence: a new multidisciplinary paradigm. IEEE Trans. Syst. Man Cybern. B Cybern. - Part A: Systems and Humans **35**(1), 1–6 (2005)

22. Ribino, P., Oliveri, A., Lo Re, G., Gaglio, S.: A knowledge management system based on ontologies. In: Proceedings of the International Conference on New Trends in Information and Service Science, pp. 1025–1033. IEEE (2009)
23. Russomanno, D.J., Kothari, C.R., Thomas, O.A.: Building a sensor ontology: a practical approach leveraging iso and ogc models. In: Proceedings of the International Conference on Artificial Intelligence, pp. 637–643 (2005)
24. Sakakibara, Y.: Recent advances of grammatical inference. Theoret. Comput. Sci. **185**(1), 15–45 (1997)
25. Sheth, A., Henson, C., Sahoo, S.S.: Semantic sensor web. IEEE Internet Comput. **12**(4), 78–83 (2008)
26. Sheth, A., Perry, M.: Traveling the semantic web through space, time, and theme. Internet Comput. **12**(2), 81–86 (2008)
27. Stevenson, A., Cordy, J.R.: Grammatical inference in software engineering: an overview of the state of the art. In: Proceedings of the 5th International Conference on Software Language Engineering, pp. 204–223. Springer, Berlin (2012)
28. Stocker, M., Rönkkö, M., Kolehmainen, M.: Making sense of sensor data using ontology: a discussion for road vehicle classification. In: R. Seppelt, A.A. Voinov, S. Lange, D. Bankamp (eds.) Proceedings of the (iEMSs) International Congress on, Environmental Modelling and Software (2012)
29. Tapia, E., Intille, S., Larson, K.: Activity recognition in the home using simple and ubiquitous sensors. In: Ferscha F.M. A. (ed.) Proceedings of Pervasive 2004, vol. vol. LNCS 3001, pp. 158–175. Springer, Berlin (2004)
30. Valiant, L.G.: A theory of the learnable. Commun. ACM **27**(11), 1134–1142 (1984)

Gait Analysis Using Multiple Kinect Sensors

Gabriele Maida and Marco Morana

Abstract A gait analysis technique to model user presences in an office scenario is presented in this chapter. In contrast with other approaches, we use unobtrusive sensors, i.e., an array of Kinect devices, to detect some features of interest. In particular, the position and the spatio-temporal evolution of some skeletal joints are used to define a set of gait features, which can be either static (e.g., person height) or dynamic (e.g., gait cycle duration). Data captured by multiple Kinects is merged to detect dynamic features in a longer walk sequence. The approach proposed here was been evaluated by using three classifiers (SVM, KNN, Naive Bayes) on different feature subsets.

1 Introduction

Gait analysis is a biometric technique that aims at identifying people by their walking style. Gait recognition techniques can be categorized into three categories [9], involving the use of computer vision methods, floor sensors, or wearable sensors.

The main advantage of computer vision approaches is that the identification can be performed without physical contact between subjects and data acquisition devices. Various chapters have looked into processing the images captured by standard video cameras and analyzing the human silhouette [13, 24] in order to extrapolate characteristic features. Computer vision techniques for gait recognition can be classified as model-free and model-based ones. In model-based approaches, a set of body parameters are obtained by fitting a body model to the person captured in each frame. In contrast, model-free approaches do not utilize a model for people but the entire

G. Maida · M. Morana (✉)
University of Palermo, Viale delle Scienze, ED. 6, 90128 Palermo, Italy
e-mail: gabriele.maida@unipa.it

M. Morana
e-mail: marco.morana@unipa.it

S. Gaglio and G. Lo Re (eds.), *Advances onto the Internet of Things*, 167
Advances in Intelligent Systems and Computing 260, DOI: 10.1007/978-3-319-03992-3_12,

shapes of silhouettes or the whole motion of human bodies are used [11, 22]. Model-based approaches are view-invariant and scale-independent, but usually require high quality gait sequences and more computing time than model-free approaches.

The use of floor sensors and wearable sensors can provide more accurate feature detection than computer vision approaches, but the former are expensive because of the use of force plates installed on the floor [15, 17], while the main disadvantage of the latter is that they are more intrusive.

Our approach falls into the computer vision category, in which a person's gait is captured by a camera. The proposed system aims to unobtrusively identify people in an Ambient Intelligence (AmI) scenario [7] by using an array of Kinect sensors to detect gait features, and then recognize the person by his walking style.

We started from the OpenNI 2.0 APIs [2] which provide an efficient skeleton detection method that makes it possible to represent a human body as a set of connected joints. By analyzing joint positions and their spatio-temporal evolutions, it is possible to extrapolate static and dynamic features. In order to preserve the pervasiveness of the system, the Kinect sensors are coherently connected to a miniature fanless computer with reduced computation capabilities.

This chapter is organized as follows: relevant research is presented in Sect. 2, while the system architecture proposed here is described in Sect. 3. The experimental scenario is discussed in Sect. 4, followed by our conclusions in Sect. 5.

2 Related Work

Over the years, the issue of gait recognition has been addressed in different ways in various works. The first attempt at automatic gait recognition was probably the one reported in [5], where a camera was used to capture light sources mounted on selected joints of walking people.

Gait recognition techniques based on silhouette analysis have been proposed in various works. Typically, most of them involve the following phases: subject detection, silhouette extraction, feature extraction, feature selection and classification [23]. In [25], a spatio-temporal silhouette analysis is performed to detect a sequence of static body poses. For each frame of the sequence, a background subtraction procedure is used to extract the binary silhouette. Then, a principal component analysis is applied to compute the predominant components of gait signatures, and a classification technique is employed to identify the person.

Lee [12] describes a gait appearance feature vector based on moments obtained from silhouettes. The whole body is segmented into regions, and for each region, an ellipse is fitted to the visible portion of the foreground object in order to extract a set of moment-based region features. The Mahalanobis distance between feature vectors is used as a measure of similarity, and a Nearest-Neighbor approach is used to rank a set of training sequences according to their distances, by means of a query sequence.

The authors of [11] proposed a spatio-temporal gait representation called Gait Energy Image (GEI). To preserve temporal information, the motion pattern within a gait cycle is represented in a single image. GEI needs less storage space and computation time, but it is still view-dependent and performs better when a side view is used. A possible way to solve the view dependency issue is to use multiple cameras with overlapping fields of view.

The authors of [21] extended the concept of GEI by introducing the concept of Gait Energy Volume (GEV), that uses averaged reconstructed voxel volumes instead of temporally averaging segmented silhouettes. Such 3D data is reconstructed using depth images captured by a frontal viewpoint of the Microsoft Kinect sensor.

In [20], an approach that utilizes both gait and face recognition is presented. The authors demonstrate that the integration of face and gait recognition provides better performance than the use of a single technique alone. Face recognition usually works better with front-parallel images, while a gait recognition technique, based on silhouette analysis, performs more efficiently on side-view sequences.

The authors of [18] use the Kinect SDK provided by Microsoft to obtain a 3D virtual skeleton. A single Kinect sensor is placed to give a side view of the walking path in order to acquire video sequences of walking people. In each frame, the skeleton is converted into a vector containing static features (i.e., the height of the subject and the length of certain body parts). Two dynamic features are also computed along the walk (i.e., step length, and speed). Such feature vectors are then classified using different classifiers.

In [14], a module for the management of an office environment is described, using Microsoft Kinect as the primary interface between the user and the AmI system. In particular, a fuzzy classifier is trained to recognize some simple gestures (such as open/closed hands) in order to produce a set of commands, opportunely structured by means of a grammar, which are used to control the actuators of the AmI system.

3 System Overview

In recent years, the availability of an ever-increasing number of cheap and unobtrusive sensing devices has piqued the interest of the scientific community into producing novel methods for understanding what is happening in the environment, based on the raw measures acquired. Due to the heterogeneity of the data captured, an information fusion mechanism [8] is usually required to address a specific goal, such as that of understanding what the user is doing.

In this chapter, we address the issue of gait recognition using an array of Kinect sensors to unobtrusively identify people in an Ambient Intelligence scenario.

In particular, the approach proposed here relies on the analysis of certain features extracted by means of the OpenNI 2.0 APIs [2], which provide a real-time representation of the human body as a set of connected joints. Moreover, multiple Kinects are used to increase the acquisition range provided by a single Kinect in order to include a longer walk (e.g., in a hallway before the user enters the office). By analyzing joint

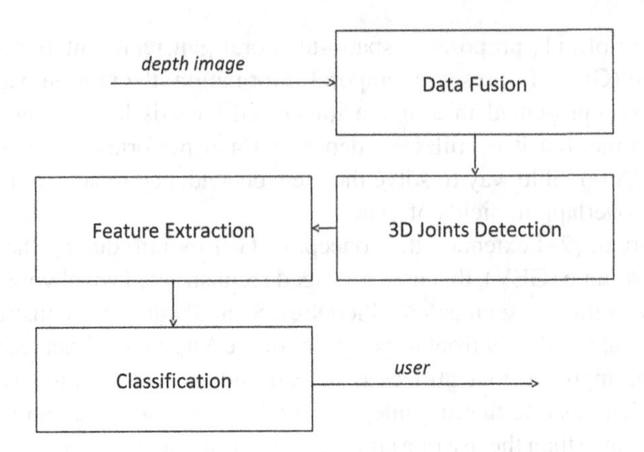

Fig. 1 System overview

positions and their spatio-temporal evolutions, it is possible to extrapolate a set of static and dynamic gait features. Finally, the classifiers are trained using the data collected and then they are used to recognize users' walks. The training set contains ten repetitions of a walk per user. If a walk is not recognized, it will be marked as "unknown".

Our system architecture (see Fig. 1) consists of four components: a *Kinect Data Fusion* step to merge data from multiple Kinects, a *3D Joints Detection* step to obtain human skeletons from depth maps, a *Feature Extraction* step to detect gait features, a *Classification* step to identify people according to the extracted features, and a *Calibration* step which is performed the first time the Kinects are placed into the environment.

3.1 Multi Kinect Architecture

As reported in [3], we found that Kinect is the most suitable device for pervasive AmI tasks, both in terms of cost and functionality, since it is equipped with ten input/output components which allow the device to perceive and interact with the surrounding environment.

The core of the Kinect is represented by the vision system, composed of an RGB camera with VGA standard resolution (i.e., 640×480 pixels), an IR-projector that shines a grid of infrared dots over the scene and an IR-camera for capturing the infrared light. The information obtained from projected dots is used to create three-dimensional depth maps of the observed scene (i.e., pixel values represent distances). The other components include four microphones (three on the right side and one on the left side), a led indicator that shows the state of the Kinect, a motor that allows you to control the tilt angle of the camera (30° upward or downward), and a 3-axis accelerometer that measures the position of the sensor.

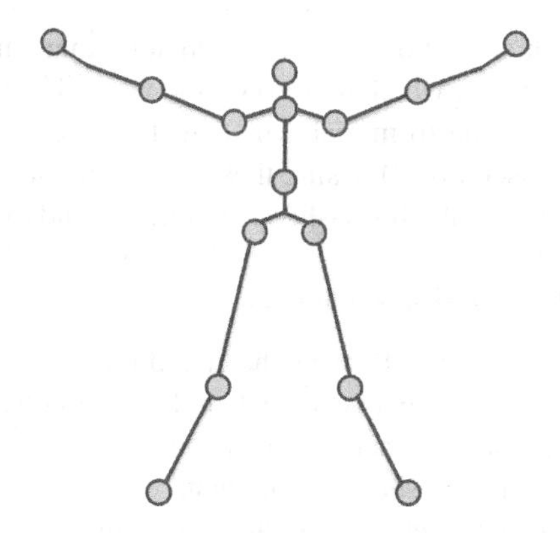

Fig. 2 The 15 joints of the human body: head, neck, torso, shoulders, elbows, hands, hips, knees, feet

A Kinect can see people standing between 0.5 and 3.5 m from the sensor, but we chose to extend this range by putting two Kinects at a distance of approximately 3 m from each other, in order to have about 0.5 m of overlapping frames acquired by both devices during a walk. In the *Calibration* step, an object is positioned at a point visible from both Kinect sensors, in order to find the roto-translation matrix representing the transformation between the two coordinate systems. Such conversion matrix is then used in the *Kinect Data Fusion* step.

Note that it is possible to add more devices by repeating the *Calibration* process, so that any Kinect addition will increase the captured walk sequence of about 3 m, thereby allowing for a better analysis of dynamic features (e.g., the walking speed). We verified that two Kinects appear to be adequate given the cyclical nature of certain features (e.g., gait cycle duration).

Kinect Data Fusion is the core of the Multi Kinect Architecture. During this step, each frame from a different device is ordered in the right position and converted into a single coordinate system, so that the entire walk can be considered as captured by a single virtual Kinect. This may be done by using the timestamp related to each frame and applying the roto-translation matrix for the conversion of the coordinate system. This step is essential, especially when the person is visible from both Kinects. Such new virtual gait frames are used in the *3D Joints Detection* step to detect the joint positions of the person and then in the *Feature Extraction* step to detect gait features.

3.2 Gait Features

In [16], a gait cycle is defined as the time interval between two successive floor contacts of the same foot. During a gait cycle, two distinct periods can be observed, namely a *swing phase*, when a leg is moving forward, and a *stance phase*, when both feet are touching the ground [26].

We chose to use both static body-shape parameters, that can not change during a walk, and dynamic features, which depend on the walk. The OpenNI 2.0 skeleton detection method is able to perform real-time detection (i.e., to find the 3D coordinates) of 15 body joints (see Fig. 2). Using these points and their evolution during a walk, we are able to extract static body-shape parameters and dynamic features from the acquired sequences.

The considered static body-shape parameters are:

- *person height*: vertical distance between head and feet;
- *torso length*: vertical distance between neck and torso joints;
- *shoulder width*: horizontal distance between shoulders;
- *arm length*: distance between shoulder and hand joints;
- *leg length*: vertical distance between hip and foot joints.

The considered dynamic features are listed below:

- *gait cycle duration*: time interval between successive floor contacts of a foot;
- *walk speed*: total walk time divided by total walk distance;
- *step width*: maximum distance between feet during the gait cycle;
- *footstep frequency*: total walk time divided by number of footsteps;
- *arm swing frequency*: total walk time divided by number of arm swings.

where *total walk time* is the difference between the first and the last frame timestamps acquired. The static body-shape parameters are extracted during the stance phase, whilst the dynamic features are extracted by analyzing the evolution of the joints during an acquired walking sequence. Once the whole walking sequence has been acquired, the features extracted are used to train a classifier in the *Classification* step.

3.3 Feature Classification

We used three different classifiers to evaluate the system's performance with different sets of selected features. In particular, the following classifiers were used: a Multi-Class Support Vector Machines, a K-Nearest Neighbor and a Naive Bayes Classifier.

Support Vector Machines (SVM) [19] are supervised learning models with associated non-probabilistic algorithms used to analyze data for binary classification and regression. A multi-class SVM is a net of SVMs able to classify instances into more than two classes.

The K-Nearest Neighbor decision rule [4] assigns to an unclassified observation the most common class amongst its k closest samples in a reference set (where k is a positive integer).

Naive Bayes [27] is a kind of probabilistic classifier based on Bayesian networks that assigns a new observation to the most probable class, assuming that the attributes are conditionally independent given the class value.

Table 1 The feature sets used for the experiments

Static features (SF)	Dynamic features (DF)	Mixed set 1 (MS1)	Mixed set 2 (MS2)
Person's height	Gait cycle duration	Person's height	Torso length
Torso length	Walk speed	Shoulder width	Arm length
Shoulder width	Step width	Leg length	Gait cycle duration
Arm length	Footstep frequency	Walk speed	Footstep frequency
Leg length	Arm swing frequency	Step width	Arm swing frequency

In order to find the most discriminating features we tested the system by using a single feature at a time and evaluating the recognition rate. The features that yielded the highest recognition rate in the test (about 25–30 %) were included in the *MS1* set, whereas the remaining features were included in the *MS2* set. The various feature sets that were tested for the evaluation of the system, reported in Table 1, include the *SF* set consisting of static features only, the *DF* set consisting of dynamic features only, and two mixed sets (i.e., *MS1* and *MS2*).

3.4 System Ontology

Ontology [10] is a formalism to share and re-use knowledge among different systems. It represents the conceptualization of the relevant entities and their relations in a common vocabulary.

Our system ontology is shown in Fig. 3. The subdivision among system modules, data types and devices is sketched. System modules are the components that manipulate data from devices. *DataType* is subdivided into *RawData*, that represents data which has not been elaborated, and *Symbolic*, representing data containing information.

There are two types of *SystemModules*, namely *TranslationModule* and *UnderstandingModule*. *TranslationModule* can be subdivided into three types called *DataFusionModule* that transforms raw data (*DepthMap*) in other raw data (new *DepthMap*), *JointExtractionModule* which transforms *DepthMap* in symbolic data (*Joint*) and *FeatureExtractionModule*, that obtains other symbolic data (*Feature*) from joints. The *UnderstandingModule* classifies symbolic features to detect the user who is performing a walk.

In our system two kinds of *Devices* are used, called, *Sensor* (Kinect), to acquire *RawData* from the environment, and *Node*, which is the device where the sensor is installed (e.g., Kinect needs a computer to work).

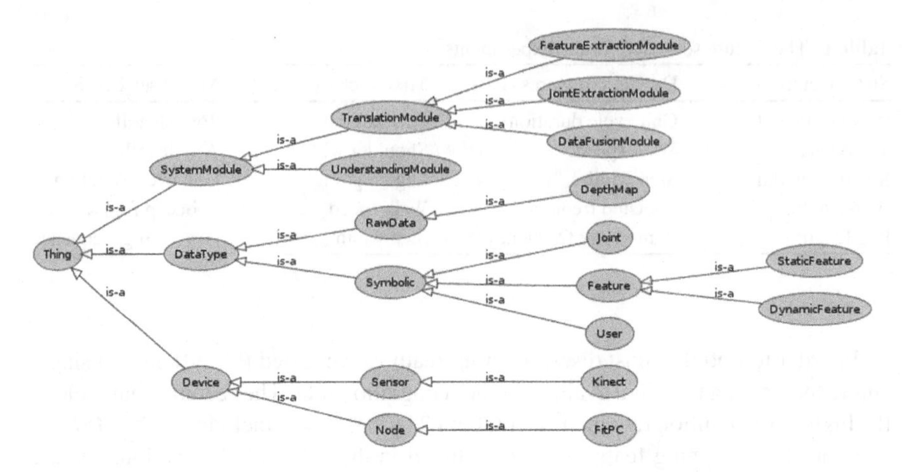

Fig. 3 System ontology

4 Experimental Results

The sensory component in our AmI architecture is implemented through a Wireless Sensor and Actuator Network (WSAN), whose nodes are equipped with off-the-shelf sensors (i.e., for outdoor temperature, relative humidity, ambient light exposure and noise level) [6].

The gait recognition module was evaluated by performing a number of tests on gait data collected at our Department.

In particular, two Kinects were placed on an office hallway at a distance of approximately 3 m from each other (see Fig. 4). Ten subjects (three women and seven men in the height range from 160 to 185 cm) were asked to walk at their normal speed down a path of approximately 8 m. The Kinect sensors captured the user's walks from a frontal-view with a frame rate of 30 fps. Each person repeated the walk ten times, so a total of 100 walking sequences were collected.

The main goal of our tests was to find the best feature set for our Gait Recognition System. Thus, we trained three classifiers with the sets described in Table 1 with all of features together. In order to evaluate the couple *classifier/feature set*, we computed the classification rate of the system by using the Leave-One-Out-Cross Validation method [1], then, for each test, 99 sequences were used for training and the remaining sequence was used for validation.

The accuracy values obtained are reported in Table 2. It is noticeable that static features (*SF*) are more relevant than the dynamic ones (*DF*), since they provided a better recognition rate for each classifier. However, increasing the number of features is not enough to improve the performance. In fact, the recognition rate decreases when all of the features were used. On the other hand, by selecting the most relevant features (i.e., the *MS1* set composed of *person's height*, *shoulder width*, *leg length*,

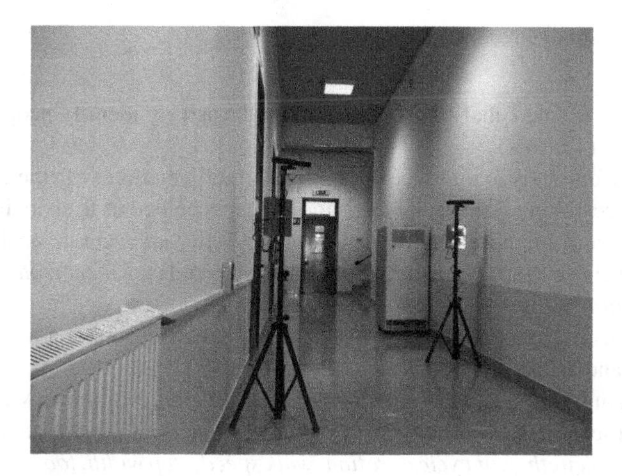

Fig. 4 Kinect sensors placed in the office hallway

Table 2 Accuracy obtained for each classifier and feature set

Classifier	All features (%)	SF (%)	DF (%)	MS1 (%)	MS2 (%)
SVM	54	79	38	73	47
KNN	68	79	40	**92**	58
Naive Bayes	83	75	50	86	66

walk speed and *step width*), we obtained the best results achieving a recognition rate of 92 % with the KNN classifier.

As far as classifiers are concerned, the Naive Bayes classifier was found to be more stable, yelding the highest overall classification rate for each feature set (from 50 % by using dynamic features only, to 86 % by using the *MS1* set), whereas the SVM classifier gave the worse results (from 38 % by using dynamic features only, to 79 % by using the static set). The KNN classifier provided good results with all sets, and in particular with the *MS1* set.

The overall system was tested using C language for the *Kinect Data Fusion* and *3D Joint Detection* steps and MATLAB for the *Feature Extraction* and *Classification* steps. A prototype of the activity recognition module was implemented connecting the Kinect to a miniature fanless PC (i.e., a fit-PC2i with Intel Atom Z530 1.6GHz CPU and Linux OS with kernel 2.6.32), that guarantees real-time processing of the observed scene with minimum levels of obtrusiveness and low power consumption.

5 Conclusion

The research presented analyzes a system to unobtrusively identify people by using multiple Kinect sensors.

We used an ontology to interface our gait recognition module with the entire ambient intelligence system. The use of the same concepts helped us to share information with other system modules (e.g., user presence). After an accurate analysis of the ontology, we chose to structure our system in four modules: *Kinect Data Fusion*, *3D Joints Detection*, *Feature Extraction* and *Classification*.

We collected our gait dataset by placing two Kinect sensors in a hallway in our Department and asking ten people to walk ten times.

By analyzing joint positions and their spatio-temporal evolutions, we have been able to detect several features, namely a *person's height*, *torso length*, *shoulder width*, *arm length*, *leg length*, *gait cycle duration*, *walk speed*, *step width*, *footstep frequency* and *arm swing frequency*. The features are then used to train a classifier to recognize the user performing the walk.

We carried out a number of experiments to evaluate the feature sets and classifiers by using the 100 walking sequences we captured. As a result of our tests we found that a subset of features composed by *person's height*, *shoulder width*, *leg length*, *walk speed* and *step width* was sufficient to correctly identify a person with a recognition rate of 92 %.

Acknowledgments This work has been partially supported by the PO FESR 2007/2013 grant G73F11000130004 funding the SmartBuildings project.

References

1. Arlot, S., Celisse, A.: A survey of cross-validation procedures for model selection. Stat. Surv. **4**, 40–79 (2010)
2. consortium, O.: OpenNI. http://www.openni.org/
3. Cottone, P., Lo Re, G., Maida, G., Morana, M.: Motion sensors for activity recognition in an ambient-intelligence scenario. In: IEEE International Conference on Pervasive Computing and Communications Workshops (PerCom Workshops), pp. 646–651 (2013). doi:10.1109/PerComW.2013.6529573
4. Cover, T., Hart, P.: Nearest neighbor pattern classification. IEEE Trans. Inf. Theory **13**(1), 21–27 (1967)
5. Cutting, J.E., Kozlowski, L.T.: Recognizing friends by their walk: gait perception without familiarity cues. Bull. Psychon. Soc. **9**(5), 353–356 (1977)
6. De Paola, A., Gaglio, S., Lo Re, G., Ortolani, M.: Sensor9k : A testbed for designing and experimenting with wsn-based ambient intelligence applications. Pervasive Mob. Comput. **8**(3), 448–466 (2012). http://dx.doi.org/10.1016/j.pmcj.2011.02.006
7. De Paola, A., La Cascia, M., Lo Re, G., Morana, M., Ortolani, M.: User detection through multi-sensor fusion in an ami scenario. In: Information Fusion (FUSION), pp. 2502–2509 (2012)
8. De Paola, A., La Cascia, M., Lo Re, G., Morana, M., Ortolani, M.: Mimicking biological mechanisms for sensory information fusion. Biol. Inspired Cogn. Architectures **3**(0), 27–

38 (2013). doi:10.1016/j.bica.2012.09.002. http://www.sciencedirect.com/science/article/pii/S2212683X12000527
9. Gafurov, D.: A survey of biometric gait recognition: approaches, security and challenges. In: Annual Norwegian Computer Science Conference, pp. 19–21 (2007)
10. Gruber, T.R., et al.: A translation approach to portable ontology specifications. Knowl. Acquisition **5**(2), 199–220 (1993)
11. Han, J., Bhanu, B.: Individual recognition using gait energy image. IEEE Trans. Pattern Anal. Mach. Intell. **28**(2), 316–322 (2006)
12. Lee, L., Grimson, W.E.L.: Gait analysis for recognition and classification. In: Proceedings of Fifth IEEE International Conference on Automatic Face and Gesture Recognition, pp. 148–155 (2002)
13. Liu, Z., Sarkar, S.: Simplest representation yet for gait recognition: averaged silhouette. In: Proceedings of the Seventeenth International Conference on Pattern Recognition (ICPR'04), vol. 4, pp. 211–214 (2004)
14. Lo Re, G., Morana, M., Ortolani, M.: Improving user experience via motion sensors in an ambient intelligence scenario. In: Pervasive and Embedded Computing and Communication Systems (PECCS), pp. 29–34 (2013)
15. Middleton, L., Buss, A., Bazin, A., Nixon, M.: A floor sensor system for gait recognition. In: Fourth IEEE Workshop on Automatic Identification Advanced Technologies, pp. 171–176 (2005). doi:10.1109/AUTOID.2005.2
16. Murray, M.P., Drought, A.B., Kory, R.C.: Walking patterns of normal men. J. Bone Joint Surg. **46**(2), 335–360 (1964)
17. Orr, R.J., Abowd, G.D.: The smart floor: a mechanism for natural user identification and tracking. In: CHI'00 Extended Abstracts on Human Factors in Computing Systems, pp. 275–276. ACM Press (2000)
18. Preis, J., Kessel, M., Werner, M., Linnhoff-Popien, C.: Gait recognition with kinect. In: Proceedings of the First International Workshop on Kinect in Pervasive Computing (2012)
19. Scholkopf, B., Smola, A.J.: Learning with Kernels: Support Vector Machines, Regularization, Optimization, and Beyond. MIT Press, Cambridge (2001)
20. Shakhnarovich, G., Lee, L., Darrell, T.: Integrated face and gait recognition from multiple views. In: Proceedings of IEEE Conference on Computer Vision and Pattern Recognition, vol. 1, pp. 436–439 (2001)
21. Sivapalan, S., Chen, D., Denman, S., Sridharan, S., Fookes, C.: Gait energy volumes and frontal gait recognition using depth images. In: IEEE International Joint Conference on Biometrics (IJCB), pp. 1–6 (2011)
22. Tao, D., Li, X., Wu, X., Maybank, S.J.: General tensor discriminant analysis and gabor features for gait recognition. IEEE Trans. Pattern Anal. Mach. Intell. **29**(10), 1700–1715 (2007)
23. Wang, J., She, M., Nahavandi, S., Kouzani, A.: A review of vision-based gait recognition methods for human identification. In: Digital Image Computing Techniques and Applications (DICTA), pp. 320–327 (2010)
24. Wang, L., Tan, T., Hu, W., Ning, H.: Automatic gait recognition based on statistical shape analysis. IEEE Trans. Image Proc. **12**(9), 1120–1131 (2003)
25. Wang, L., Tan, T., Ning, H., Hu, W.: Silhouette analysis-based gait recognition for human identification. IEEE Trans. Pattern Anal. Mach. Intell. **25**(12), 1505–1518 (2003)
26. Yamauchi, K., Bhanu, B., Saito, H.: Recognition of walking humans in 3d: initial results. In: Proceedings of the IEEE Computer Society Conference on Computer Vision and Pattern Recognition Workshops (CVPR), pp. 45–52 (2009)
27. Zhang, H.: The optimality of naive bayes. In: Proceedings of the Seventeenth International Florida Artificial Intelligence Research Society Conference (2004)

3D Scene Reconstruction Using Kinect

Marco Morana

Abstract The issue of the automatic reconstruction of 3D scenes has been addressed in several chapters over the last few years. Many of them describe techniques for processing stereo vision or range images captured by high quality range sensors. However, due to the high price of such input devices, most of the methods proposed in the literature are not suitable for real-world scenarios. This chapter proposes a method designed to reconstruct 3D scenes perceived by means of a cheap device, namely the Kinect sensor. The scene is efficiently represented as a composition of superquadric shapes so as to obtain a compact description of environment, however complex it may be. The approach proposed here is intended to be used as a novel processing module of a well-established cognitive architecture for artificial vision. Experimental tests have been performed on real images and the results look very promising.

1 Introduction

Over the last 40 years, the issue of automatically recognizing real-world objects has been investigated by a considerable body of research related to different fields, from computer vision to neuroscience. The techniques proposed therein can be roughly classified as those recognizing the objects contained in a scene in a 2D or 3D space. Both 2D and 3D object recognition still present challenges for the computer science community since the same object usually looks very different according to its orientation, scale and more generally to the acquisition conditions.

A further distinction can be made between "full object recognition" and "recognition by parts" approaches. In many cases the latter is preferred since a complex object can be described as a combination of simpler primitives which can be related to each other by logical relations (e.g., *above*, *below*, *larger*, *smaller* and so on).

M. Morana (✉)
University of Palermo, Viale delle Scienze, Edificio 6, 90128 Palermo, Italy
e-mail: marco.morana@unipa.it

S. Gaglio and G. Lo Re (eds.), *Advances onto the Internet of Things*, 179
Advances in Intelligent Systems and Computing 260, DOI: 10.1007/978-3-319-03992-3_13,

In this chapter, we describe a framework for efficiently representing a 3D scene as a combination of superquadric curves. In particular, the object is perceived by means of a cheap device containing both an RGB camera and a depth sensor, namely the Microsoft Kinect. A volumetric analysis is then performed to discard noisy data and the object is reconstructed by estimating a set of best-fitting superquadrics.

The chapter is organized as follows: related works are outlined in Sect. 2, whilst the system architecture proposed here is described in Sect. 3. Experimental results are detailed in Sect. 4, and conclusions are discussed in Sect. 5.

2 Related Work

A mutual relationship exists between scene reconstruction and object recognition processes. The reason for this is that, in order to reconstruct a scene it is useful to break the scene down into objects. Then, once a description of the scene has been provided, it is possible to recognize the observed objects by classifying their descriptors.

Several systems for 3D object representation have been proposed over the last few years. The main challenge of such approaches is to obtain satisfactory results not only in a controlled testing environment, but also in complex scenarios with unconstrained conditions, e.g., a home environment or an office. In many cases, range images, i.e., 2D images in which each pixel contains the distance between the sensor and a point in the scene, are preferred to the RGB ones since they generally provide a better discriminable data representation.

Since range images are more robust in the face of changes in environment conditions, a number of works have focused on how they should be processed.

In [13], an approach for the direct recovery of a set of volumetric models, i.e., superquadrics, from unsegmented range data is presented. The method is divided into two stages: model-recovery and model-selection. During the first stage, several seeds are placed at random points in the input image, and for each seed, a model is iteratively built and allowed to grow. Finally, those models which produce the simplest and most accurate approximation of the input data are selected.

A technique for part-level object recognition using superquadrics is presented in [12]. The system is based on interpretation trees [10] and can handle flexible articulated objects, i.e., human figurines, that cannot be perfectly modeled by superquadrics.

In [15], a framework is described for extracting some 3D primitives (i.e., spheres, cylinders, cones) from range data captured by a laser scanner.

Several systems provide good results, although high quality range sensors are needed to obtain high resolution input images. Since range sensors are usually very expensive, most of the methods proposed so far have not been suitable for extensive use in real-world scenarios. For this reason, our proposal involves the use of a cheap device containing both an RGB camera and a depth sensor.

The method proposed here is intended to be used as a novel processing module of the framework presented in [4, 5]. In their work, the authors describe a cognitive architecture for an artificial vision system, in which an effective internal representation of the environment is built up by means of processes defined over a suitable intermediate level, the *conceptual level*, that acts between the sensory data, the *subsymbolic level* and the linguistic *symbolic level*. In particular, the conceptual level is characterized by a conceptual space whose dimensions are the parameters of the 3D geometric primitives, i.e., superquadrics, which constitute the scene. The aim of this work is to provide a more efficient technique for reconstructing 3D objects by means of the Kinect sensor.

Microsoft Kinect is based on the hardware reference design and the structured-light decoding chip provided by PrimeSense, an Israeli company which also provides a framework, OpenNI [16], that supplies a set of APIs to be implemented by sensor devices and middleware components.

The core of the Kinect is represented by the vision system composed of an RGB camera with VGA standard resolution (i.e., 640×480 pixels), an IR projector that shines a grid of infrared dots over the scene and an IR camera that captures the infrared light. The factory calibration of the Kinect makes it possible to establish the exact position of each projected dot against a surface at a known distance from the camera. The deformation of this dot pattern against the scene is captured to derive depth images of the observed scene, and capture the objects' position in a three-dimensional space.

Even though Kinect has only been on the market for a couple of years, it has attracted the attention of a number of researchers, thanks to the availability of open-source and multi-platform libraries that reduce the cost of developing new algorithms. A survey of the sensor and corresponding libraries is presented in [3, 11]. In [1], an approach based on RANSAC (Random Sample Consensus) [9], an algorithm for robustly fitting models in the presence of many data outliers, is described. The authors proposed a solution for 3D object localization using superquadrics to model image data captured by the Kinect. Because it is easy to use, the Kinect sensor has also been successfully adopted as an input device for gesture [14] or activity [6] recognition systems in ambient intelligence scenarios.

3 System Overview

In this section a description of the system is given, explaining both the basis of superquadric shapes and the reconstruction technique proposed here.

3.1 Superquadrics

The term *superquadrics* was first used by [2] to define a family of geometric shapes that includes superellipsoids, superhyperboloids of one piece, superhyperboloids of two pieces and Supertoroids.

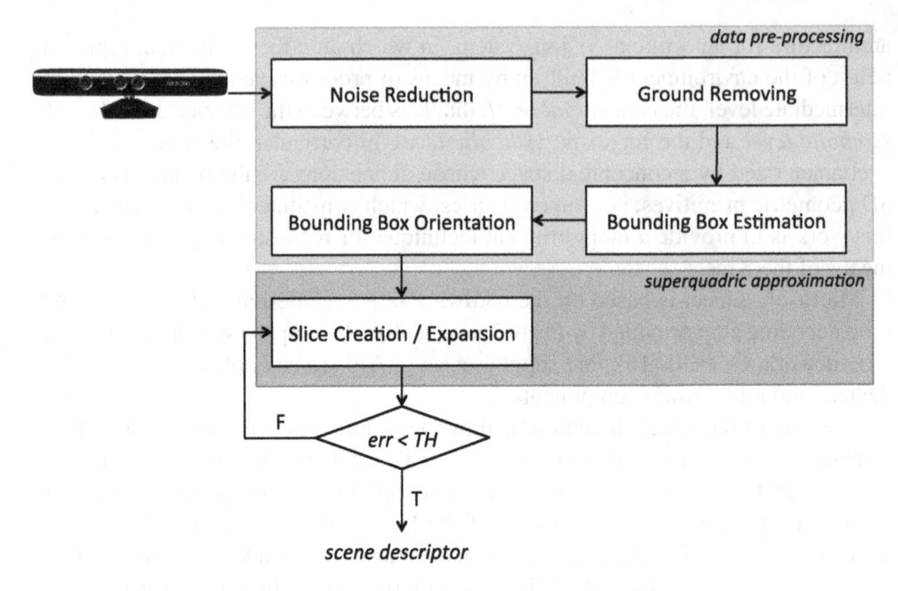

Fig. 1 System overview

The explicit form of a superquadric is given by the equation:

$$S_p(\eta, \omega) = \begin{bmatrix} x(\mathbf{p}; \eta, \omega) \\ y(\mathbf{p}; \eta, \omega) \\ z(\mathbf{p}; \eta, \omega) \end{bmatrix} = \begin{bmatrix} a_1 \cos(\eta)^{\varepsilon_1} \cos(\omega)^{\varepsilon_2} \\ a_2 \cos(\eta)^{\varepsilon_1} \sin(\omega)^{\varepsilon_2} \\ a_3 \sin(\eta)^{\varepsilon_1} \end{bmatrix} \tag{1}$$

where $-\pi/2 \leq \eta \leq \pi/2$ and $-\pi \leq \omega \leq \pi$.

The elements of the vector $\mathbf{p} = (a_1, a_2, a_3, \varepsilon_1, \varepsilon_2)$ are the parameters of the superquadric. In particular, a_1, a_2, a_3 represent the size of the model along the **X**, **Y**, **Z** axes, and ε_1, ε_2 control the shape of the model. More specifically, ε_1 is the squareness parameter in the north-south directio , while ε_2 is the squareness parameter in the east-west direction (see Fig. 2).

The inside-outside equation of the superquadric in implicit form is:

$$F(x, y, z) = \left[\left(\frac{x}{a_1} \right)^{\frac{2}{\varepsilon_2}} + \left(\frac{y}{a_2} \right)^{\frac{2}{\varepsilon_2}} \right]^{\frac{\varepsilon_2}{\varepsilon_1}} + \left(\frac{z}{a_3} \right)^{\frac{2}{\varepsilon_1}} \tag{2}$$

where $F(x, y, z)$ assumes a value equal to 1 when the point (x, y, z) is a superquadric boundary point, a value less than 1 when it is an inside point, and a value greater than 1 when it is an outside point.

In order to model a superquadric in a general position, six additional parameters are needed. In particular, p_x, p_y, p_z define the translation of the model relative to

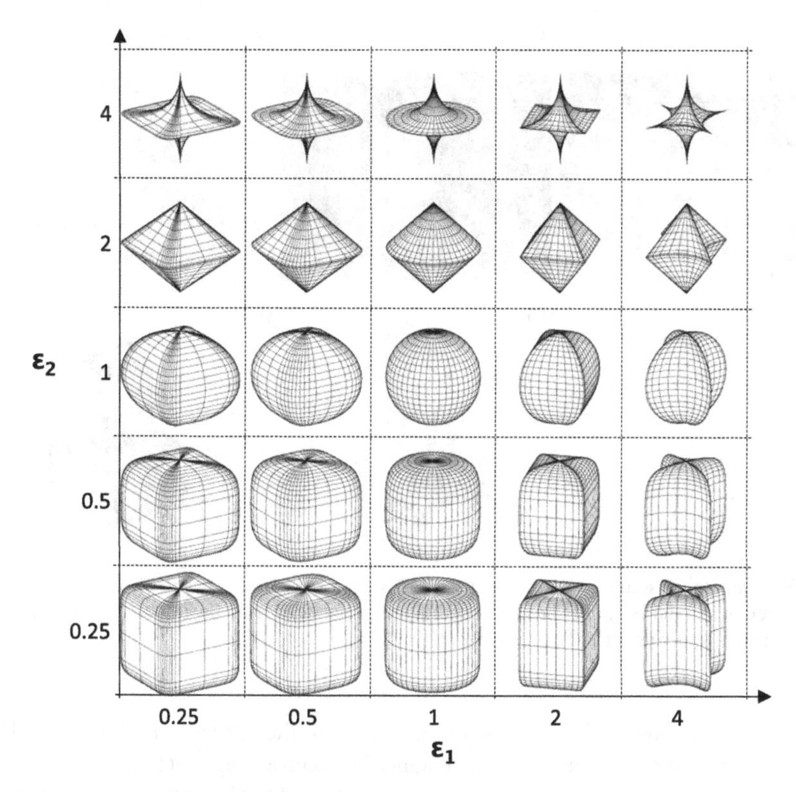

Fig. 2 Shapes obtained with ε_1, ε_2 in the range [0, 4]

the origin of the coordinate system, while the orientation in space is expressed by means of the angles ϕ, θ, ψ.

Thus, the model parameter vector **p** in the general position is:

$$\mathbf{p} = \left(a_1, a_2, a_3, \varepsilon_1, \varepsilon_2, p_x, p_y, p_z, \phi, \theta, \psi\right) \tag{3}$$

3.2 Scene Reconstruction

The method proposed in this chapter aims to reconstruct 3D scenes captured by the Kinect as a composition of some superquadric shapes. As previously discussed, research in the literature has addressed this problem by processing the images made by traditional range cameras or stereo vision systems. Here, in order to obtain a more detailed data representation, we directly process the 3D point cloud captured by the Kinect.

In order to correctly approximate the object some data pre-processing is required. Firstly, the whole set of 3D points (Fig. 3c) is analyzed to reduce the noise related to

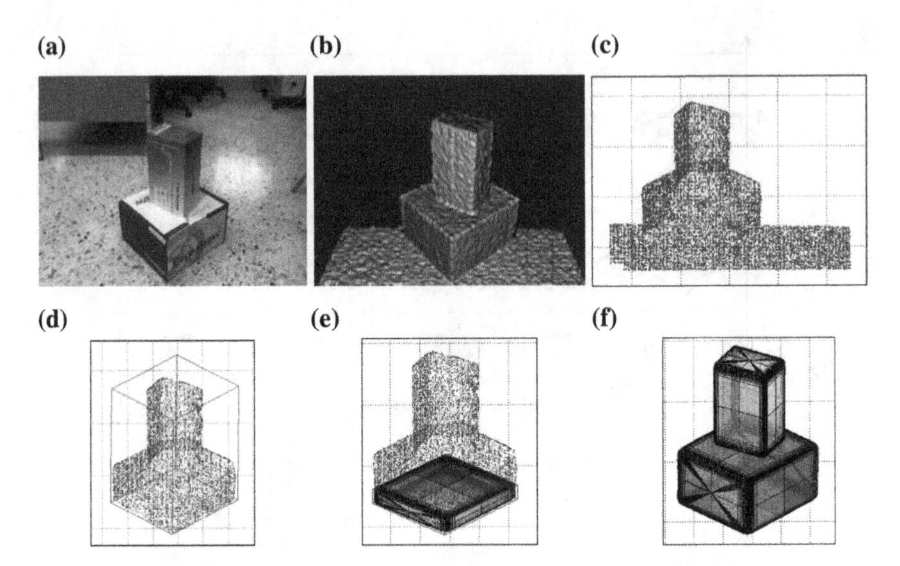

Fig. 3 Example of some processing steps. RGB image (**a**), depth image (**b**), point cloud (**c**), ground removing and bounding box (**d**), superquadrics obtained from the initial slice (**e**), optimal approximating superquadrics (**f**)

the acquisition process, i.e., points not belonging to the object or to the scene. The noise reduction method computes the distance between a couple of points, discarding those points whose distance is above a given threshold. The maximum distance is dynamically computed according to the mean distance measured for the considered point cloud.

Once the set of points has been filtered, a ground removing algorithm is applied to separate the object from the plain it lies in. The algorithm, based on RANSAC, computes the plane defined by 3 randomly chosen points and evaluates the number of inliers for that plane. This process is repeated for a certain number of iterations and the best plane, that is the plane with the greater number of inliers, is selected as *ground*.

Next, an overall bounding box BB_O is estimated for the whole set of points (Fig. 3d) and, in order to correctly break up the object into slices, the point cloud is rotated to the angle needed to arrange the bounding box parallel to the 3D axes.

As shown in Fig. 1, the superquadric approximation process is based on an iterative procedure for the creation and expansion of slices of 3D points.

The creation of a slice consists in the selection of a set of 3D points in a randomly chosen direction. For example, a slice of height H in the z-direction is created by selecting the (x, y, z) points of the cloud in the range $z_{\min} \leq z \leq z_{\max}$, where $z_{\max} - z_{\min} = H$.

In order to find the superquadrics that best approximates the point cloud contained in each slice, both the scale parameters a_1, a_2, a_3 and the form factors $\varepsilon_1, \varepsilon_2$ need to be defined. In particular, the size of the superquadric is estimated according

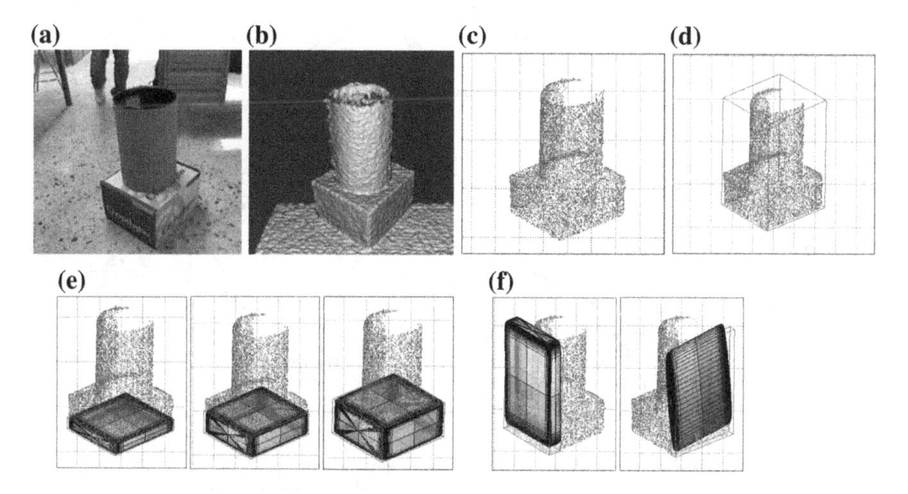

Fig. 4 Example of the slice creation and expansion process. RGB image (**a**), depth image (**b**), point cloud (**c**), ground removing and bounding box (**d**), superquadrics obtained from a correct slice selection (**e**), superquadrics obtained from an incorrect slice selection (**f**)

to the dimensions of the minimal bounding box BB that fits the set of 3D points contained in a slice, that is $a_1 = BB_x/2$, $a_2 = BB_y/2$, $a_3 = BB_z/2$. The form factors are computed by applying the RANSAC algorithm to search for the couple $(\varepsilon_1, \varepsilon_2)$ in the range $[0, 1]$ that best fits the input points. The remaining parameters $(p_x, p_y, p_z, \phi, \theta, \psi)$ are computed according to the position and orientation of BB. Note that the dimensions of BB are dependent on the number of 3D points effectively discovered in each slice region. In fact, since the Kinect is able to capture only those points belonging to the object surfaces, it usually happens that no points are selected in a particular direction.

Once the superquadric has been computed, the fitting error, i.e., a measure of how well the current model fits the points of the slice, is computed according to the least-squares minimization of the superquadric inside-outside function (Eq. 2) proposed in [17].

During the expansion step the size of the slice is increased in the chosen direction, e.g., the z-direction in the example given above. Then the fitting error e_i at the step i is compared with a threshold TH and the current slice is expanded until $e_i > TH$.

Once a slice can not be expanded any further, the method continues the processing of the remaining point cloud by iterating the slice creation-expansion steps until the whole scene has been analyzed.

Figure 4 shows the processing steps involved in the approximation of an object composed of a box and a cylinder. In particular, the images in Fig. 4e show the superquadrics obtained from the selection of a correct slice, while an example of slice selection along two incorrect directions is shown in Fig. 4f.

Once the object has been approximated, it can be fully described by the whole set of parameters of the approximating superquadrics.

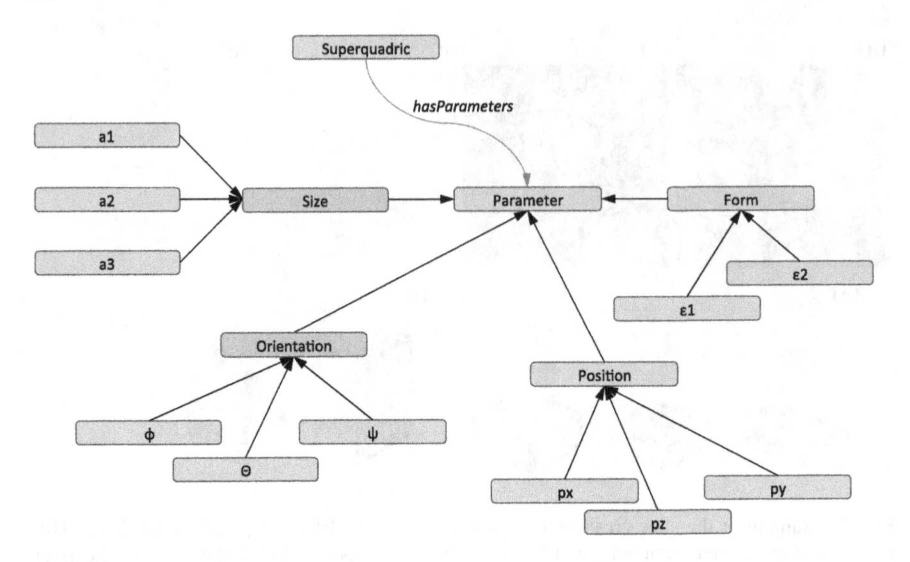

Fig. 5 The ontology representing the superquadrics

Information gathered through the reconstruction process is represented by means of the ontology shown in Fig. 5. As discussed above, the **Superquadric** shape is defined by a set of **Parameters** that capture properties related to the size, form, orientation and position of the curve, i.e, the object. Thus, the **Size** parameters a_1, a_2, a_3, the **Form** parameters ε_1, ε_2, the **Orientation** parameters ϕ, θ, ψ and the **Position** parameters p_x, p_y, p_z fully describe the **Superquadric** in the 3-D space.

4 Experimental Results

The proposed architecture has been designed to address a specific application scenario involving the management of indoor environments, e.g., offices or homes [8]. The main characteristic of such environments is that their interior design is usually based on a number of objects (e.g., chairs, desks, bookcases) that can be successfully represented as a composition of simple shapes (e.g., parallelepipeds, spheres, cylinders). In the AmI architecture adopted, a Wireless Sensor and Actuator Network (WSAN), whose nodes are equipped with off-the-shelf sensors (i.e., outdoor temperature, relative humidity, ambient light exposure and noise level) [7] is used to monitor the whole environment, while the Kinect sensor is used to detect specific objects placed within the office.

In order to evaluate the accuracy of the proposed scene reconstruction module in a real world scenario, several tests were performed on data captured by means of a Kinect device. In particular, we wanted to understand how some objects' properties

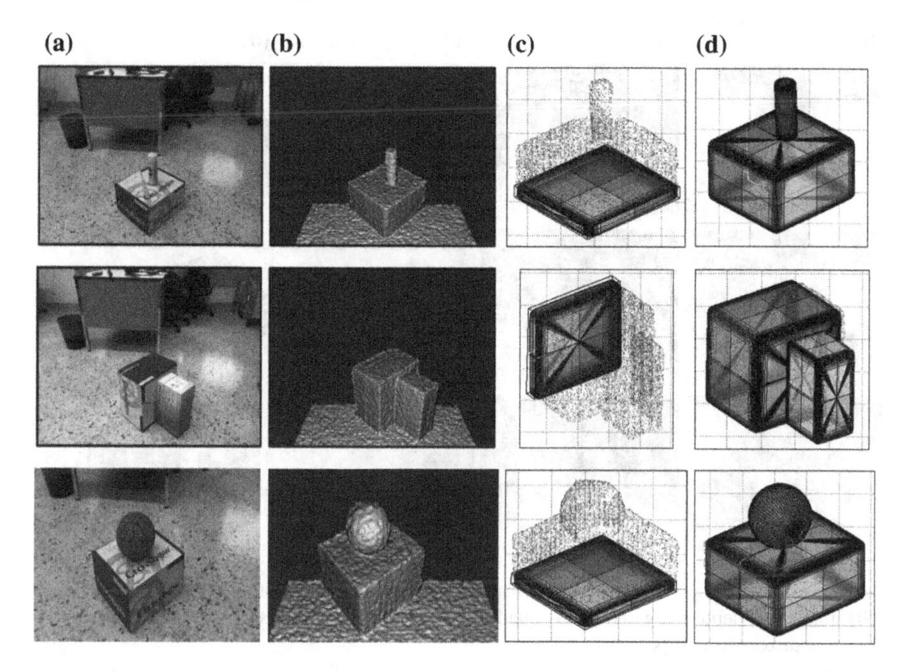

Fig. 6 Superquadric approximation of three simple scenes consisting of a single composite object. RGB image (**a**), depth image (**b**), superquadrics obtained from initial slice selection (**c**) and (**d**), optimal approximating superquadrics

(e.g., size, position, material) would eventually affect the overall performance of the proposed method.

Tests were conducted on 3D objects that can be broken down into different configurations of adjacent cubes, parallelepipeds, cylinders or spheres. These basic shapes are obtained by limiting the possible values of ε_1, ε_2, so the same approach could thus easily be extended to more complex shapes by considering different values of the ε_1, ε_2 parameters.

Some significant examples of scene reconstructions are shown in Figs. 6 and 7. The set of tests shown in Fig. 6 is oriented to observe how the system deals with objects that can be approximated with two simple superquadric shapes. The first row shows some images related to the reconstruction of a scene consisting of a spray placed above a box. This test serves to evaluate the ability of the proposed approach in approximating small noisy objects, such as the spray. The second row shows the reconstruction of two adjacent boxes. This kind of test was performed to evaluate how efficiently partial occlusions are managed. The third row shows the reconstruction of a scene consisting of a ball placed above a box. This test allowed us to demonstrate that symmetric and partially occluded objects, i.e., the ball, can be successfully processed.

The reconstruction of three more complex scenes is shown in Fig. 7. The difficulty associated with these scenes is mainly represented by the limited amount of free space

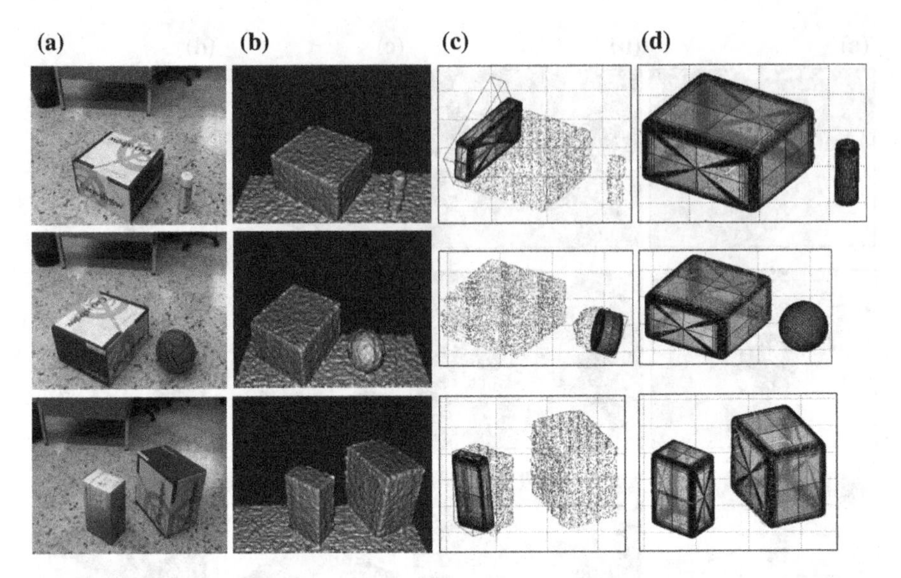

Fig. 7 Superquadric approximation of three more complex scenes consisting of two distinct objects. RGB image (**a**), depth image (**b**), superquadrics obtained from initial slice selection (**c**) and (**d**), optimal approximating superquadrics

between the three pairs of captured objects. For example, the top row shows that the smaller object (i.e., the spray) is detected and correctly approximated even though it is close to a bigger object characterized by a larger number of points.

The method proposed was been tested on about 30 scenes with different levels of complexity. For each scene, the whole process was run 10 times, obtaining an average reconstruction rate of 84 %.

The prototype was implemented connecting the Kinect to a personal computer (i.e., 2.5 GHz dual-core Intel Core i5, 4 GB of RAM and Unix OS) running MATLAB. The average scene reconstruction takes about 1–2 min.

From the analysis of the experimental results it emerges that some constraints need to be satisfied during the acquisition process. In particular, we noticed that three sides of the object should always be visible from the Kinect's point-of-view. This requirements has to be met to correctly drive the bounding box estimation process. Otherwise, it would not be possible to determine the scale parameters and consequently the form factors.

Moreover, some objects cannot be correctly captured by the Kinect because of to the material they are made of (see Fig. 8). For example, reflecting objects cause the IR ray to be reflected and lost, whilst transparent objects are not-correctly reconstructed since the ray is distorted when passing through them.

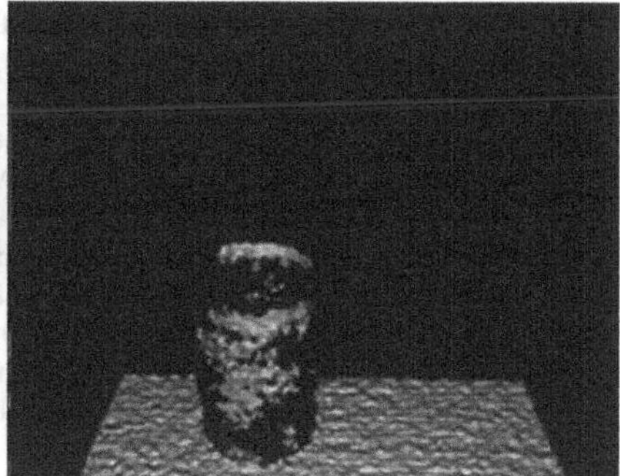

Fig. 8 Example of a misreconstructed object

5 Conclusions

This chapter describes a system for the automatic reconstruction of 3D objects captured by means of the Kinect sensor.

As compared to other solutions for object reconstruction from range images or stereo vision systems, the goal here was to demonstrate that composite objects can be efficiently reconstructed and represented by using inexpensive devices.

The experimental results demonstrate that the quality of the images provided by the Kinect is good enough to obtain satisfactory results, even under partially constrained conditions.

We are already working on improving the decomposition module in order to be able to reconstruct a greater set of composite objects, and that is, to consider a wider range of superquadric shapes. Moreover, once we have tested the effectiveness of our approach, we are planning a more efficient implementation of the prototype to speed up the reconstruction time.

Acknowledgments This work has been partially supported by the PO FESR 2007/2013 grant G73F11000130004 funding the SmartBuildings project.

References

1. Afanasyev, I., Biasi, N., Baglivo, L., Cecco, M.D.: 3D object localization using superquadric models with a kinect sensor. Technical report: Mechatronics Department, University of Trento, Italy (2011). http://www.ing.unitn.it/afanasye
2. Barr, A.: Superquadrics and angle-preserving transformations. Comput. Graph. Appl. IEEE **1**(1), 11–23 (1981). doi:10.1109/MCG.1981.1673799
3. Borenstein, G.: Making Things See: 3D Vision with Kinect, Processing, Arduino, and MakerBot. Make: Books. O'Reilly Media Inc., Sebastopol (2012)
4. Chella, A., Frixione, M., Gaglio, S.: A cognitive architecture for artificial vision. Artif. Intell. **89**(1–2), 73–111 (1997). doi:10.1016/S0004-3702(96),00039-2

5. Chella, A., Frixione, M., Gaglio, S.: Understanding dynamic scenes. Artif. Intell. **123**(1–2), 89–132 (2000). doi:10.1016/S0004-3702(00),00048-5
6. Cottone, P., Lo Re, G., Maida, G., Morana, M.: Motion sensors for activity recognition in an ambient-intelligence scenario. In: IEEE International Conference on Pervasive Computing and Communications Workshops (PerCom Workshops), pp. 646–651 (2013). doi:10.1109/PerComW.2013.6529573
7. De Paola, A., Gaglio, S., Lo Re, G., Ortolani, M.: Sensor9k : a testbed for designing and experimenting with wsn-based ambient intelligence applications. Pervasive Mob. Comput. **8**(3), 448–466 (2012). http://dx.doi.org/10.1016/j.pmcj.2011.02.006
8. De Paola, A., Lo Re, G., Morana, M., Ortolani, M.: An intelligent system for energy efficiency in a complex of buildings. In: Sustainable Internet and ICT for Sustainability (SustainIT), pp. 1–5 (2012)
9. Fischler, M.A., Bolles, R.C.: Random sample consensus: a paradigm for model fitting with applications to image analysis and automated cartography. Commun. ACM **24**(6), 381–395 (1981). doi:10.1145/358669.358692. http://doi.acm.org/10.1145/358669.358692
10. Grimson, W.E.L.: Object Recognition by Computer: The Role of Geometric Constraints. MIT Press, Cambridge (1990)
11. Kean, S., Hall, J., Perry, P.: Meet the kinect: An Introduction to Programming Natural User Interfaces, 1st edn. Apress, CA (2011)
12. Krivic, J., Solina, F.: Part-level object recognition using superquadrics. Comput. Vision Image Underst. **95**(1), 105–126 (2004). doi:10.1016/j.cviu.2003.11.002
13. Leonardis, A., Jaklic, A., Solina, F.: Superquadrics for segmenting and modeling range data. IEEE Trans. Pattern Anal. Mach. Intell. **19**(11), 1289–1295 (1997). doi:10.1109/34.632988
14. Lo Re, G., Morana, M., Ortolani, M.: Improving user experience via motion sensors in an ambient intelligence scenario. In: Pervasive and Embedded Computing and Communication Systems (PECCS), pp. 29–34 (2013)
15. Marshall, D., Lukacs, G., Martin, R.: Robust segmentation of primitives from range data in the presence of geometric degeneracy. IEEE Trans. Pattern Anal. Mach. Intell. **23**(3), 304–314 (2001). doi:10.1109/34.910883. http://dx.doi.org/10.1109/34.910883
16. PrimeSense: Openni. http://www.openni.org/
17. Solina, F., Bajcsy, R.: Recovery of parametric models from range images: the case for superquadrics with global deformations. IEEE Trans. Pattern Anal. Mach. Intell. **12**(2), 131–147 (1990). doi:10.1109/34.44401

Sensor Node Plug-in System: A Service-Oriented Middleware for Wireless Sensor Networks

Giuseppe Di Modica, Francesco Pantano and Orazio Tomarchio

Abstract Sensors populate our environment in a pervasive way. You may find them at home, in your car, in streets, or even in your smartphone. They are usually employed to measure various kind of phenomena, and can serve very specific purposes such as monitoring, surveillance, prediction, controlling. In the IoT vision, the potential represented by the huge amount of *raw* data that millions of sensors produce every day need to be transformed into a more exploitable *knowledge*. In order to keep up with the pace at which raw data are being produced today, we need new solutions that combine tools for data management and services capable of promptly structuring, aggregating and mining data even at the time they are produced. This chapter discusses some of the issues related to the management of sensors and sensor data, and proposes a solution to face these issues. The proposed solution is a middleware to be deployed on top of physical sensors, capable of abstracting away sensors' proprietary interfaces, and offering them to third party applications in an as-a-Service fashion for an immediate and universal use. The viability of the approach was finally tested on real-life use case scenarios.

1 Introduction

Today we are witnessing to a widespread diffusion of physical sensor devices which are capable of capturing data from the environment they are deployed in and offering such data to (even remote) elaboration units for further mining and computation.

G. Di Modica · F. Pantano · O. Tomarchio (✉)
Department of Electrical, Electronic and Computer Engineering, University of Catania, Catania, Italy
e-mail: Orazio.Tomarchio@dieei.unict.it

G. Di Modica
e-mail: Giuseppe.DiModica@dieei.unict.it

F. Pantano
e-mail: Francesco.Pantano@dieei.unict.it

S. Gaglio and G. Lo Re (eds.), *Advances onto the Internet of Things*,
Advances in Intelligent Systems and Computing 260, DOI: 10.1007/978-3-319-03992-3_14,

Not to talk about the millions of personal handheld devices (smartphone, tablets) which manufacturers equip with sensors able to capture a wide variety of phenomena. If on the one hand this sensors' explosion is perfectly in line with the expectation of the Internet of Things (IoT) vision [10], on the other one the potential provided by the huge stream of data that can be generated by sensors worldwide is not yet thoroughly exploited.

Though specifications and standards have been trying to define data models and protocols for sensors and sensor networks [13], still proprietary sensor networks are not connected to each other, nor they are connected to a globally accessible information network. Data produced by sensors are very often not structured and application-oriented, in the sense that they serve just the specific purpose of the application that makes use of them. The IoT vision of worldwide deployed sensors, universally searchable and usable over the Internet by any end-consumer, is far from being realized.

On the developer end, applications that extract data from sensors must be written in compliance with the proprietary data interface; further, mining the extracted data is a complex operation as proprietary data meaning must first be accurately interpreted. Relieving the sensor applications' developer from the burden of extracting and interpreting sensor data is with no doubt a key step towards the fulfillment of the IoT vision. In this respect, the service-oriented approach [8, 15] along with the Cloud technology [2] provides adequate abstractions for application developers. We believe that integrating heterogeneous sensors and sensor network technologies with Cloud platforms over the Internet would pave the way for the development of new useful IoT applications.

The SeNSori research program, funded by the Italian Ministry of the Economic Development, pursues the ambitious objective of defining a new paradigm for the development of services exposing the functionality of sensors and sensor networks, capable of interacting to third party applications, and that can be adequately composed with each other to serve the needs of any application domain, thus following an IoT-inspired approach. The target scope of the research program is the control of energetic flows in civil and home environments, but its approach to the sensor's issues is general enough to be potentially adopted in any applicative domain. In particular, in the context of the program, a prototype of a middleware for wireless sensor networks was implemented and tested against some energy-saving use cases [4]. Basically, the middleware takes care of abstracting the sensor physical layer and presenting sensors as services to be easily accessed and composed. In the remainder of this reading the main technical features of the middleware are discussed.

The chapter is structured as follows. Section 2 discusses the motivation and objectives of the SeNSori project. Section 3 presents the middleware, delves into the technical details of the components that the middleware is made of, and finally describes one of the use case against which the middleware was tested. Section 5 draws the conclusion of the work.

2 SeNSori: Sensor Node as a Service for Home and Buildings Energy Saving

SeNSori is an undergoing research program funded by the Italian Ministry of the Economic Development.

Main objective of the program is the definition and implementation of an "ambient intelligent" platform for the overall control of energetic flows in civil and home environments. This ambitious program investigates on how the sensor node specifically, and the sensor network in its dinamicity, may be regarded as the result of the composition of multiple services which can be produced/consumed through the platform. The platform, which we will refer to as the Sensor Node as a Service (SNS) platform, will enable a new paradigm for building sensor-fed services for the management of home and office environments, which are conceived to maximize the ambient's quality perceived by humans and to minimize the overall energy consumption as well. Long term objective of the program is to leverage on the platform to foster the proliferation of an ecosystem of services which any consumer may access in order to satisfy their need for controlling and managing energetic flows in any civil environment. Further, in the purpose of adding more value to the implementation of final services that will be delivered to end consumers, the research program also intended to stimulate the participation of the sensor industry's stakeholders in the implementation and management of added-value services; in this regard, the platform also offers the possibility to integrate natively implemented services to third party services.

The technological challenge launched by the research program is played at different levels: from the wireless sensor networks deployed in the environment to be controlled, up to the intelligent systems committed to reasoning on the gathered data. The main technological issues faced by the program are synthesized as follows:

- Definition of a virtualization layer capable of abstracting sensors' functionality independently of their proprietary technology;
- Definition of a basic set of tools for the management of virtual sensors;
- Definition of a service-oriented interface to sensors and the respective offered functionalities;
- Definition of a set of high-level services to support the environmental governance;
- Definition of a mash-up layer for the integration of third party services and technologies.

The result of the scientific program is the implementation of a platform's software prototype, whose services and involved actors are depicted in Fig. 1. In the perspective proposed by the figure, services are categorized according to the actor that is in charge of using them. On the one end it is depicted the end user in the role of owner or responsible for the environment to be governed. On the other one it is the stakeholder which, leveraging on the basic services offered by the platform, will provide end users with added-value services.

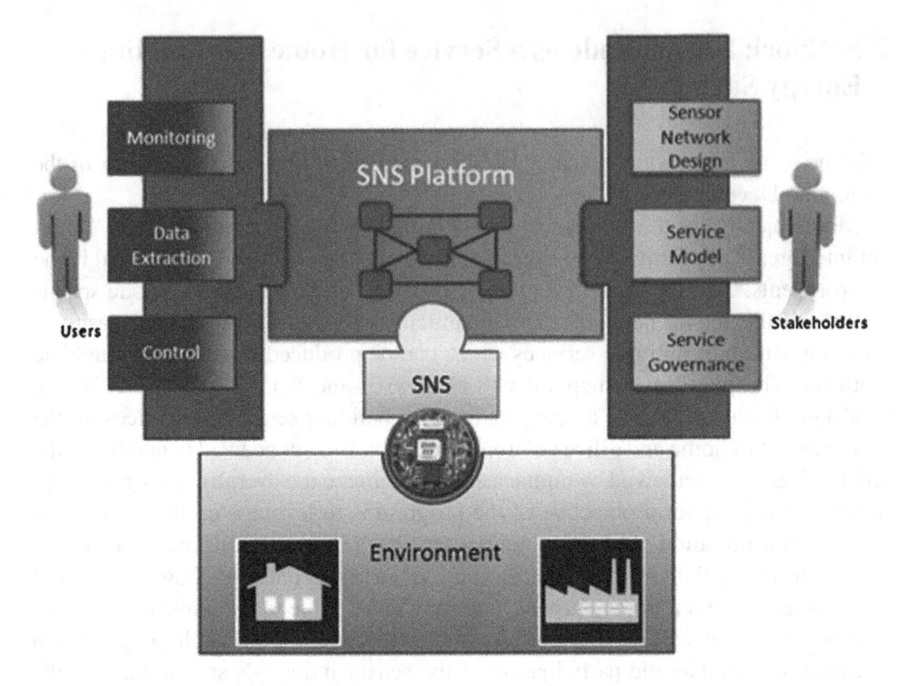

Fig. 1 SNS: services and actors

In the program's aim, the following stakeholders are called to contribute to the implementation of the ecosystem of sensor-oriented services:

- *Service provider*. Implements and pushes sensors that are searchable according to the service's capabilities or to the capabilities of the associated device(s).
- *Sensor device manufacturer*. Produces physical sensor device and provides the description of their capabilities according to the platform's specification. The sensor can be hot-plugged into the platform, searched and invoked by any appliance.
- *Sensor networks' designer and developer*. Design and implement a sensor network and their control system through the design functionality offered by the platform
- *Installer*. Provides a turn-key sensor-based control system, with a minimum and configurable set of services meeting the end user's needs.

The research work carried on by the scientific program is articulated into three different lines. The first line focuses on the definition of an infrastructural model for monitoring and controlling energetic flows in home and office environments. This line's objective is to conceive a novel model of sensor node which embeds the structural and informational aspects of a distributed, service-oriented system. The second line works on the definition of a software middleware acting as a bridge between the layer of sensor networks and the network of services exposed by the platform. The middleware will have to be universally adaptable to any underlying sensor technology on the one hand, and will have to offer flexible and powerful APIs to services' developers on the other one. The definition of a semantic model to

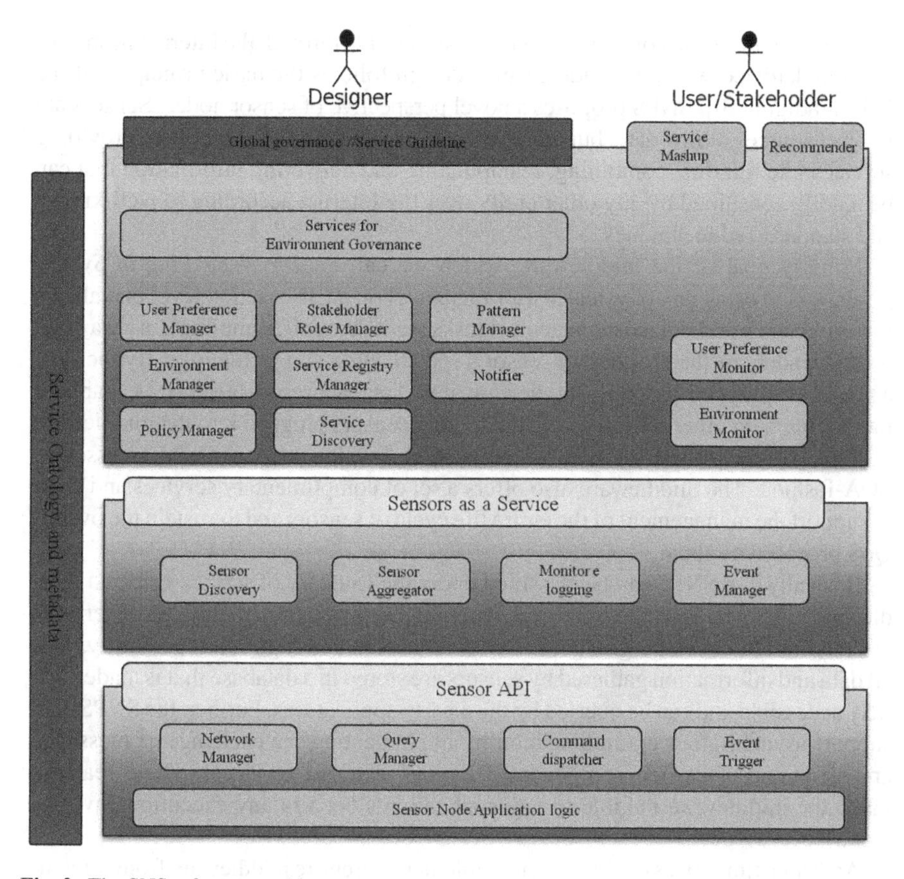

Fig. 2 The SNS reference architecture

represent the sensors' domain of information is part of this line of work too. Finally, the third line is in charge of defining a service-oriented framework that will have to offer both high-level sensor management services and tools to implement ad-hoc services according to a pattern that will grant to the end user a high level of QoS in terms of security, reliability, accessibility and usability. All the aspects regarding the service governance are also investigated in this line. Figure 2 illustrates the reference platform architecture. The three depicted layers respectively reflect the objectives of each of the above discussed lines of work.

3 SNPS: An OSGi Middleware for Wireless Sensor Networks

In this section we give a detailed overview of the platform's middleware layer and the components it is made of. The middleware has been devised to lay on the physical layer of wireless sensors, to abstract away the sensors' specific features, and to turn sensors into smart and composable services accessible through the Internet in an easy and standardized way. The middleware's design follows the basic principles of the IoT paradigm [10]. SNPS proposes a novel perspective of sensor nodes. Sensors are not just sources of raw data, but are conceived as smart objects capable of providing services like filtering, combining, manipulating and delivering information that can be readily consumed by any other entity over the Internet according to well-known and standardized techniques.

Primary goal of the middleware, which we call *Sensor Node Plug-in System (SNPS)*, is to bring any physical sensor (actuator) on an abstraction level that allows for easier and standardized management tasks (switch on/off, sampling), in a way that is independent of the proprietary sensor's specification and technology. By the time a sensor is "plugged" into the middleware, it will act as a resource/service capable of interacting with other resources (be them other sensors plugged into the middleware or third party services) in order to compose high-value services to be accessed in SOA-fashion. The middleware also offers a set of complimentary services and tools to support the management of the entire life cycle of sensors and to sustain the overall QoS provided by them.

Basically, the SNPS can be classified under the category of service-oriented middlewares [15]. In fact, the provided functionality are exposed through a service-oriented interface which grants for universal access and high interoperability. Yet, all data and information gathered by sensors are stored in a database that is made publicly accessible and can be queried by third party applications. Further, the SNPS also support asynchronous communication by implementing the exchange of messages among entities (sensors, components, triggers, external services). These features make the middleware flexible to any application's need in any execution environment.

At design time it was decided not to implement the entire middleware from scratch. A scouting was carried out in order to identify the software framework that best supported, in a native way, all the characteristics of flexibility and modularity required by the project. Eventually, the OSGi framework [14] was chosen. The OSGi framework implements a component-oriented model, which natively supports the component's life cycle management, the distribution of components over remote locations, the seamless management of components' inter-dependencies, and the asynchronous communication paradigm.

The SNPS middleware was then organized into several components, and each component was later implemented as a software module (or "bundle") within the OSGi framework. Figure 3 depicts the architecture of the middleware and its main components.

The overall architecture can be broken down into three macro-blocks:

- Sensor Layer Integration
- Core and related Components
- Web Service Integration

In the following we provide a description of each macro-block.

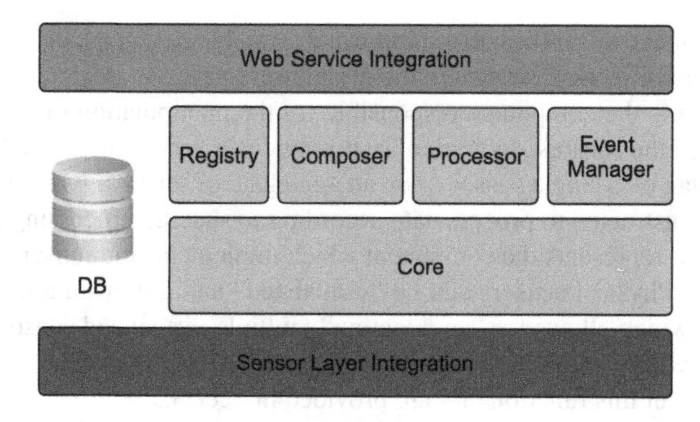

Fig. 3 SNPS architecture

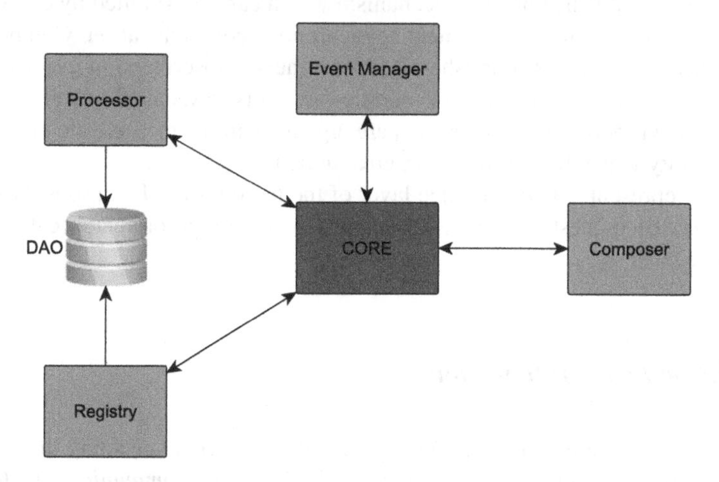

Fig. 4 Core and related components

3.1 Core and Related Components

Figure 4 depicts the connections among the software components of the SNPS architecture. The responsibility of each component is described in the following:

- **Core.** It is where the business logic of the Middleware resides. The Core acts as an orchestrator who coordinates the middleware's activities. Data and commands flowing forth and back from the web service layer to the sensor layer are dispatched by the Core to the appropriate component.
- **Registry.** It is the component where all information about sensors, middleware's components and provided services are stored and indexed for search purpose. As for the sensors, data regarding the geographic position and the topology of the managed wireless sensor networks are stored in the Registry. Also, each working component needs to signal its presence and functionality to the Registry, which

will have to make this information public and available so that it can be discovered by any other component/service in the middleware.

- **Processor.** It is the component responsible for the manipulation of the data flow coming from the sensors. In particular, it provides a service to set and enforce a sampling plan on a single sensor or on an aggregate of sensors. Also, this component can be instructed to process data according to specific processing templates.
- **Composer.** It represents the component which implements the sensors' composition service. Physical sensors can be "virtualized" and are given a uniform representation which allows for "aggregating" multiple virtualized sensors into one sensor that will eventually be exposed to applications. An insight and practical examples about this functionality are provided in Sect. 3.5.
- **Event Manager.** It is one of the most important components of the middleware. It provides a publish/subscribe mechanism which can be exploited by every middleware's component to implement asynchronous communication. Components can either be producers (publishers) or consumers (subscribers) of every kind of information that is managed by the middleware. This way, data flows, alerts, commands are wrapped into "events" that are organized into topics and are dispatched to any entity which has expressed interest in them.
- **DAO.** It represents the persistence layer of the middleware. It exposes APIs that allow service requests to be easily mapped onto storage or search calls to the database.

3.2 Sensor Layer Integration

The Sensor Layer Integration (SLI) represents the gateway connecting the middleware to the physical sensors. It implements a *bidirectional communication channel* (supporting commands to flow both from the middleware to the sensors and from the sensors to the middleware as well and a *data channel* (for data that are sampled by sensors and need to go up to the middleware).

The addressed scenario is that of wireless sensor networks implemented through so called Base Stations (BS) to which multiple sensors are "attached". A BS implements the logic for locally managing its attached sensors. Sensors can be wiredly or wirelessly attached to a BS, forming a network which is managed according to specific communication protocols, which are out of our scope. The SLI will then interact just with the BS, which will only expose its attached sensors hiding away the issues related to the networking.

The integration is realized by means of two symmetrical bundles, which are named respectively *Middleware Gateway bundle (iMdmBundle)* and *WSN Gateway Bundle (iWsnBundle)*. The former lives in the middleware's runtime context, and was thought to behave as a gateway for both commands and data coming from the BSs and directed to the middleware; the latter lives (runs) in the BS's runtime context, and forwards commands generated by the middleware to the BSs. Since the middleware and the BSs may be attached to different physical networks, the communication between the two

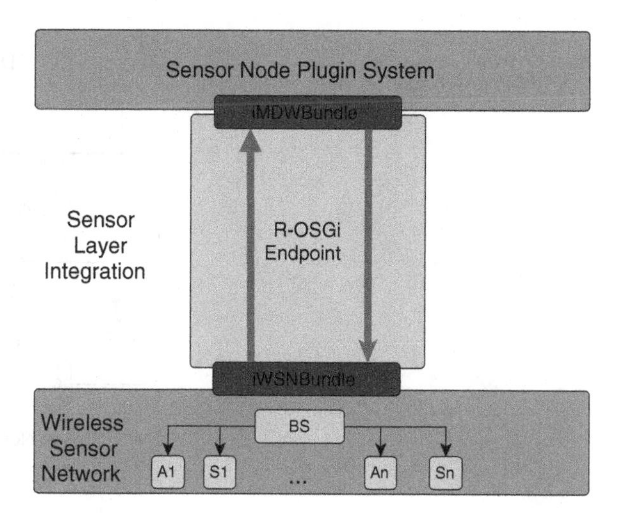

Fig. 5 OSGi bundles implementing the sensor layer integration

bundles is implemented through a remote "OSGI Context", which is a specific OSGi's features allowing bundles living in different runtime contexts to communicate to each other's. In Fig. 5 the two bundles and their respective runtime contexts are shown.

The SLI was designed to work with any kind of BS, independently of the peculiarity of the sensors it manages, with the aim of abstracting and uniforming the access to sensors' functionality. Uniforming the management of the sensors' life cycle does not mean giving up the specific capabilities of sensors. Physical sensors will maintain the way they work and their peculiar features (in terms, for instance, of maximum sampling rate, sampling precision, etc.). But, in order for sensors (read base stations) to be pluggable into the middleware and be compliant to its management logic, a minimal set of requirements must be satisfied: the *iWsnBundle* to be deployed on the specific BS will have to interface to the local BS' logic and implement the functionality imposed by the SNPS middleware (switch on/off sensors, sample data, run sampling plan) by invoking the proprietary base station's API.

3.3 Web Service Integration

As depicted in Fig. 6, the *OSGi bundle Wrapper* exports the functionality of the SNPS middleware to a Web Service context.

SNPS services can be invoked from any OSGi compliant context. On the other hand, making the SNPS accessible as a plain Web Service will make its services profitable for a great number of applications in several domains. The functionality implemented by the SNPS' bundles have been packaged into the following categories of services:

- Search for sensors;

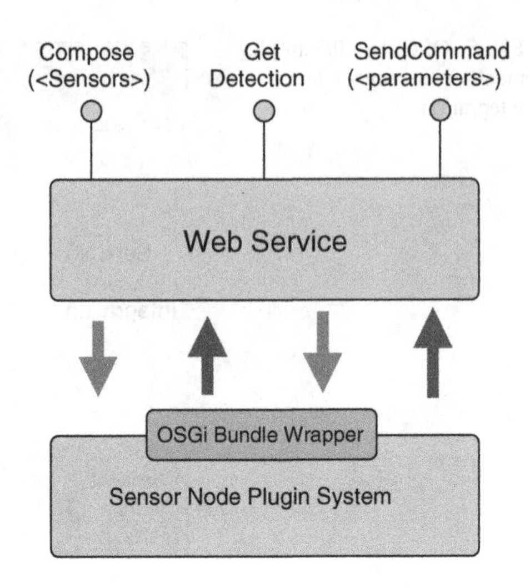

Fig. 6 Wrapping and exposing SNPS as a Web service

- Retrieve sensors capabilities and sensors data;
- Compose sensors;
- Send commands to sensors (enable/disable, set a sampling plan).

3.4 SNPS Data Model

The SNPS data model is one of the most interesting features of the middleware. Goals like integration, scalability, interoperability are the keys that drove the definition of the model at design time. The objective was then to devise a data model to structure both sensors' features (or capabilities) and data produced by sensors. The model had to be rich enough to satisfy the multiple needs of the middleware's business logic, but at the same time had to be light and flexible to serve the objectives of performance and scalability. We surveyed the literature in order to look for any proposal that might fit the middleware's requirement. Specifications like SensorML and O&M [1] seam to be broadly accepted and widely employed in many international projects and initiatives. SensorML is an XML-based language which can be used to describe, in a relatively simple manner, sensors capabilities in terms of phenomena they are able to offer and other features of the specific observation they are able to implement. O&M is a specification for describing data produced by sensors, and is XML-based as well. XML-based languages are known to be hard to treat, and in many cases the burden for the management of XML-based data overcomes the advantage of using rigorous and well-structured languages. We therefore opted for a solution that calls on a reduced set of terms of the SensorML specification to describe the sensor capabilities, and makes use of a much lighter JSON [7] format to structure the data produced by sensors. An excerpt of what a description of sensor capabilities look like is depicted in Fig. 7.

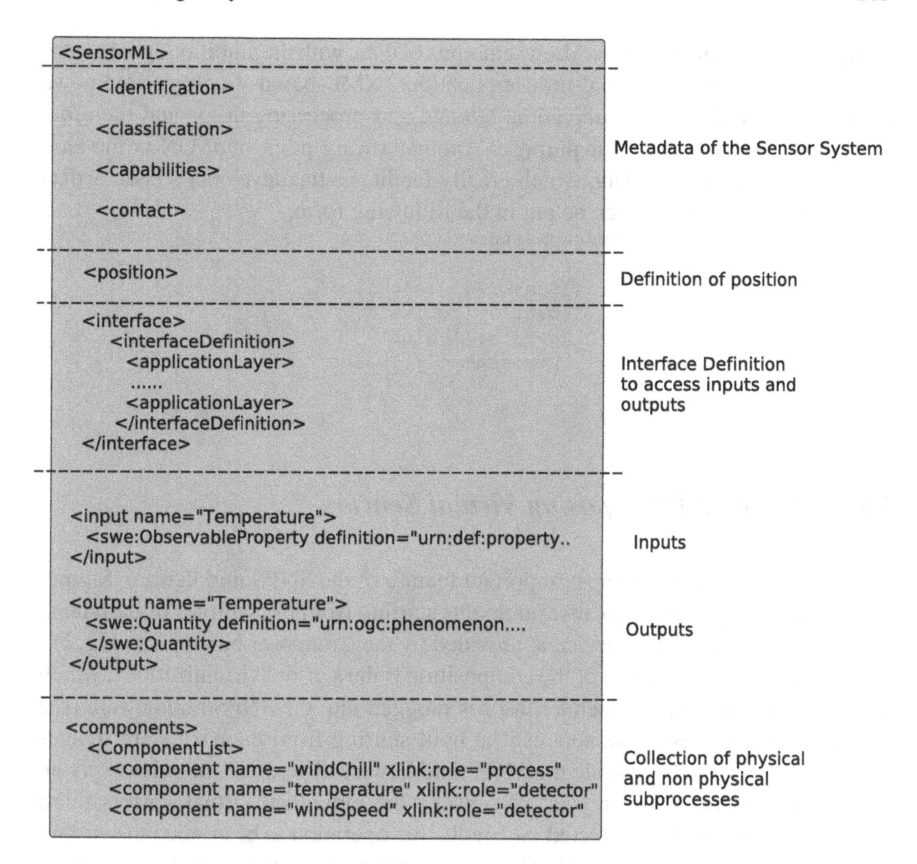

Fig. 7 Description of sensor capabilities in SensorML

This is the basic information that must be attached to any sensor before it is plugged into the middleware. Among others, it carries data regarding the phenomena being observed, the sampling capabilities, and the absolute geographic position. When the sensor wakes up, it sends this information to the middleware, which will register the sensor to the Registry bundle, and produce its *virtualized image*, i.e., a software alter-ego of the physical sensor which lives inside the middleware run-time. The virtual sensor has a direct connection with the physical sensor. Each interaction involving the virtual sensor will produce effects on the physical sensor too. It is important to point out that all virtual sensors are treated uniformly by the middleware's business logic.

Furthermore, SensorML is by its nature a process-oriented language. Starting from the atomic process, it is possible to build the so-called *process chain*. We exploited this feature to implement one of the main service provided by the SNPS, i.e., the sensors' composition service (see Sect. 3.5 for more details). This service, in fact, makes use of this feature to elaborate on measurements gathered by multiple sensors.

As regards the definition of the structure for sensor data, JSON was chosen because it ensures easier and lighter management tasks. The middleware is designed to handle (sample, transfer, store, retrieve) huge amounts of data, with the ambitious goal to also satisfy the requirements of real-time applications. XML-based structures are known to cause overhead in communication, storage and processing tasks, and therefore they do not absolutely fit our purpose. Another strong point of JSON is the ease of writing and analyzing data, which greatly facilitates the developer's task. A data sampled by a sensor will then be put in the following form:

```
Sensor_Measure:
{
        ''SensorId'':''value'',
        ''data'':''value'',
        ''type'':''value'',
        ''timestamp'':''value''
}
```

3.5 Building and Composing Virtual Sensors

Sensor Composition is the most important feature of the SNPS middleware. Simply said, it allows to get complex measurements starting from the samples of individual sensors. The composition service is provided by the Composer bundle (see Fig. 2).

An important prerequisite of the composition is the sensor "virtualization", which is a procedure performed when a sensor is plugged into the SNPS middleware (see Sect. 3.2). Aggregates of sensors can be built starting from their software images (virtual sensors) that live inside the SNPS middleware. Therefore, in order to create a new composition (or aggregate) of sensors, the individual virtual sensors to be combined need to be first selected. Secondly, the operation to be applied to sensor's measurements must be specified. This is done by defining the so-called *Operator*, which is a function that defines the expected input and output formats of the operation being performed. The final composition is obtained by just applying the Operator to the earlier chosen virtual sensors. By that time, a new virtual sensor (the aggregate) is available in the system, and is exposed as a new sensor by the middleware. Figure 8 depicts the structure of an aggregate of sensors.

Let us figure out a practical use case of sensor composition. Imagine that there are four temperature sensors available in four different rooms of an apartment. An application would like to know about the instant average temperature of the apartment. A new sensor can be built starting from the four temperature sensors applying the average operator, as depicted in Fig. 9.

In this specific case, the input sensors are homogeneous. The middleware also provides for the composition of heterogeneous sensors (e.g., temperature, humidity, pressure, proximity), provided that the operator's I/O scheme is adequately designed to be compatible with the sensors' measurement types.

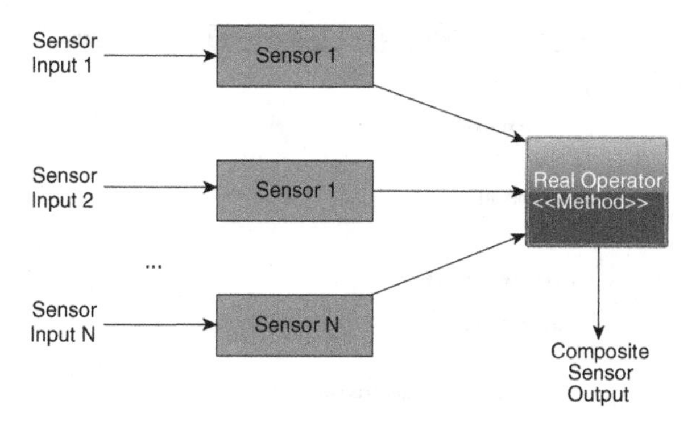

Fig. 8 Sensors composition

Fig. 9 Average operator

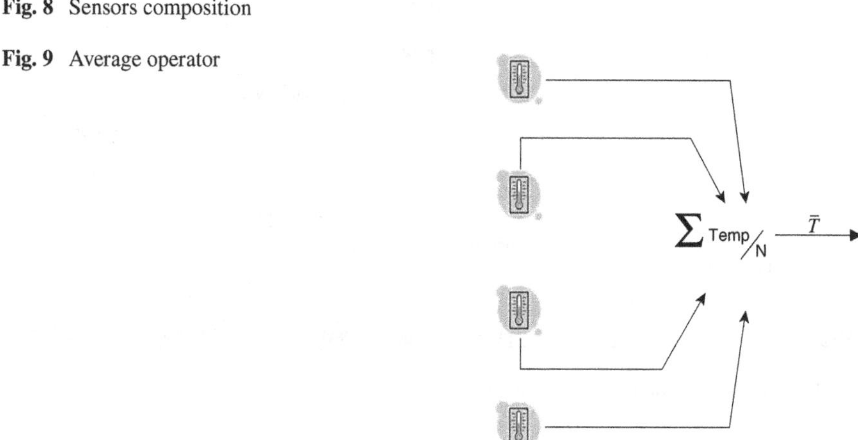

3.6 Use Case Scenario

A prototype of the middleware has been implemented and its functionality have been tested. In this section we provide some insight on a real use case that we set up in order to prove the effectiveness of the implemented mechanisms. In particular, here we focus on what we believe is the most important middleware's provided service, which is the sensors composition service.

The sensors composition process puts emphasis on the semantics of the operation, rather than relying on the simple measure. In this regard, it has been developed a use case, in order to emphasize the power and importance of the aggregation of sensors. Let us imagine an apartment in which for every room there is a temperature sensor; it may be interesting to detect the average temperature of the apartment and, if necessary, use this value to perform further processing.

Let us consider this use case as divided into two distinct phases: sensors composition and aggregate sensor inquiry.

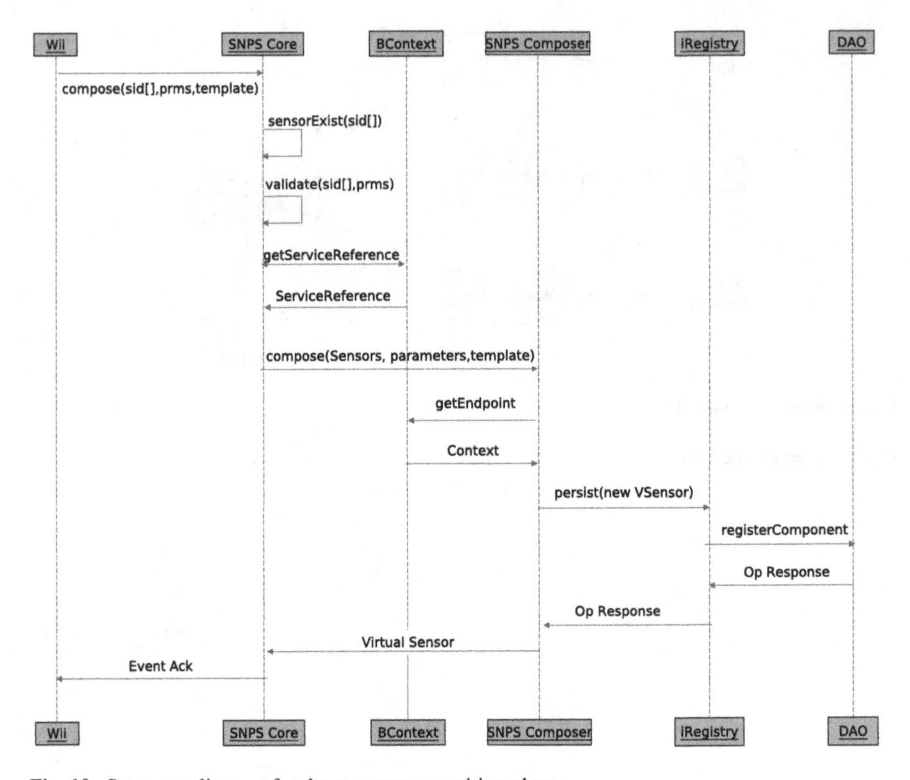

Fig. 10 Sequence diagram for the sensor composition phase

Phase 1: Sensors composition.
In the first stage, we are going to consider the following actors:

- *Web Integration Interface (Wii)*. It represents the entity generating the composition request;
- *Composer*. It generates the new virtual sensor from simple temperature sensors;
- *Registry*. It registers the sensor;
- *Core*. It orchestrates the composition task among the middleware components.

The operations carried out in the scenario, depicted in Fig. 10, are the following:

1. The Wii propagates the request to the Core of the platform.
2. The Core retrieves the images of the selected sensors and perform a two-steps validation:
   - Verification of the existence of the sensor images in memory;
   - Validation of the operator to be applied to the sensors;
3. The Core invokes the composition service provided by the Composer;
4. The Composer generates the new (virtual) aggregate sensor;
5. The Registry registers the new sensor;

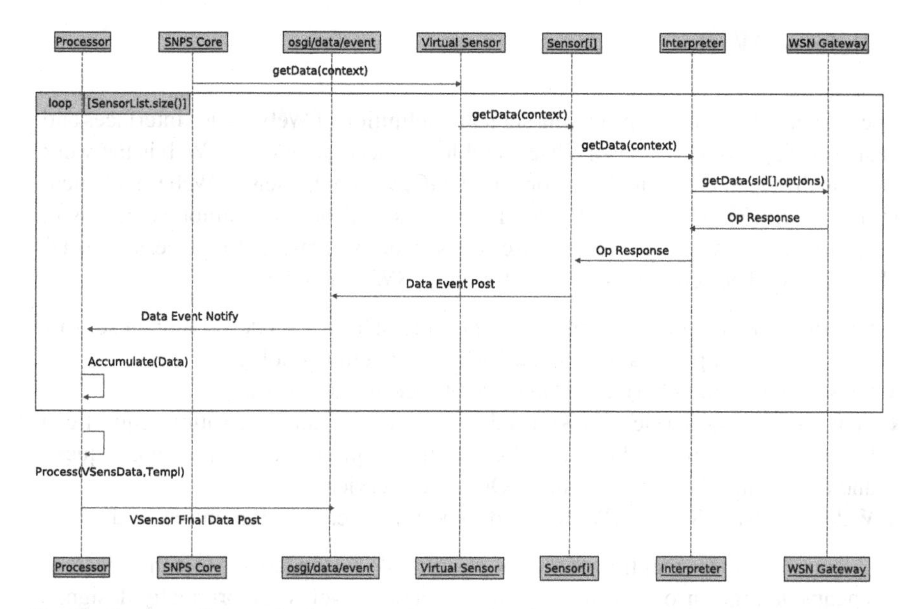

Fig. 11 Sequence diagram for the aggregate sensor inquiry

6. The Core generates a "registration" event for the new sensor, according to the Publish/Subscribe paradigm;

Phase 2: Aggregate sensor inquiry.

The steps made in this phase, described in Fig. 11, are the following ones:

1. The Wii propagates the request to the Core;
2. The Core, after selecting the aggregate sensor, invokes the get-data operation;
3. The virtual sensor image invokes, for each composing sensor, a service provided by the Sensor Layer Integration (SLI);
4. The SLI propagates the request to the gateway (at WSN Level), which is able to interact directly with the Base Station, which maps the command into a direct command to each physical sensor;
5. After getting the data, the SLI generates a Data response event, which the aggregate sensor is able to collect;
6. Finally, the aggregate sensor applies the operator to the previously collected data, and generates an event on the topic of interest;
8. The Processor records the measurement.

4 Related Work

The most notably effort in providing standard definition of Web service interfaces and data encodings to make sensors discoverable and accessible on the Web is the work done by the Open Geospatial Consortium (OGC) within the Sensor Web Enablement initiative [1, 13]. The role of the SWE group is to develop common standards to determine sensors capabilities, to discover sensor systems, and to access sensors' observations. The principal services offered by SWE include:

- Sensor Model Language (SensorML): provides a high level description of sensors and observation processes using an XML schema methodology
- Sensor Observation Service (SOS): used to retrieve sensors data.
- Sensor Planning Service (SPS): used to determine if an observation request can be achieved, determine the status of an existing request, cancel a previous request, and obtain information about other OGC web services
- Web Processing Service (WPS): used to perform a calculation on sensor data.

A common misconception of the adoption of SWE standards is that they, instead of encapsulating sensor information on application level, were originally designed to operate directly on a hardware level. Of course, supporting interoperable access on the hardware level has some advantages and comes very close to the "plug and play" concept. Currently, some sensor systems such as weather stations and observation cameras already offer access to data resources through integrated web servers. However, besides contradicting the view of OGC's SWE of uncoupling sensor information from sensor systems, the downside of this approach arise when dealing with a high number of specialized and heterogeneous sensor systems, and in resource-limited scenario (as typical WSNs) where communication and data transportation operations have to be highly optimized. Even a relatively powerful sensor gateway is not necessarily suitable as a web server: in many cases it may typically be networked via a low-bandwidth network and powered by a battery and so it has neither the energy or bandwidth resources required to provide a web service interface.

The need for an intermediate software layer (middleware) derives from the gap between the high-level requirements from pervasive computing applications and the complexity of the operations in the underlying WSNs. The complexity of the operations within a WSN is characterized by constrained resources, dynamic network topology, and low level embedded OS APIs, while the application requirements include high flexibility, re-usability, and reliability to cite a few. In general, WSN middleware helps the programmer develop applications in several ways: it provides appropriate system abstractions, reusable code services and data services. It helps the programmer in network infrastructure management and providing efficient resource services.

Some research efforts have been done on surveying the different aspects of middleware and programming paradigms for WSN. For example, [6] analyzed different middleware challenges and approaches for WSN, while [16] and [12] analyzed programming models for sensor networks.

As an example of different approaches, we cite here TinyDB [9], a query processing system based on SQL-like queries that are submitted by the user at a base station where the application intelligence resides. Enabling dynamic re-configuration is one of the main motivations for component-based designs like the RUNES middleware [3]. Finally, operating systems for WSNs are typically simple, providing basic mechanisms to schedule concurrent tasks and access the hardware. In this respect, a representative example is TinyOS [17] and the accompanying nesC language.

Very recently, in order to provide high flexibility and to add new and advanced functions to WSN middleware, the service-oriented approach has been applied to sensor environments [8, 11]. The common idea of these approaches is that, in a sensor application, there are several common functionalities that are generally irrelevant to the main application. For example, most services will have to support service registries and discovery mechanisms and they will also need to provide some level of abstraction to hide the underlying environments and implementation details. Furthermore, all applications need to support some levels of reliability, performance, security, and QoS. All of these can be supported and made available through a common middleware platform instead of having to incorporate them into each and every service and application developed.

In this context, the OSGi technology [14] defines a standardized, component/service oriented, computing environment for networked services. Enabling a networked device with an OSGi framework adds the capability to manage the life cycle of the software components in the device from anywhere in the network without ever having to disrupt the operation of the device. In addition, the service oriented paradigm allows for a more smooth integration with Cloud platforms and for advanced discovery mechanisms also employing semantic technologies [5].

5 Conclusion

The proliferation of physical sensors and handheld devices equipped with sensors is responsible for the production of huge amount of raw data which today's tool for data management are not able to keep up with. In the IoT vision, sensors and their respective sampled data are useful not just for the environment they were designed for, but should feed an ecosystem of services accessible world wide over the Internet. To implement such a scenario a new perspective of sensors and sensor networks must be adopted, which shifts the role of an individual sensor from a mere physical device producing raw data to a powerful service capable of undertaking some data manipulation tasks and of interacting to other sensors/services for serving more ambitious objectives in wider domains. In this chapter we have discussed of a middleware which faces some of the issues underlying this view. A prototype of the middleware was developed and tested in a typical problem of energy-aware control in civil and home environments, but its approach to sensor abstraction makes the middleware deployable in other application domains.

Acknowledgments This work has been partially funded by the Italian project "Sensori" (Industria 2015 - Bando Nuove Tecnologie per il Made in Italy) - Grant agreement n. 00029MI01/2011.

References

1. Botts, M., Percivall, G., Reed, C., Davidson, J.: OGC sensor web enablement: overview and high level architecture. In: Nittel, S., Labrinidis, A., Stefanidis, A. (eds.) GeoSensor Networks, Lecture Notes in Computer Science, vol. 4540, pp. 175–190. Springer, Berlin (2008)
2. Buyya, R., Yeo, C.S., Venugopal, S.: Market-oriented cloud computing: Vision, hype, and reality for delivering it services as computing utilities. In: 10th IEEE International Conference on High Performance Computing and Communications, 2008. HPCC '08, pp. 5–13 (2008)
3. Costa, P., Coulson, G., Gold, R., Lad, M., Mascolo, C., Mottola, L., Picco, G.P., Sivaharan, T., Weerasinghe, N., Zachariadis, S.: The runes middleware for networked embedded systems and its application in a disaster management scenario. In: Fifth Annual IEEE International Conference on Pervasive Computing and Communications (PerCom 2007), IEEE Computer Society, pp. 69–78 (2007)
4. Di Modica, G., Pantano, F., Tomarchio, O.: SNPS: an OSGi-based middleware for Wireless Sensor Networks. In: Workshop on Cloud for IoT (ESOCC 2013 workshops), Communications in Computer and Information Science, vol. 393. Malaga (Spain) (2013)
5. Di Modica, G., Tomarchio, O., Vita, L.: A P2P based architecture for Semantic web service discovery. Int. J. Soft. Eng. Knowl. Eng. **21**(7), 1013–1035 (2011)
6. Hadim, S., Mohamed, N.: Middleware: Middleware challenges and approaches for wireless sensor networks. IEEE Distrib. Syst. **7**(3), 1 (2006)
7. IEEE Network Working Group: JavaScript Object Notation (JSON) (2006). http://www.ietf.org/rfc/rfc4627.txt?number=4627
8. Issarny, V., Georgantas, N., Hachem, S., Zarras, A., Vassiliadist, P., Autili, M., Gerosa, M.A., Hamida, A.B.: Service-oriented middleware for the Future Internet: state of the art and research directions. J. Internet Serv. Appl. **2**(1), 23–45 (2011)
9. Madden, S.R., Franklin, M.J., Hellerstein, J.M., Hong, W.: Tinydb: an acquisitional query processing system for sensor networks. ACM Trans. Database Syst. **30**(1), 122–173 (2005)
10. Miorandi, D., Sicari, S., Pellegrini, F.D., Chlamtac, I.: Internet of things: vision, applications and research challenges. Ad Hoc Netw. **10**(7), 1497–1516 (2012)
11. Mohamed, N., Al-Jaroodi, J.: A survey on service-oriented middleware for wireless sensor networks. Serv. Oriented Comput. Appl. **5**(2), 71–85 (2011)
12. Mottola, L., Picco, G.P.: Programming wireless sensor networks: Fundamental concepts and state of the art. ACM Comput. Surv. **43**(3), 19:1–19:51 (2011)
13. OGC: Sensor Web Enablement (SWE) (2013). Available at http://www.opengeospatial.org/ogc/markets-technologies/swe/
14. OSGi Alliance: Open Service Gateway initiative (OSGi) (2013). Available at http://www.osgi.org/
15. Papazoglou, M.P., van den Heuvel, W.J.: Service Oriented Architectures: approaches, technologies and research issues. VLDB J. **16**(3), 389–415 (2007)
16. Sugihara, R., Gupta, R.K.: Programming models for sensor networks: A survey. ACM Trans. Sen. Netw. **4**(2), 8:1–8:29 (2008)
17. TinyOS community: TinyOS (2013). Available at http://www.tinyos.net/

Toward the Next Generation of Sensors as a Service

Dario Lombardo, Vito Morreale and Giuseppe Li Calsi

Abstract This chapter presents a novel approach for service oriented architectures applied to the Sensors-as-a-Service paradigm. This work illustrates a flexible, scalable programming model of applications based on a service platform. The major contribution of this work is a new idea for service model, SNS model, compliant to sensor, where service is designed to use the information coming from sensor, and it is possible to their integration. The service model allows to optimize information that are coming from service or from sensor versus a new concept Service SNS.

1 Introduction

One of the most known and well-established development and deployment approaches to software systems is Service oriented architecture (SOA). It indicates a specific type of distributed system, in which the entities that constitute it are precisely the services. An emerging trend in recent years makes various types of Web-resident sensors, instruments, imaging devices, and repositories of sensor data, discoverable, accessible and controllable via the World Wide Web. It would be interesting to combine the two different aspects, i.e. service-orientation and sensor-as-a-service, into a unique services platform conceived, specialized, and optimized for sensors. This chapter aims at this goal, i.e. a new idea to develop a platform for ease of use and the integration of sensors as services.

D. Lombardo (✉) · V. Morreale
Engineering Ingegneria Informatica spa, Viale Regione Siciliana, 7275, 90146 Palermo, Italy
e-mail: dario.lombardo@eng.it

V. Morreale
e-mail: vito.morreale@eng.it

G. Li Calsi
demetriX srl, Piazza Stazione, 60, 90044 CARINI, PA
e-mail: giuseppe.licalsi@demetrix.it

S. Gaglio and G. Lo Re (eds.), *Advances onto the Internet of Things*, 209
Advances in Intelligent Systems and Computing 260, DOI: 10.1007/978-3-319-03992-3_15,

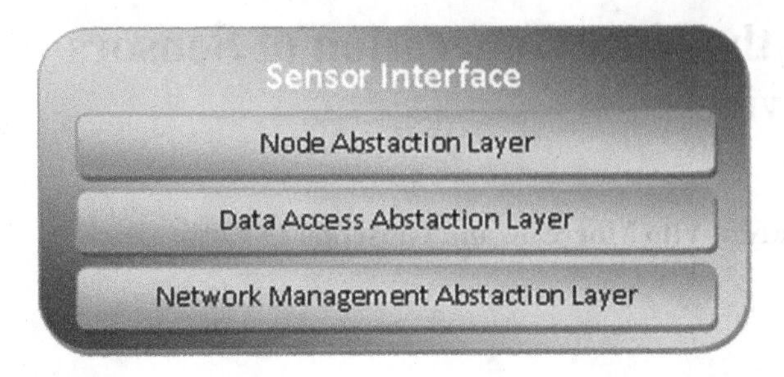

Fig. 1 Sensor interface

This chapter is organized as follows: first a brief state of the art is presented (Sect. 2); then our solution is presented: Model SNS Service (Sect. 3), SNS platform (Sect. 4).

2 Service Oriented Architecture for Sensor Networks

One major reason for the increasing interest in wireless sensor networks (WSN) in the last few years is their potential usage in a wide range of application domains. This interest brings to the identification of the new paradigm of Sensors as a Service, to employ the same concept of service at sensor with the suitable correction for this case. The appropriate levels of abstraction [1–4] contribute to obtain the required specifications (Fig. 1):

- The *Node Abstraction Layer* handles information on sensory capabilities of sensor nodes providing mechanisms for uniform access by the higher levels.
- *Data Access Abstraction Layer* is related to collect data from networks sensory and move to higher levels according to the specifications
- *Network Management Abstraction Layer* provides an interface to the functionality most closely related to the management system of the network (setting, state nodes, etc.)

At the application level, each sensor may be modelled according to a service-oriented design [5–7], i.e. as including a set of actions that may be demanded to the sensor, with the help of services called by means of the framework, where services were developed. Those actions concern primarily querying the sensor in order to acquire the physical quantities under monitoring, and configuring the behaviour of the sensor by adding the new functionalities offered by specialized services.

The logical model of the sensor has to be agnostic with respect to its physical characteristics and to the measured quantity, thus allowing a wide range of flexibility whenever new sensors are added. There are some necessary features that a Service-Oriented Middleware for WSN would be required to implement. They include functional and non-functional requirements. In many application domains specific functions are essential since they derive from the application logic itself.

A middleware common to all applications has to include all major core services, in order to avoid that these should then be integrated in each application developed for the specific domain. The common requirements for such services are [8–10]:

1. Runtime support for the delivery of services and their execution
2. Support to consumers for the discovery of registered services
3. Transparency for client applications
4. Abstraction to hide the heterogeneity of environment sensors
5. Configurable services
6. Support to the mechanisms of self-organization
7. Interoperability with a wide range of devices
8. Efficient management of large volumes of data with high throughput communication
9. Cooperative processing of tasks should lead to more precise results and new application fields
10. Sensor networks require security mechanisms that are adaptive to environmental conditions
11. Support for the requirements of QoS
12. Support for integration with other software systems
13. Sensor nodes must perform tasks of network maintenance
14. All algorithms and protocols must be energy optimized.

Finally, service-oriented approaches shape the sensors as a series of services, allowing then to adopt the common and powerful paradigm of service-oriented architecture for building applications. The major strong point of the service-oriented middleware is the ease with which you can add advanced features simply by implementing new services.

3 SNS Service Model

The SNS model, see Fig. 2, is defined to represent SNS service with its structure and its attributes. The SNS model would like to convey a possible representation of sensor as a service. The model's goal is to create an abstraction of the sensor, the physical layer, considering it as a service, and a software level, for interacting with sensor like if it was a service, by adopting the whole service-oriented paradigms.
In the SNS Service model a connection between service and sensor is expressed formally, where Service SNS finds an integration at high level. It is possible to represent this formalization:

$$\begin{aligned} Service\,SNS = {} & Sensor(data, dependence) \\ & + Service(data, operation, dependence) \end{aligned}$$

The SNS service model is an optimized conjunction of service and sensor, where the information related to both are represented. The SNS service model is composed

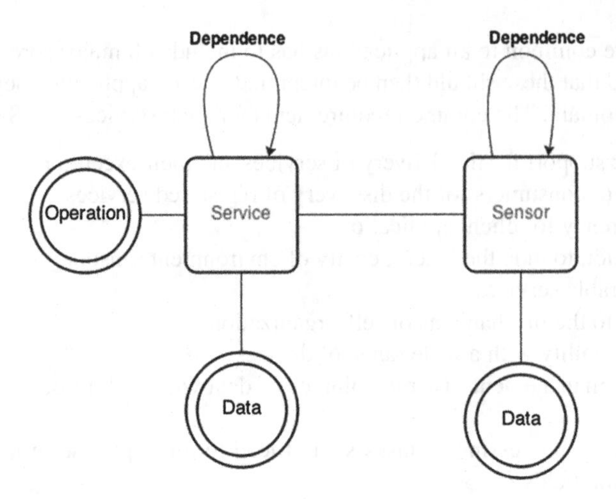

Fig. 2 SNS service model

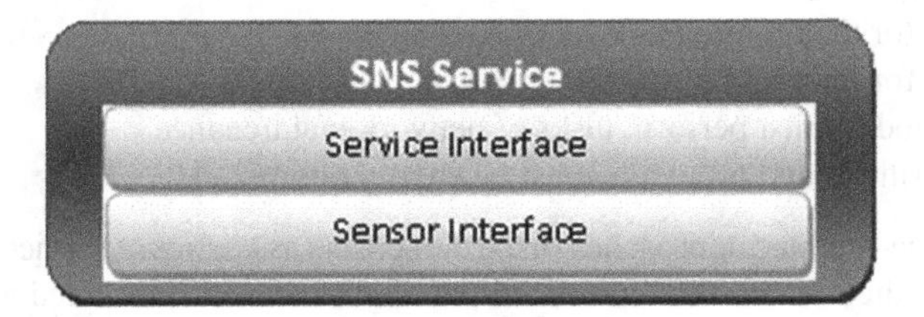

Fig. 3 SNS service

of a service-related part and a part more specific to the wrapped sensor. The service part includes information about operations and data related to them, whereas the sensor part specifies only the data to gather from the sensor; however a dependence relationship between service and sensor can be specified.

The *operation* attribute specifies what the service can do and its interaction with the external world, while the *data* attribute represents what information the service/sensor manage. The above-mentioned *dependence* concerns with the connection between services and sensors and allows to wrap a combination of several services or sensors with a SNS service.

4 Service Platform

This work proposes a standardized configuration of SNS service that strictly depends on its description. SNS service (Fig. 3) will exhibit an unique Web service interface through which client services and systems can exploit sensor's services.

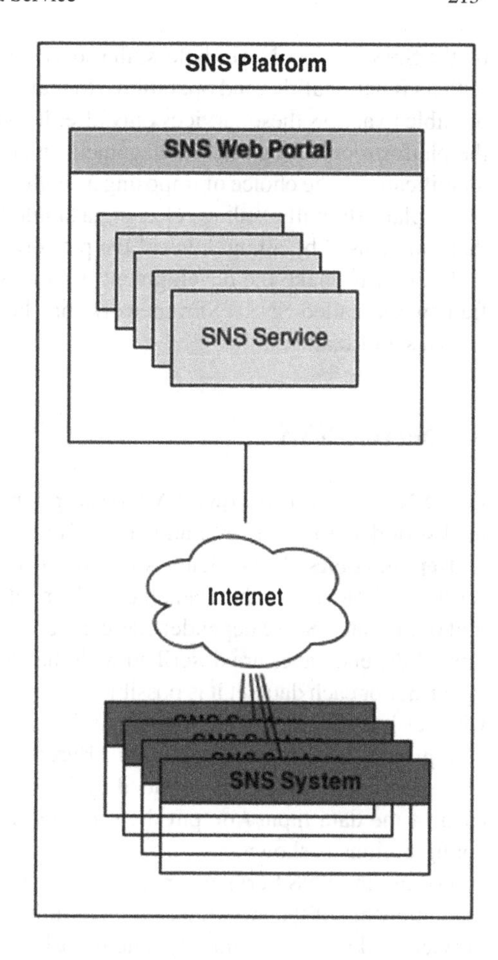

Fig. 4 SNS platform

The SNS service has available all data necessary for its operation: such data come from the data supplied by sensors on network or other SNS services. So that SNS service can be properly configured by the user or application supposed to use it.

A SNS service (Fig. 3) is a software component that can be plugged into the platform to offer only one interface, to manage indifferently services and sensors. The SNS service, Fig. 4, will access to the sensors managed by the SNS platform through the sensor interface, being able to achieve measurements or perform any operation specified in the interface.

The information that the SNS service can exchange varies from service to service. Each SNS service makes accessible its description, by which it explains the dependencies (data sources), the operations and the data to be known, so that it can offer its services. The configuration of each SNS services will exposure a standard WS management interface, that it is fixed by platform. This choice allows to have a single interface shared by all SNS services, providing the description of the SNS service, to carry out the basic operations and the configuration and the operations offered by the SNS service. Nevertheless, the service will expose WS customized interface with its functionalities and operations, to make available outside of the platform. It is possible to access these services only after having carried out the configuration from the platform or through the management interface of the service, and obtaining its identification. The choice of imposing a single interface for managing all the services of the platform will result in releasing a single library for managing all services and that can be used by all modules of the platform.

In order to make the development of new SNS services as fasted as possible, a framework, called SNSWSFramework, has been defined and released, which contains a set of feature.

4.1 Service SNS

Each SNS service has its own XML description that contains the list of dependencies, the list of data to configure and list of offered operations.

Dependencies: Dependencies can be of two types: "sensors" and other "SNS services". For each of them a name is given, alonh with the type of measure and the unit of measure. Some dependencies can be selected among several alternatives. This type of dependencies are described with the element $< dependencyChoices/ >$.

data: For each data set it is possible to indicate the name, the unit of measurement, whether it is required to configure and, in case it is not present, its default value.

Operations: For each operation offered by the service, this field provides the name, the list of input data provided by the operation and the list of output data. For each of the data input / output the same information provided for the data service configurations is shown.

For use the SNS Service is necessary configure. The SNS Service configuration, in relationship of the own description, permit to build a new instance of specific SNS Service, and it is stored into definite repository.

Configure SNS services: Each SNS service needs to know in advance the information to access the data required for its operation. This information varies for each user and should be saved in an appropriate user configuration. When the user wants to use the services offered by SNS service, he or she has to provide the ID of this associated configuration. The information that can be configured for each service are of two types: dependencies and data.

The configuration management can be done through the management interface of the service using one of the methods exposed for configuration management:

- editServiceConfiguration: to edit a configuration already existing on the system, replacing the old one with the new one, by providing the identifier of the old configuration and the element <serviceConfiguration /> with the new one; it returns the identifier of the new configuration in case of success;
- getServiceConfiguration: to retrieve an existing configuration, by taking the identifier of the configuration as input and getting back an element of type <serviceConfiguration /> with the corresponding configuration.

Configuring dependencies: SNS Services offer functionalities to get data from sensors of the user or other SNS services configured by the user. In order to access this data, the service needs to know the required parameters that it is possible to distinguish according to the type of dependencies. The dependencies that need to be configured for the service SNS are indicated in its description; for each of them the type of measurement returns is indicated, and the unit of measure. To complete the configuration of dependency, it is necessary to specify a data source that returns a data consistent with the description. As previously mentioned, the dependencies are of two types, i.e. sensors and other SNS services, whereas configuration parameters of the dependence are:

1. *sensors dependencies*: to specify the data needed to access the instance of the SNS system that manages the sensors (address on which the service responds) and the identifier of the sensor;
2. *dependencies on other SNS services*: to specify the data needed to access the management interface of the service, the name of the operation of interest, the *id* of the configuration to be used;

Configuration data: The SNS services may require some parameters that have to be configured for their correct operation. These are also indicated in the description of the service and for each of them the necessary information to their configuration is given.

5 Conclusion

This chapter describes a new approach for developed services connected with a sensor. It defines a structure of an abstract service and the list of requirements for this service model. The information generated by a service and a sensor is handled by SNS platform: the source information is different every time for every sensor, but the interface to interact with the platform is uniform.
This work allows to be compliant with a service approach to distributed systems (such as sensor networks). The resulting API and platform are released according to the open source rules.

Acknowledgments This work is partially supported by the Italian Ministry of Economic Development through the project SeNSori ("SeNSori– Sensor Node as a Service for Home and Buildings Energy Saving").

References

1. Girolami, M., Lenzi, S., Furfari, F., Chessa, S.: SAIL: A Sensor Abstraction and Integration Layer for Context Awareness. University of Pisa, Pisa (2008)
2. Escolar, S., Isaila, F., Calderón, A., Sánchez, L.M., Singh, D.E.: SENFIS: a Sensor Node File System for increasing the scalability and reliability of Wireless Sensor Networks applications. J. SuperComputing (2009)

3. Escolar, S., Carretero, J., Isaila, F., García, F.: Deconstructing the wireless sensor networks architecture. In: IEEE Symposium of Embedded Systems [IES' 2006], Antibes, Juan- Les-Pins, Francia (October 2006)
4. Suba, F., Prehofer, C., van Gurp, J.: Towards a Common Sensor Network API: Practical Experiences. Nokia Research Center, Helsinki (2008)
5. Golatowski, F., Blumenthal, J., Handy, M., Haase, M., Burchardt, H., Timmermann, D.: Service-oriented software architecture for sensor networks. Institute of Applied Microelectronics and Computer Science, University of Rostock, Rostock (2003)
6. http://www.opengeospatial.org/ogc/markets-technologies/swe
7. Amundson, I., Kushwaha, M., Koutsoukos, X., Neema, S., Sztipanovits, J.: Oasis: a service-oriented middleware for pervasive ambient-aware sensor networks. Technical report, Institute for Software Integrated Systems Department of Electrical Engineering and Computer Science Vanderbilt University, Nashville, Tennessee, USA (2006)
8. Anastasi, G.F., Bini, E., Romano, A., Lipari, G.: A service-oriented architecture for qos configuration and management of wireless sensor networks. In: 15th IEEE International on Emerging Tecnology and Factory Automation, Bilbao (2010)
9. Anke, J., Muller, J., Spieß, P., Chaves, L.W.F.: A service-oriented middleware for integration and management of heterogeneous smart items environments. In: Processing of 4th MiNEMA Workshop, Lisbon (2006)
10. Caporuscio, M., Raverdy, P-G., Moungla, H., Issarny, V.: ubisoap: a service oriented middleware for seamless networking. In: 6th International Conference on Service Oriented Computing: ICSOC, Stockholm (2008)

Advances in Internet of Things as Related to the e-government Domain for Citizens and Enterprises

Francesco Beltrame and Virginia Dagostino

Abstract This work deals with the role of Internet and, specifically, of Internet of things with its advanced performances (i.e. Cloud technology), in the *e-government* domain. It is based on the multi-year direct experience of the authors committed to develop the complex digitalization process of the Italian Public Administration with the objective of yielding ever improved services delivery, with respect to quality and time, to citizens and enterprises. The underlying rationale is that since the Public Administration, in Italy, but in most developed countries as well, acts as a critical switching node for about 50 % of the Gross National Product, its internal machinery improvement achieved through a largely diffused advanced and well suited ICT solutions adoption, will strongly affect in a positive manner the national competitiveness, leading, in the end, to a better scenario for employment and wealth. Apps, open data and Cloud are keywords offered to the reader as basic seeds for the understanding of a proposed model and corresponding solution to deal with the complex problem of setting up both single and multi-Public Administration business processes. In both instances, a novel logical and technological model is introduced and discussed, strictly driven by the end user needs, on the front office side, i.e. by the real demand and not by a theoretical, often generic, technology offer.

1 Introduction

A Public Administration (PA) unable to accomplish its various missions as determined by law, constitutes a serious deadlock to the development of any country, and keeps far from it also potentially interested foreign investors, which do not feel

F. Beltrame (✉)
University of Genova, Genova, Italy
e-mail: francesco.beltrame@unige.it

V. Dagostino
RINA S.p.A., Genova, Italy
e-mail: virginia.dagostino@rina.org

S. Gaglio and G. Lo Re (eds.), *Advances onto the Internet of Things*,
Advances in Intelligent Systems and Computing 260, DOI: 10.1007/978-3-319-03992-3_16,

properly protected by a too slow and inefficient machinery, difficult to understand from the outside world.

The main objective of this work is that of analyzing and defining strategical criteria in order to realize a knowledge information system of prototypal nature designed for the PA considered as a whole, resulting from a new modeling of organizational processes, based on a complete:

- document flow "de-materialization"
- introduction of activity and business process powerful workflow tools
- introduction of a semantic level able to ease document organization and an information smart access from citizens and enterprises, including the full tracking versus time of their instances to the various reference PA over the territory
- development and use of Cloud technologies for PA front-office and back-office applications.

The system is conceived to yield a series of services to the various end user categories, among which access to the citizen digital record and the enterprise digital record, support to the open publication of public data as produced by a given PA and to their integration with data from other PAs and, more in general, with other open data, as ruled in the various countries and, for example, in Italy by the Digital Administration Code (CAD), according to the laws n. 82/2005 and n. 235/2010 and their further modifications and integrations.

Following a deep analysis of the information system status currently installed at the various hosting PAs, it is evident the need of re-thinking the nature and architecture of such systems. The main reasons are related to the obsolescence of the running systems and to the architectural model previously adopted, such as the client-server one. Such considerations do not allow an easy evolution of such systems towards new and more performing solutions. Furthermore, the expected increase of data as generated by document and information de-materialization in each single PA and those expected from the opening of the internal processes from and to the citizens (trend known with the denomination of *big data* or *data deluge*), will lead to the point of the need of reconsidering also the tools supporting such data persistency to be able to manage data amount of the order of peta or exabyte for each single PA [4].

To such extent, it is worth to mention some technologies potentially capable to optimize the use of processing and storage resources:

- **virtualization**: it allows to disentangle the applications from the underlying hardware, yielding in an optimal resource usage at run time, reducing management and maintenance costs, as well as, if properly monitored, energy consumption, also making easier disaster recovery procedures, service quality and operational continuity
- **Cloud technologies**: it is based on virtualization techniques usage and it allows resource sharing among different administrative entities (*tenant*), yielding to each of them the feeling of having available infinite resources and, at the same time, making possible cost attribution only as a function of their real use. Such an approach keeps also into account the adoption of programming languages based on parallel

and logic-functional paradigms, which better fit the dynamic and distributed nature of the new applications on the server-side

- **no SQL databases**: this is an approach to heterogeneous data storage and access, promoted in 2003 by Google researchers and also adopted by other relevant Internet providers, which removes the limits in performance and scalability of current solution of commercial databases as applied to *big data* contexts.

The realization of the prototypal system should adopt the most innovative techniques both for the development of software modules (*apps*), also by taking into proper account the guidelines on "reuse by parts", and for the development of user interfaces highly usable, by taking advantage of the interaction features as offered by mobile devices.

Another important aspect for system success, will be its independence from the application domain, in order to obtain a product easily usable in many PA sectors (portability and scalability of the prototypal solution). Of course, the strategic design criteria do not refer only to technological profiles, but also to the normative ones, thanks to a careful analysis of the various constrains which limit a full digitalization of the administrative process, area internationally known as e-government, in order to guarantee that when the system is realized and in place, it will fit the numerous current laws and ready to implement also new tools which will be in turns later regulated, until even to prepare new proposals "de jure condendo".

It is worth to mention that the PA innovation need meets a relevant connection item of the Horizon 2020 European Union (EU) program for two main reasons:

- reason number one is that, for the first time, Horizon 2020 joins, inside the same framework, Research&Development and Innovation (until FP7, included, the situation has been different)
- reason number two is due to the observation that in Horizon 2020 the content titles for Research&Development and for Innovation are named in its third part only: Societal Needs, to represent as the new knowledge production (Research& Development part) and Innovation (which includes larger and more differentiated dimensions, such as the anthropological, the social and the economic ones) is mostly driven (see quantitative aspects about the resource distribution over the three aforementioned chapters) from the *demand* of providing adequate solutions to concrete needs of citizens and enterprises and *not* from the offer of even advanced technological solution looking for problems. *In fact, the main need is to yield value to citizens and enterprises of the various countries through a more efficient PA thanks to the use of advanced ICT technologies, such as Cloud, in order to provide services in such a manner to be law-compliant and economically sustainable.*

For what it concerns the aspects related to new ICT technologies, it is necessary to keep into account the activity lines of the European Research Area (ERA) and for the industrial and technological leadership in the ICT domain:

- **development, diffusion and functioning of ICT based electronic infrastructures**: the PA need allows to make available either for the PA itself or for other institution data related to the various activities (with respect to the privacy-law constraint), also historical data in order to perform analyses and studies from various disciplines researchers, either scientific and humanistic ones, achieving an incentive effect for the industrial context development

- **next generation processing: advanced information systems and technologies:** the ways of application development, particularly the server-side ones, since they must be compliant with specific maintenance requirements, laws, technical rules and certifications, reuse and migration, will allow the experimentation of paradigms different from object-orientation, in order to better take advantage of the parallel and distributed features of Cloud (for example, parallel and logic-functional languages)
- **Future Internet: infrastructures, technologies and services:** the approach to shared development and reuse of applications starting from open specifications is coherent with the concept of Future Internet [3] as currently promoted.

2 Qualification of the Socio-Economic Relevance of the Proposed Solution and of its Positive Fallouts for End Users Resources Use and Optimization

The strategic criteria which constitute the basis of the envisaged solution reflect the vision of a logic based on three words: LAW-ECONOMY-TECHNOLOGY, capable to offer the best guarantees about the final prototype either for the research partners or for the industrial ones, because it calls for new knowledge production (and therefore research activity) able to optimize as efficiency regards the crucial PA bottlenecks. Such an optimization, will in turns make more competitive each country (the PA, each year, does intermediate relevant amount of the Gross National Product), according to a regulated framework ruled by the various laws which are in force in the various countries.

The envisaged solution can be synthesized in the development and use of Cloud technologies for the PA front-office and back-office.

In the following, the Cloud as applied to e-government is described, in order to understand its potential for the qualification of the socio-economic relevance of the matter presented in this work.

The Cloud concept has various definitions in the literature. Considering the authority of the source, the USA NIST definition is considered:

> Cloud computing is a model to access through Internet to a pool of assemblable processing resources (networks, servers, storage, applications and services) which can be easily allocated at the time of need, and similarly can be also easily released when no more needed, leaving to the providers the operational management task of such resources [5].

One of the main point of strength of Cloud is the one of being capable to provide almost unlimited ICT capacities directly available through Internet, therefore from any point where a network connection is present and almost with any device: desktop, laptop, tablet, smartphone and other intelligent devices. Another fundamental point of strength arises from the fact that are just the service providers to be responsible for the hardware and software resources, of their maintenance, of their updating, security and of other essential resources. To Cloud clients, in the case study PAs, made free from the management duties, will be leaved only the task of identifying the services and the tools useful for their specific needs. In this way, the single client (end user) can focus the activities on its own core business leaving (like in outsourcing) to the service provider all the ICT support activities. Furthermore, the needed ICT services can be acquired in a fashion directly related to the specific needs of any specific moment. Well known examples of Cloud provided applications are Facebook, Amazon, eBay. For example, a given PA, will not need to acquire, maintain and manage large infrastructures for its information/knowledge systems, since it will have the possibility to acquire all or almost all the needed ICT services from a Cloud provider. Even today, many private enterprises of different size take advantage of Cloud services, and, by acting in this way, they realize consistent savings and improve their economic yield. For PAs, instead, still cultural and legislative barriers are present to use such service approach for what it concerns the localization and management of data of sensitive nature. In technologically advanced countries, such as the Republic of Korea (ROK), the USA, the United Kingdom (UK), Japan, Finland, programs to include Cloud as part of their ICT architecture are running, while instead some services have been already migrated on Cloud platforms provided by private enterprises. This has been accomplished to allow for the administrations and for other public institutions to reduce management costs and to obtain other improvements for the provisioning of qualified services to citizens and enterprises.

The service Cloud offering is relatively new and it is undergoing a great development. This fact implies that near to the numerous benefits as offered by this new way of conceiving ICT services, are still present, in view of including such services in the information/knowledge PA systems, limitations and problems which is important to tackle and solve. Particularly, large amount of work has to be carried out in terms of specifically customized applications at design and development level in order for them to be used in a Cloud context. For such reasons, new knowledge production is needed to study, experiment and evaluate both the requirements and the opportunities as offered by the Cloud, inside the framework of the processes correspondent to the services provided by PAs, particularly on the front-office side, the one more perceivable by citizens and enterprises and on which, even in recent years, less has been invested with respect to the back-office side.

Nowadays, PAs of developed countries are subject to heavy cost reduction policies, while, at the very same time, to the search of ways to improve the efficacy and efficiencies of their units. The Cloud is a key element for the PA modernization, by obtaining, at the same time, the result of improving its own efficiency while reducing costs. In the USA, the official Government web portal, USA.gov, has been moved onto a platform named Enterprise Cloud, produced by Terremark. USA.gov is one of the USA more visited Government website, with more than 100 million visitors per day, and it has been designed to function as access point to the information available on the public USA website. However, the online traffic underwent a very high degree of variation under particular cases. Before the porting on the Cloud of the portal, in case of high traffic peaks, service delays or interruptions were frequent. Enterprise Cloud has been the answer to this problem. The effective porting required only 10 days, and the Cloud trial has been carried out in one weekend. Martha Doris, the *Deputy Associate Administrator of The Office of Citizen Services*, estimated that the porting onto the Terremark Cloud platform allowed a cost cut of about 90 %, by improving, at the same time, system performances and flexibility. As a consequence of such positive results, the DATA.gov portal porting onto the same Cloud platform is in progress. In the United Kingdom (UK), it has been decided to move some services over the Cloud. During 2011, it has been announced the creation of a private Cloud infrastructure named G-Cloud. G-Cloud is designed to include not only the Government data-center, but also to yield tools supporting working activities, to implement a private Government wiki and a mail server for the UK municipalities. It is currently running a study to identify applications and tools useful for the Government and available over the Cloud, which could be shared with other departments and public institutions. According to the running plan, 80 % of the UK public departments will make use of the G-Cloud platform, leading to saving of about 3.2 billion pounds per year on the expenses for ICT resources. Further to economic reasons, G-Cloud has also other motivations. There is an improvement on security, since all the information would be hosted in the Government private Cloud. Other services, mainly those of vital importance, would be separated by G-Cloud. Other benefits include the possibility of implementing good practices for what it regards energy efficiency. In Japan, the Ministry of Internal Affairs and of Communications (MIC) developed a plan named Hatoyama, introduced in 2009. The Hatoyama plan has the objective to create new ICT markets to act as a stimulus for Japan economy. As part of the Hatoyama plan, the Government is designing a Cloud national infrastructure, named Kasumigaseki Cloud, to be realized by 2015. Kasumigaseki Cloud will provide the platforms for the shared functions at the Government departments, in order to harmonize systems and processes, by reducing processing time. As a consequence, new investments in the order of various thousands of billions of yen are envisaged and, at the same time, the creation of 300.000–400.000 new jobs. In summary, the USA are trying to save money by using the resources made available by the Clouds of the private sector, discharging over them the costs related to the investment, operating management and updating of the ICT devices. Instead, the UK Government decided to create its own Cloud structure maintaining their ICT still bounded to the PA hands, but, at the same time, gaining the advantages offered by the Cloud. Japan, a country where high technology is largely diffused, produced a very ambitious plan to create new ICT markets and to use the Cloud either in the PA as well as in other sectors. On the other side, one of the barrier to a pervasive diffusion of Cloud services is given by the lack of reliable access points to the network. From this point of view, Japan is advanced with respect to the other countries, since wireless Internet connections are available almost everywhere.

From the described examples, it appears evident the tendency toward an ever increasing Cloud adoption, and that the Cloud will be most likely a dominant service model in the next years. Nowadays, the acquisition of very large Government ICT structures is not more necessary and not more useful, since it is possible to acquire services on the basis of the need in any moment according to a centralized fashion. Since many Government structures or public institutions use the same class of applications, it is possible to consider the development of common ICT solutions based onto Cloud services. In this way, it could be possible to put in place strategies to allow reduction of the costs related to investment on new ICT devices, to their maintenance, to their updating and to their energy consumption. The Cloud could be therefore fundamental in the future for the PA ICT services.

For example, by looking at the future for ICT investment in the Italian PA, a potential useful model could be represented by a mixed solution between the UK G-Cloud and the Kasumigaseki of Japan. In fact, it is necessary to properly consider the legislative limitations on the possible localization of data of sensitive nature, which forces to use the already available ICT infrastructure, and, versus time, to include all the other public entities such as hospitals, schools, municipalities, by implementing a *Hybrid Cloud*. The public will have access to information through services made available for the citizens over a Public Cloud, while the Private Cloud will be used by the PA. Standard applications should be developed for similar institutions (e.g., municipalities). Furthermore, different approaches should be studied for institutions which have different missions, both for the nature of the institution itself and for reason related to data security (e.g., Ministry of Justice). It could also be possible to study set of services offered on virtual shops, where the end users could select the application type for their needs in terms of cost, performances and reliability, of course always being law requirement compliant. USA, UK and Japan, envisage in the Cloud the direction towards which integrate and evolve their information/knowledge systems, and are investing quite large amount of resources in this strategy.

From a research point of view, it has to be considered that while several tools and services on the Cloud are already available at a sufficiently robust and mature stage, there are also various interesting investigation directions related to the development of intelligent applications for workflow management either inside a given PA or for what it regards interoperability and information and data exchange among offices belonging to different PAs. As matter of fact, it is possible to make available for the citizens a centralized point to access information managed by various PA departments as, for example, it happens in the USA with the aforementioned USA.gov portal. Furthermore, it could be of interest, among the various potential research perspectives, the one of developing applications and adaptive workflows for processing procedures and documents inside a single PA or even to process complex procedures and documents involving more PAs.

3 Functional and Performance Requirements of the Proposed Solution

Starting from the state of the art on Cloud, workflow management, ICT applications running at the PA of some of the most advanced countries in this domain, and from the analysis of the availability of some commercial products for the PAs, the most important scientific and technological needs for the PAs have been identified.

Such collected needs have been then set inside frameworks of concrete operational activities of the PAs. In such pictures, some of the requirements that future research should fulfil have been identified. As matter of fact, the collected needs cover a larger domain with respect to that outlined by the pictures and correspondent requirements as reported in the following text. A more deep step of analysis is envisaged in subsequent phases, where further requirements could be derived from the expressed needs in conjunction of their inclusion into further applicative situations.

3.1 Needs List

From the studies carried out, the following needs can be identified (B).

B1. Improved application interoperability thanks to the adoption of open standard

a. *at data (secured) access level*
b. *at applications interoperability level.*

B2. Systems migration support to allow applications migration over Cloud platforms as offered by different providers, also with regard to disaster recovery functions, service maintenance, provider substitution.

B3. Service availability coherent with the operational needs on the basis of shared service levels and security and legislative constraints.

B4. Security

a. *data security (no-repudiation, high availability, compliance to laws and technical rules which are into force)*
b. *identification mechanisms (of people) and authentication mechanism (of documents) set according to a federated context, through trust mechanisms among different administrative domains and roles and rights delegation on resources*
c. *compliance to running rules of networking solutions related to virtualization.*

B5. Adoption of environments and methodologies Cloud-specific for applications test and development

a. *apps should be driven by user needs (PA personnel, citizens, enterprises), not from the hosting Cloud. Specific reference is made to current and possible research activities on self-configuring and, more in general, auto-configuring*, capable to look for an equivalent service to that present on another Cloud provider, in order to ensure transparency of it for the PA*
b. *languages with native support for parallelism and the analysis of large amount of data.*

B6. Support to persistency and shared access of large amount of data, in the order of peta- and esabyte.

B7. Business process and adaptive workflow systems over the Cloud

a. *business process should dynamically adapt to the variable resources availability and to legislative changes*
b. *business process could be distributed also on different Clouds of different administrations, during its execution, citizens and enterprises should be able to track in real time their instance status. Furthermore, each of the involved PAs should be able to monitor the overall process efficiency as a basis to the performance evaluation of its responsible managers (performance cycle).*

B8. Semantic engines development for

a. *research, identification and indexing of services which can be assembled in new applications on demand*
b. *structured and un-structured information semantic integration*
c. *easy composition of services (where the description can also be provided in natural language), to be eventually made automatic in the future.*

B9. To adopt Cloud solutions for Open Data [6]. The advantages of such solutions include

a. *data publication made more easy*
b. *data presentation additional modalities*
c. *data interoperability.*

B10. Definition of innovative techniques to allow metadata generation and use to overcome

a. *the problem of existing legacy systems integration*
b. *privacy problems in data use.*

B11. Specific application frameworks

a. *management, execution and monitoring of PA processes*
   - *processes involving a single PA*
   - *processes involving multiple PAs*
b. *citizens/enterprises PA interaction*
   - *processes related to citizens PA interaction*
c. *health*

- *healthcare application tight constraints, as: sensitive data (privacy), need of sharing in an easy and trusted way the data among the different involved actors.*

The listed needs (B) in the following are inserted inside typical operational PA contexts, and from them some of the main requirements will be derived to be pursued in the research activities to be carried out.

3.2 Operational Scenarios

The operational scenarios described in this paragraph represent the context for the realization of the aforementioned listed needs, and for the subsequent research requirements identification. The relationship between needs and requirements is intermediated by the environment in which they have been inserted and, in any case, the requirement is proposed as a solution strategy for the need and, as such, is not the only way for the satisfaction of the need itself, but, more properly, the one considered as the most adequate on the basis of the status of the art and of its advancement perspectives both in terms of basic/applied research and of technological infrastructures.

3.2.1 Business Process Inside a Single PA

Figure 1 depicts how any Public Administration (PA) conducts its missions (as established by the law) by providing performances to citizens. Performances, in turns, are realized through the involvement of various cooperating services to achieve the objective.

According to the always increasing digital vision of the PA, services are operated through *apps*, they exchange data/informations (*open data*) and are carried out according to a pre-ordered sequence, i.e. service orchestration. Such services could use *apps* available on the PA Cloud and/or on other public Cloud (*Hybrid Cloud*). The procedures of services applicative cooperation to achieve the desired performance could give rise to the specification of a workflow of activities, which requires a correspondent document flow. Final result will be the performance provided to citizens or to another PA according to a controlled procedure. Availability and possibility to have access from anywhere and through any device to documents is ensured by the digital nature of the documents themselves. Document digitalization allows storage, preservation, access, operational continuity, for which, nevertheless, it is necessary to provide, for each PA, a detailed plan, according to specific technical rules.

The workflows of activities are normally automatic through the use of web-based applications and web services if they allow to

- control and coordinate the workflow of activities
- manage the resources to be utilized for the performance

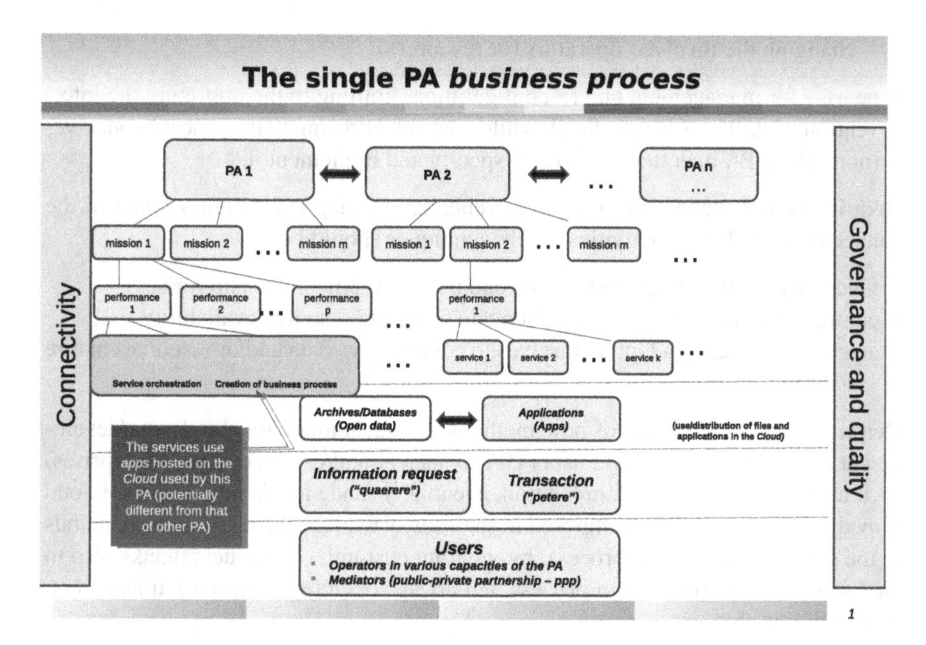

Fig. 1 PA missions, performances and services

- generate the document flow and monitor its status
- monitor and track the status of the entire performance while running
- act to overcome eventual advancement barriers during performance provision to the end user.

The future proposed framework envisages a large presence of mobile devices, such as smartphones and tablets, which will make use of *apps* available over the Cloud, assembled at the usage moment according to the specific user needs, both in the case of PA people or citizens and enterprises. Full realization of a workflow of activities fully compliant with PA needs is still today not feasible due to limitations in the status of art in relationship with what is represented in the following requirements.

Requirement 1-Definition of a new generation of adaptive workflow of activities to implement PA business processes. Such workflows should:

- be distributed over the Cloud
- include activities supported by *apps* over the Cloud and *open data*
- be able to learn from the use
- be able to re-organize the process of performance providing
  - changing the services execution order (while respecting the logic dependence constraints among them)
  - substituting those services that could constitute problem for an optimal execution of the performance with other services properly identified

– changing the involved operators (or resources)

- be easy for management and re-configuration, limiting to the minimum the intervention of skilled ICT personnel, while, instead, allowing to the process conceiver (normally a PA operator) to directly specify and implement it.

Requirement 2-Definition of service orchestration adaptive techniques and of the connected workflow of activities. Such techniques should be:

- guided by the dynamic needs of the end user (PA, citizens or enterprises)
- configurable according to a continuously changing resource availability
- able to dynamically adapt themselves to context, law, data and/or resources nature variations.

Requirement 3-Definition of new methods and techniques for the design/development of the *apps* for the end users (PA managers and/or citizens and enterprises) such as "atomic" services composition, eventually made available by public Cloud providers. *Apps* should be composed in such a way to place the user and its demands to the center of the design process, by yielding customized products thanks also to the use of an "ad hoc" design process, but always open (*open source*), transparent, easy and re-usable.

Requirement 4-Development of methods, techniques and technologies for the deployment and operation on mobile devices, such as smartphones, of hardware embedded *apps*. Such *apps* should allow the interaction among different devices in a transparent way with respect to manufacturers, configuration, device operating system, enabling the information/services exchange in a native form.

3.2.2 PA Cooperation and Multi-PA Business Process

Figure 2 depicts the cooperation among different PAs to provide a performance. According to this scheme, the information/knowledge system enlarges from a PA to another one, becoming, in the end, multi-PA. The multi-PA business processes include activities carried out in different PAs, and it is therefore necessary to provide what it is outlined in the following requirements.

Requirement 5-Definition of new methods, techniques and technologies for the distribution of the workflows of activities over different Cloud platforms in order to serve different PAs. Such methods, techniques and technologies should allow the realization of:

- workflows of activities distributed over different and interoperable platforms
- coordination among distributed activities
- interoperability at data (open and big) level and at application level (data exchange among services carried out by different PAs)
- transparent and easy cooperation among processes executed over different PA Clouds
- monitoring and supervision of the entire process from each single PA

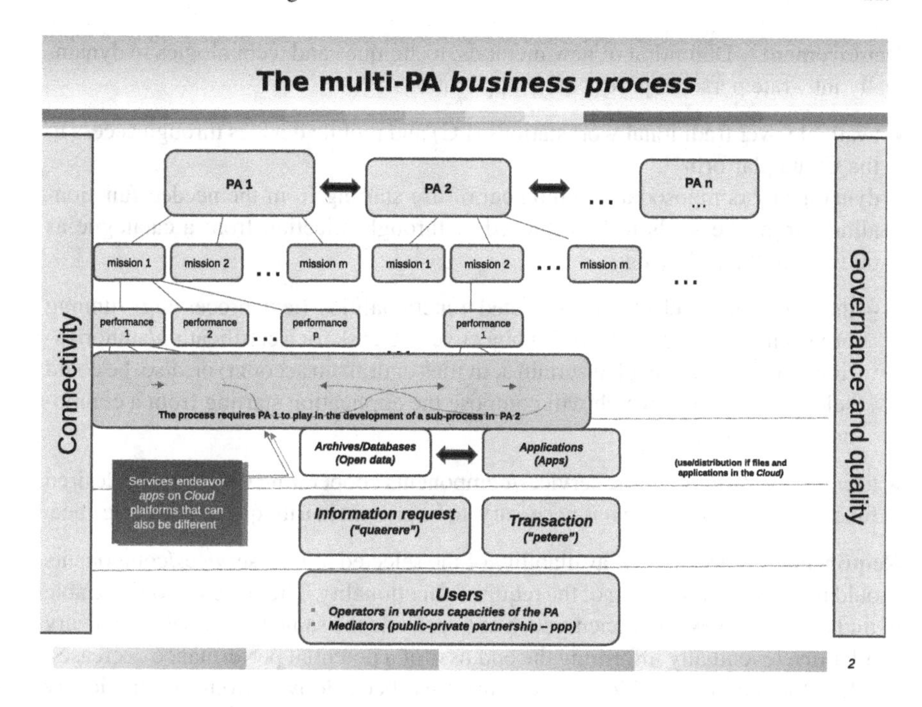

Fig. 2 The multi-PA business process envisages a cooperation among services of different PAs to provide the performance to the end user

- workflow flexibility to face real time variations
- components and workplan (or of part of it) re-usability (activities, supporting *apps*).

3.2.3 New Generation *Apps* for Personal Productivity

The availability of advanced and new services, made possible by the presence of a Cloud provider always devoted to new products development, allows to overcome the classical concepts of application software for personal productivity. Each system user (PA operators, citizens, enterprises) will benefit a customized *app* (or application, if he/she operates through a traditional PC terminal) specifically conceived for its duty. The *app* will be dynamically composed (on the fly) starting from basic services jointly assembled in order to provide only those functionalities required versus time (LEGO-like approach, a catalogue of *apps* which can be dynamically composed and re-used). New services will be, progressively, made available and they will be transparently (to the user) integrated inside the personal productivity applications, yielding new functionalities, easiness of use and enabling, in the end, a greater productivity. Such considerations are summarized in the following requirements.

Requirement 6-Definition of new methods, techniques and technologies to dynamically integrate personal productivity applications:

- available over traditional workstations (PC) and mobile devices through access to the Cloud platform
- dynamically composed at the moment of use starting from the needed functionalities, from the goals to be achieved or through selection from a catalogue as outlined in the following:
  - the user could select the interested functionalities (text processing, running project monitoring) or the target objective (e.g., ask for a certification/authorization, fill-in a module, plan certain activities or draft an account) or, also, he could select the modules which will compose the application starting from a components catalogue
- to compose various atomic services/components cooperating to realize the desired functionalities and, when it is necessary, to be present in a unique graphic interface.

Requirement 7-The lack of availability of the selected atomic services/components should not stop the providing of the required functionality. The system should be able to find other services/components such as to provide the same functionality or a very similar one (eventually informing the end user of a potential performance decrease).

The aforementioned 1–7 requirements have been derived from the previously listed needs as described in the points B1–B10. Further requirements will be necessary as the result of a successive needs analyses (e.g., from B7 to B10) and from their fine tuning.

4 Specification of the Technological Innovation Gap to be Filled and of the Level of Novelty and Originality of the Knowledge to be Produced by the Proposed Solution

During the last 5 years, the Cloud concept deserved the attention of the market and of the users, mainly, mass-market and small-business. As previously reported, the Cloud model is based on the demand providing of processing resources or data persistency starting from an homogeneous and shared pool of resources among various users, resulting in an infinite resource availability illusion [2]. Nowadays, the offered services according to the Cloud model are structured over three levels:

1. the lowest level, named IaaS (*Infrastructure as a Service*), yields basic services, such as process execution (*jobs*) or Virtual Machine (VM) and storage capacity for user files and data persistency, eventually with interfaces for network customization, firewall and other basic ICT resources. The first service of this kind has been introduced by Amazon, with its Amazon Web Services (AWS). The user has full control (as administrator) of its own servers (even if virtualized) and he needs to install all the required software to run its application (operating

systems, databases, application servers). On the contrary, the user does not have control neither on the physical infrastructure nor onto the selection of the physical resources supporting the virtualized layer, including, in many instances, the control onto applications for geo-localization
2. the second level, named PaaS (*Platform as a Service*), is a platform offering in a native manner (both expanding a IaaS and/or as autonomous solution) functionalities and resources needed for the implementation and execution of applications (such as no SQL databases or parallel computational models). Examples of this type are MS-AzureTM and Google App Engine. Over this platform, with heavy constraints on the syntax and on the behavior of the applications, the user-developer can create, maintain and develop applications needed by the end user. The developer does not concern about maintenance and evolution of the development platform, rather been focused onto the applicative logic and onto the interaction with the end user
3. the highest level is named SaaS (*Software as a Service*), and it provides (shared) access to applications from the end user. Examples of this nature are GMailTM or FacebookTM. The client does not own the application and therefore does not need to install, maintain or update such applications, but he cannot also customize, extend or evolve that application for its own specific needs. As matter of fact, the client gets only the right to the use of the application, which will paid according to its use or through some form of indirect remuneration (such as the possibility to receive commercial messages in the mass-market version).

All the Cloud currently commercially available on the market undergo three main limitations, which constitute barriers for their immediate adoption inside the PA context: the so-called *vendor lock-in*, the *compliance* to PA laws and regulations and the still present systems *non-interoperability*. The concept of *vendor lock-in* means the risk deriving from the adoption of proprietary non-standard solutions, which binds PA to a single provider, without the possibility to find easily on the market better quality or more convenient solution as substitutes, unless spending high costs for migration [7]. The term *compliance* makes reference to the problem of certification or guarantee that certain services or functionalities required by PA are compliant with laws and regulations which are into force in the various countries. This fact is highly debated inside the European and global ICT market, essentially due to the fact that each Member State or Nation has its own set of law and regulation constraints, sometime in contradiction among themselves. In this context, the market reacted with the identification and adoption of international certifications (e.g., PCI or ISO 27001/2), which should prove the adoption of attention and quality level in service provision. Such certifications are, sometime, used by single PAs for the selection of providers or in calls as proxy of regulation respect: such solution has the limitation to leave to the PA the duty of requirements settings to be regulations-compliant. This fact is easy to achieve for large institutions with properly qualified legal support, but less useful for small entities not fully aware of the various implications as deriving from the use of today technologies. In some European Union Member States (e.g., Germany [1]), the PA issued, through specific national organisms for this purpose, guidelines for ensuring conformity of ICT vendors to national or local regulations: in such a way, single PAs are made free from the duty to decide the operational modalities to respect the laws into force and, at the same time, a unique and homogeneous indication to enterprises is given on how to refer and to take into proper account current regulations necessary to invest in solutions which can be adopted by more than a single PA and, therefore, more economic. Finally, are considered *non-interoperable* those technological choices on programming languages, Application Programming Interfaces (API), application stack or non-standard data formats which prevent the migration of a given application toward a new provider or the re-use of the application or of its data from another PA. The kind of application which should be developed, and which, obviously does not exist at commercial level, is based on the implementation of intelligent and adaptive workflow able to use either already available services or also new ones as support to the processing of workflow inside a department of a given PA, or even between departments of different PAs. Various technological gaps and races need to be tackled in order to implement such a system. In fact, consolidated models and tools to design and manage adaptive workflow do not exist. Furthermore, the most innovative identified strategy is the concept of a workflow able to automatically compose the services available to the end user for the achievement of his goal, even if this last one is going to be changed at run time. For this reason, it is necessary to call for new knowledge production as the result of original research activities important for both the theoretical and experimental development of such models and tools, offering also the guarantee to have available the various software components (*fragments*) needed as support.

References

1. Bsi Security Recommendations for Cloud Computing Providers. https://www.bsi.bund.de/SharedDocs/Downloads/EN/BSI/Pubblications/Minimum_information/SecurityRecommendationsCloudComputingProviders.html (2011)
2. Fox, A., Griffith, R., Joseph, A., Katz, R., Konwinski, A., Lee, G., Patterson, D., Rabkin, A., Stoica, I.: Above the clouds: A Berkeley view of cloud computing. Dept. Electrical Eng. and Comput. Sciences, University of California, Berkeley, Rep. UCB/EECS 28 (2009)
3. Future Internet Public-Private Partnership. http://www.fi-ppp.eu/ (2011)
4. Gabriella Cattaneo Massimiliano Claps, S.C.M.B.: Clouds for science and public authorities SMART 2011/0055. Tech. rep., University of Zurich, Department of Informatics (2012). https://custom.cvent.com/1E8AD1B771DA4B029B78FF1784749EF5/files/fc4e90577a1243e5a22e6b0e713ea59c.pdf
5. Mell, P., Grance, T.: The nist definition of cloud computing (draft). NIST spec. publ. **800**(145), 7 (2011)
6. Open data on cloud. http://opendatasalute.cloudapp.net/ (2011)
7. The data liberation front. http://www.dataliberation.org/

Low-Effort Support to Efficient Urban Parking in a Smart City Perspective

Alessio Bechini, Francesco Marcelloni and Armando Segatori

Abstract The Internet of Things (IoT) is today considered as one of the most important enabling technologies for developing a wide variety of smart services aimed at assisting the final user in the urban environment. In this chapter, we present how IoT can be employed to develop a system for the effective and efficient management of urban parking, thus providing a small, yet relevant contribution to the implementation of a real Smart City. Our system relies on the identification of each single parking slot but, unlike other approaches proposed in the last years, it does not require dedicated sensors and/or infrastructure, thus it can be regarded as a low-cost and low-effort solution. Indeed, it collects parking data from a mobile application on the drivers' mobile devices and possibly identifies each slot by QR codes deployed on the single parking spots. The amount of data collected by the system on parking occupancy allows inferring valuable information that can be used by local governments. For instance, it will be possible to define appropriate pricing schemes so as to promote parking areas not particularly occupied. The employment of an SOA design guarantees the integration of the developed system with other existing services within a Smart City.

A. Bechini (✉) · F. Marcelloni · A. Segatori
Department of Information Engineering, University of Pisa, Largo L. Lazzarino, 52126 Pisa, PI, Italy
e-mail: a.bechini@ing.unipi.it

F. Marcelloni
e-mail: f.marcelloni@ing.unipi.it

A. Segatori
e-mail: a.segatori@for.unipi.it

S. Gaglio and G. Lo Re (eds.), *Advances onto the Internet of Things*, Advances in Intelligent Systems and Computing 260, DOI: 10.1007/978-3-319-03992-3_17,

1 Introduction

The last decade has witnessed the growth of new ICT solutions to support both pervasive monitoring and information access. Many different objects and entities found in everyday life have the possibility to take part in broadly available software applications, often involving a user interface on mobile devices. The concept of "Internet of Things" (IoT) pushes further this vision [5], making uniquely identified objects play an essential role in pervasive software applications, and not necessarily relating a "thing" to a sensor or a computing unit.

The advantages brought by IoT have been recognized in several application fields [7, 13]. In particular, the urban environment is a perfect target for IoT, and the convergence of infrastructure technologies and new ways to conceive software applications enables the development of assorted "smart" services to assist the final user in everyday life, thus building up a *Smart City*.

Parking represents a crucial problem in urban life [3] and, in a Smart City perspective, it can be efficiently addressed by dedicated applications, taking into consideration the already available services, as well as the relations with other aspects of urban living [1]. Often, good solutions fails to find widespread acceptance because of the cost of in-place deployment and maintenance, and IoT makes no exception in asking for a thoughtful economic assessment [4]. Consequently, a prime requirement for any parking support system is to keep as low as possible the effort asked by these activities.

It has been recognized that sustainability in city logistics involves parking issues as well [22]. Nowadays in urban areas car parking is a time-consuming activity that substantially impacts the everyday life of citizens. Often, finding out the proper parking place asks for wandering around for an unpredictable period, wasting both time and fuel, and moreover contributing to pollution and traffic jam. An interesting survey [17], consisting of 1,400 questionnaires and proposed to citizens of Nicosia in Cyprus and Chania in Greece, shows that about 37 % of the drivers spend more than 10 min searching for an available parking spot. Moreover, in 34 % of the cases the drivers behave "negatively" in case no available parking spot is found nearby: Some park illegally causing traffic problems, whereas others leave the area cancelling their activities.

In this setting an effective support to car parking can be given by the so-called Parking Guidance and Information (PGI) systems, whose benefits are expected to become essential to the overall sustainability of urban activities. The very first problem to solve is how to help users find the most convenient (and currently vacant, of course) parking spot. To this aim, different strategies can be applied and, following the work by Wang et al. [21], they can be categorized as follows:

Blind Search—is applied when there is no parking information. Drivers have to wander around for a vacant parking spot until one of them becomes available.

Parking Information Sharing (PIS)—is the most popular strategy adopted by smart parking systems. Drivers can check in real-time parking information for a certain area, and choose the desired place according to their preferences. This approach raises the *multiple-car-cashing-single-space* problem when the demand of drivers for vacant parking spots increases.

Buffered PIS—addresses the multiple-car-cashing-single-space problem, which may cause severe traffic congestion. It intentionally reduces the number of available parking spots shown to drivers through the published information.

Experimental results show that the blind search is the worst strategy, whereas the other two are better in terms of average driving distance, especially during rush hours.

Different types of parking areas may pose different challenges in designing a smart support software system. In particular, on-street and off-street parking should be told apart. The term *on-street parking* typically refers to public open air, roadside spaces that are usually managed by local authorities. On the contrary, *off-street parking* involves large dedicated areas (lots) located at the street level, in the underground or in specific buildings, typically with strict access control at their gates; parking lots of this kind are managed by local authorities or private companies.

In this chapter, we address the flexibility and integration issues required for a parking support system in the context of a Smart City, according to the typical IoT approach. The proposed solution supports wide area parking services, from search up to occupancy start/end, including also a booking service. Although it can even be used independently of other Smart City services, the proposed parking system is designed to be seamlessly integrated in an assorted suite of software components targeted at dealing with other aspects of urban life. The IoT vision requires the identification of the involved actors [5]: In this case, any single parking spot can be identified by different means, e.g. in-place QR codes and/or GPS positioning. This choice is driven by the need to keep low the deployment costs, developing at the same time a quick, intuitive, and user-friendly procedure for the whole parking service, payment included. In other words, the status of a parking spot is maintained and managed by means of the information obtained from users, who operate both in place (as the spot is occupied or leaved) and remotely (in searching and in booking). Parking occupancy information is kept in the system repository and used during the search phase and so crowdsourcing-based solutions [14] become unnecessary. The guidance provided by a dedicated mobile application can lead to more effective and efficient car parking, yielding fuel savings, with a positive impact on the quality of urban life. Moreover, the availability of real-time data on specific parking occupancies can be fruitfully exploited by other urban monitoring or data mining applications to better plan and manage traffic and public city services.

The proposed solution is highly flexible, and can easily manage both on-street and off-street parking. Moreover, different system instances can be integrated in an overall Smart City framework, thus taking advantage at a global level of information coming from each of them.

This chapter is organized as follows. Section 2 reports a survey of the main smart parking systems. Section 3 is dedicated to spotting out requirements for advanced smart parking solutions based on "Internet of Things" concepts. In Sect. 4 the underlying data model is developed, taking interoperability issues into particular account. Section 5 presents structural details of our proposal. In Sect. 6, we discuss some data-driven ancillary services. Final conclusions are drawn in Sect. 7.

2 Related Works

In recent years, different works have been presented about applications to automate and to improve the management of parking areas, and many of them address the optimization of the parking occupancy [6, 10]. From surveying the different architectures recently proposed for smart parking applications, it becomes evident that, according to Polycarpou et al. [17], interoperability is the prime concern in this context, especially whenever the coverage of wide geographical areas is required.

A reasonable way to organize the software infrastructure is proposed by Sujith et al. [16] who, explicitly addressing parking lots, adopt a conceptual view with three distinct layers: (i) the *Smart Parking Spot* (SPS), to represent things such as sensors, lights, and displays within a parking spot; (ii) the *Parking Management System* (PMS), i.e. the Web-based application to store data gathered by sensors, and (iii) the *Smart Parking Client* (SPC) to allow the driver/user directly interact with the sensor node placed at the parking spot, by means of a mobile application on the smartphone. Each SPS is connected to the network using the Wi-Fi available in the parking lot. When a SPS is coupled with an SPC, the corresponding parking spot is occupied and its status can be checked online.

Often the proposed smart parking solutions rely on sensors present at each parking spot. In a paradigmatic case [10], ultrasonic sensors are organized in a wireless sensor network. All sensors in a specific area, such as a garage level, or one or more streets, make up an XMesh wireless network, and data can be transmitted via a gateway. Also in this case, a mobile application can help users reach the desired parking spot. A similar infrastructure based on wireless sensor networks is proposed by Yang et al. [23], with modules that notify the web-servers about the availability of each parking spot. As usual, drivers can interact with the system through a mobile application.

Vehicle protection can be included as an additional important requirement, as in the case of SPARK (Smart PARKing scheme) [15], a system to be deployed in large parking lots. SPARK is based on Vehicle Ad Hoc Networks (VANETs) and assists drivers with assorted parking services, such as real-time navigation, anti-theft protection, and parking status information. The system model consists of On-Board Units (OBUs, provided by a trusted authority), which communicate with other cars as well as with Road-Side Units (RSUs, placed at fixed positions). When an OBU-equipped vehicle reaches the parking lot, it first communicates with the RSU to get the access authorization. Furthermore, before leaving the parking lot, drivers have to switch the OBU from sleeping to activated mode by inserting a password, otherwise the system will consider the vehicle as getting robbed, and an alert is raised.

Under some environmental conditions, sensors may be subject to interference, leading to a misreading of the parking spot state. Furthermore, because of aging effects, sensors can become inaccurate or in the worst case stop working due to failures. To overcome problems of this kind, the number and variety of the deployed sensors can be increased, and this has been done e.g. by adopting both light and vibration sensors [21] to detect vehicles.

The latest trends in the smart parking context account for the concepts of *parking reservation* and *dynamic pricing*. The correct implementation of the former functionality requires solving some particular problems. The system must guarantee that the specific spot reserved by the driver will not be occupied by other vehicles.
For *off-street* parking, several solutions can be easily put in place. For example, rollway post/barrier for each parking spot can be used to prevent unauthorized drivers from occupying it [11]. In [21], each parking lot consists of an exact number of parking spots and the current lot state is fully determined by how many spots are occupied and how many are reserved. When a driver performs a reservation via Internet, the system updates the state of the parking lot, denying access to unauthorized vehicles if no free space is available. Moreover, depending on the parking lot state and the congestion level, the system determines the parking price according to a specific schema, periodically notifying drivers.
For *on-street* parking, Geng et al. [10] underline the problem complexity: Movable gates, barriers or obstacles at the parking spot to be remotely operated can be very expensive and hard to install and maintain. They propose a different approach based on lights placed in each parking spot. A green light at a parking spot indicates that it is available, whereas a red light is used for the reserved ones. As a driver approaches the parking spot, a yellow light informs him that he can proceed with the occupation, while a blinking red light signals that another driver has already reserved that spot. In case someone illegally occupies a spot, the system also notifies the driver who made the reservation, possibly proposing an alternative location. If the original parking spot becomes available again (because the car was towed away, or simply it left), then that spot would still be kept reserved for the driver.

Recently, some different smart parking systems have been actually deployed and tested around the world, and here we just recall some cases that exhibit typical features. In Oakland, California, a system was installed in the Rockbridge Bay Area Rapid Transit (BART) station [18], giving drivers real-time parking information via Internet or Variable Message Signs (VMS). The experimentation ended up with satisfactory results, but such an experience suggested that, to recover investments and installation/maintenance costs, the system should be operated at a larger scale. Moreover, according to a survey among the participants, about 45% of the respondents were in favor of expanding the system to other BART stations, and about 70 % expressed interest in displaying other informational messages, e.g. on accidents or traffic level.
Another solution is ExpressPark,[1] deployed in Los Angeles. Among the provided services, we find PGI, dynamic pricing, and reservations. Sensors in parking spots check on the availability and send data to the parking management system. Pay stations for multiple parking spaces or single-space meters accept different payment methods, like debit/credit cards, coins, and cellular phone. The payed amount may depend upon the parking demand, the time of the day, and the length of stay. A mobile application helps drivers interact with the system, adding some useful services such as search for free spaces and PGI.

[1] http://www.laexpresspark.org

A similar smart parking system, SFPark,[2] operates in San Francisco. The infrastructure consists of 14 garages and 8,200 wireless sensors, embedded in the pavement of on-street spaces. Via wireless connections, sensors inform the SFPark data warehouse about parking availability. Similarly to ExpressPark, multi-space and single-space parking meters accept different payment methods. The actual fee depends on the time of the day, the length of stay and also on scheduled events. Meters are able to display and apply the current pricing because they are automatically updated via wireless messages.

3 Requirements Analysis

A smart parking system is expected to provide users with accurate information and to implement helping services, i.e. for the driver who asks for a convenient parking solution (hereafter, we will use the terms "end user" and "driver" interchangeably). In a Smart City context, a wide assortment of services is available, and thus we want to carry out a requirement analysis to understand what the *core* functionalities and modules of the system should be, and what other external services or data sources could be exploited, integrated, or even further supported. Trivially, real-time parking information must be a prime objective, and this is the central issue in PGI systems, because of its impact on the quality of life of citizens [3].

Apart the PGI services, the system must be able to manage the whole parking process for each single parking place. Such a process is a *transaction* that involves the driver and the supporting system. The *parking transaction* may encompass different possible services and choices for the driver. In our discussion, according to what has been pointed out also in recent related works, we consider both occupancy and booking of a parking spot. Dealing with different parking spots, either concurrently or at different times, means carrying out distinct parking transactions.

The definition of a parking transaction model is particularly important for the requirements analysis of the whole system, because it includes the fundamental actions to be supported and automatically coordinated, in concert with indications from the user. The proposed model is shown in Fig. 1: Here only the regular flow of actions is reported, whereas unexpected events or situations may determine a premature transaction commit or abort.

In the proposed model, a driver can either book a parking spot, or go straight to its occupation. In case a reservation is done, the user is expected to occupy the spot at the beginning of the reserved period. At leaving, the user must proceed to pay, and the parking spot is released. These operations do not necessarily have to be serialized, especially for automatic, off-line payment methods. It is possible to issue a reservation extension, both before and after the beginning of the spot occupancy, provided that the initial corresponding reservation already exists in the transaction.

---

[2] http://www.sfpark.org

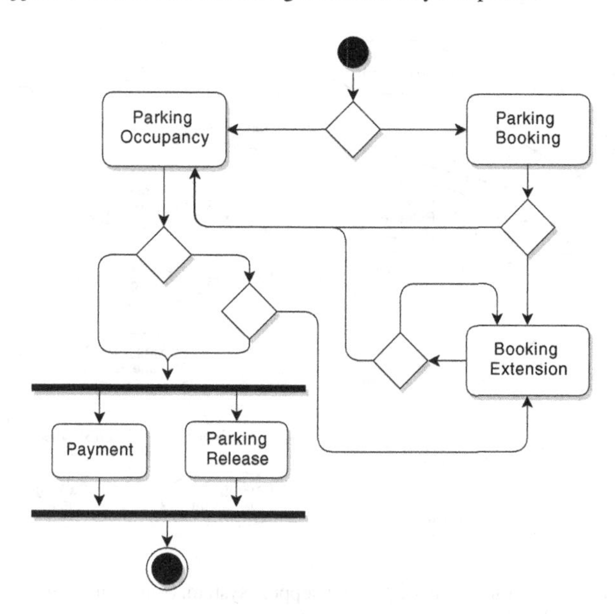

Fig. 1 The *parking transaction* corresponds to the structuring of all the actions related to the management of a single actual parking process

It is worth recalling that many setbacks may determine a deviation from the ordinary transaction flow, and the system must be able to accommodate also these situations.

Upon a precise definition of a parking transaction, we can try to sketch the basic use cases, as shown in Fig. 2. The involved actors are the *driver*, the *system administrator*, and possibly *institutional users*.

As users have to be assisted in spotting the most convenient place to park their car, the *Parking Search* function has to deliver information on the available places. In doing this, a target location must be somehow specified. For trips planned with adequate advance, it makes sense considering a *Parking Booking* service. The driver may ask for *directions* to reach the selected parking place. This service is indicated as shadowed in Fig. 2, along with others, because it does not necessarily belong to the system, but may be an external one to be integrated. Dedicated procedures must be triggered at the notification of *Parking Occupation* as well as *Parking Release*. The parking fee has to be charged via the *Payment* function, which can be directly hosted by the system or be an external service.

Users must be able to check their personal transaction data anytime (the *My Transaction Info* use case in the diagram), both for ongoing and past placements and reservations. Regarding the user data and the access control, they can be internally managed by the system but, in case of integration of the system in a broader "smart city" software support, *Login* and *Account Information* will be handled at a super-application level. Moreover, the data collected by the parking system can be analyzed by a *Statistical Service* and summarized values can be offered to institutional users.

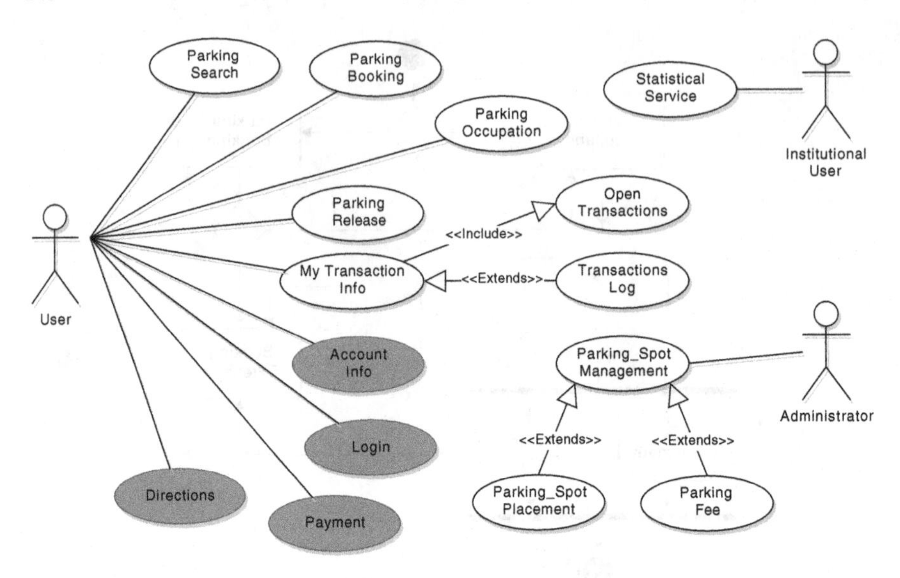

Fig. 2 Use case diagram for the smart parking support system. Some functionalities can be integrated from services already available in the Smart City

Finally, the system administrator must perform the *Parking Spot Management* tasks, possibly updating placements and fees. In a Smart City environment, management decisions can be influenced by real-time data on occupancy, traffic level, status of the road network, and also by the occurrence of events like concerts, meetings, or sport competitions nearby.

The requirements for our smart parking application do not only stem from a rough functional analysis, but they also involve the general software architecture. The required flexibility asks for a modular structure, with very loosely coupled components. Other functionalities may be developed, tested and easily integrated with the rest of the application at a later time. It must be possible to quickly (and as seamlessly as possible) integrate external services. Furthermore, some components must be able to communicate with third-party modules as well, and take part in a broader application, increasing and enhancing the functionalities and services offered to end users in the Smart City.

An additional non-functional requirement calls for a robust yet not expensive means to identify the parking spot on the field, so that the user would be able to notify the system about what precise place he/she is going to use.

4 A Model for Parking Data

The back-end of the entire system is in charge of managing a large collection of data, whose structuring influences the effectiveness and efficiency of the parking supporting functions. Besides, an appropriate data schema can also simplify the

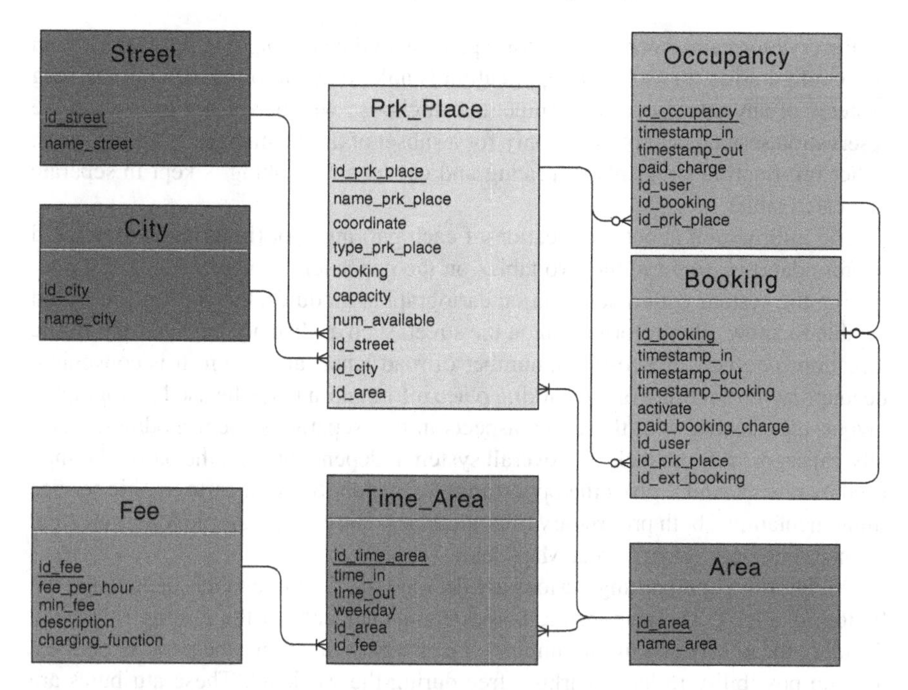

Fig. 3 The parking data model is supported by a few DB tables. Each parking spot is characterized by its own core attributes, and the corresponding information about location, booking/occupancy, and pricing, is kept in dedicated tables

control flow of the data handling procedures, thus improving their maintainability. We propose a simple yet general model for parking data, to be implemented with a few core tables. Only the basic entities and attributes are taken into account here, and ancillary data can be added aside. The model can be used along with other external modules to setup better services, or to add new ones.

The main entity of the model is the *Prk_Place*, shown in the central position in Fig. 3, and it aims at grouping up all the relevant aspects of every single parking space. A parking place is characterized by its core attributes such as the *id* to uniquely identify it, its geographical *coordinates*, and its *type* to discriminate the vehicles that can park on it (car, bus, motorcycle, camper, and so on). Note that when the *capacity* attribute holds the value 1, the parking place corresponds to a single spot, and when it is greater than one it refers to a parking lot. Thus, the system can treat both on- and off-street parking in a homogeneous way by making use of the same entity.

A parking place keeps information on how many vacant spots are currently present. Whenever a driver occupies a spot, the value of the *num_available* attribute must be decremented by one. It is not possible to park in a place whose availability is 0. On the other hand, when the driver notifies the system that he is releasing a spot, the corresponding value must be incremented by one and so the parking place becomes vacant again.

Reservations are accepted only for a place with the *booking* flag set to *true*. With this configuration, the software application is able to adapt to the different parking systems of any city. Local governments can choose whether or not to take on the reservation service, and possibly only for a subset of all the managed parking spots. Other information about place, pricing and occupancy/booking is kept in separate dedicated tables.

The information about the location of each parking spot (in terms of structured address data) is stored in the two tables on the upper left corner of Fig. 3. By integrating the system with other external cartographic modules, the application would be able to know other information on the street, such as distance, speed limit, traffic direction (one-way or two-way), number of road lanes, and so on. It is convenient decoupling the handling of the parking place information from the used cartographic layout, e.g. dealing with these two aspects in two separate system modules. Actually, this approach can make the overall system independent from the particular map used by clients, and it gives the opportunity to exploit different cartographic service implementations, both proprietary (Google Maps, Bing Maps, Apple Maps, etc) and open-source ones (Open Street Maps, etc).

Pricing rules and parking periods are detailed via the three tables depicted at the bottom of Fig. 3. The responsible bodies/companies can set the pricing rules, e.g. deciding the hourly and minimum fee (or a charging function), the parking periods, and the possibility to leave parking free during the weekend. These attributes are handled on a per-area basis, giving the opportunity to manage aggregations of parking places.

It is essential to adequately support the parking transaction model at the database level (see Sect. 3 for further details). Occupancy and booking data are associated to two specific entities (in the right upper corner of Fig. 3). Both of them have a key ID, and moreover a *timestamp_in* and a *timestamp_out* to indicate respectively start and end of the referred periods. In the *Booking* entity, the timestamps come from the indications given by the user to mark the reservation interval, whereas in *Occupancy* they are automatically communicated by the mobile application (and this is typically triggered by a QR code scan). The user who carried out the operation is identified by the corresponding ID (*id_user*). The system can store different types of ID values, e.g. the vehicle's registration number, or the user ID assigned in another module that manages the users' master data, assuming that it has been properly integrated into the system. The *Booking* entity keeps also the time instant the reservation has been committed by the user (*timestamp_booking*) and an *activate* flag to indicate whether the booking is valid or not. The extension of a reservation can be easily handled by adding a new tuple. All the records belonging both to the original reservation and to the (possibly multiple) further extensions are chained by means of the *id_ext_booking* and *id_booking* attributes values. The choice to keep trace of any extension let us precisely record how the system is actually used. Consequently, it will be possible to investigate the typical behavioral aspects of the user/system interaction, getting hints to improve the quality of the service, e.g. exploiting the data mining module described in Sect. 6.2.

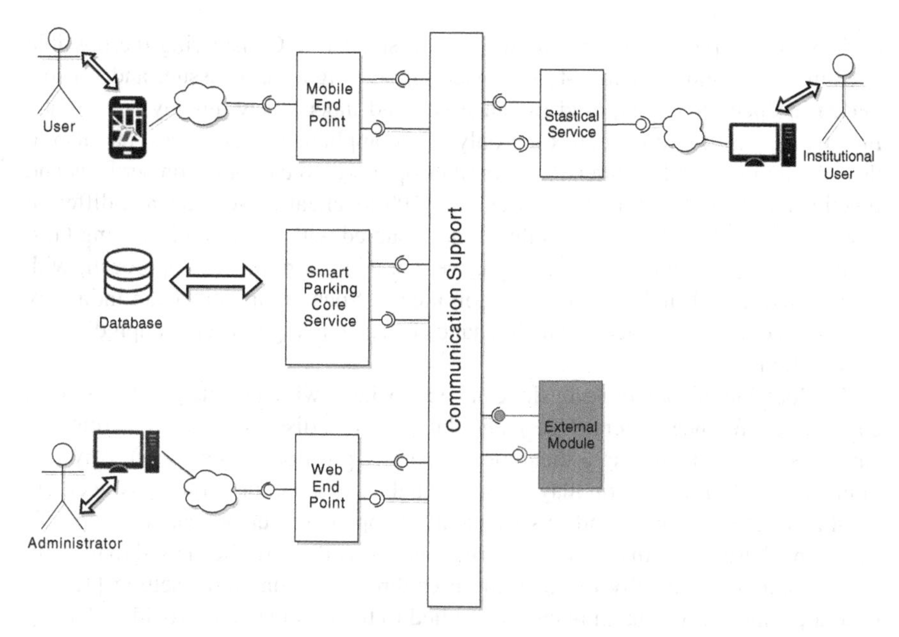

Fig. 4 Schematic view of the components of the smart parking system

5 Components for Smart Parking Support

The whole system devoted to provide the smart parking support must get together components of different types that need to exchange proper information. Figure 4 illustrates a schematic view of the main system components and their interconnection. In practice, considering the popularity of cellular/smart phones today, the interaction with the driver must necessarily be handled by a mobile application (top left corner of Fig. 4). The rest of the system can be structured taking into account the modularity and flexibility requirements expressed in Sect. 3.

Before diving into the details of the different parts of the system, a crucial point must be discussed. The whole system operation relies on the assumption that each parking spot could be identified on the field. In practice, the user must be able to drive to the chosen location, and communicate the *exact* parking spot identifier *in an extremely simple way*. After devising a satisfactory solution to this problem, we will be able to design the system components accordingly.

5.1 On-Field Identification of Parking Slots

The employment of sensors of different types or dedicated devices is a popular choice for the IoT infrastructure [5, 10], and it is suitable also for the on-field identification of parking slots. Beyond the evident benefits, problems may arise from using assorted devices that employ different communication standards. Considering the possible operating environments, long-term power supply might be an issue, and sensors placed in open air are exposed to vandalism and indeed may quickly deteriorate. Finally, the solution cost relates not only to the appliances themselves, but also to their deployment and maintenance. Another option is to capitalize on services and functions already available to the user, e.g. GPS localization services and different types of optical "readers" on mobile phones coupled with in-place identifying tags. To make the procedure robust enough, we can adopt a multi-way approach, with a concerted use of multiple identification means. Such an approach can indirectly address also fault-tolerance and maintenance facets, making it react to unpredictable malfunctions.

The location of each parking slot can be associated with a single point in a given coordinate reference system (CRS), and thus it can be discovered and identified by any GPS device (assuming a theoretically perfect positioning accuracy). Anyway, in practice GPS information may be not available, reliable, or sufficiently accurate for this specific purpose, and this frequently happens in our typical scenarios. To overcome these problems, in past experiments, QR codes have been used to identify locations, and specifically to guide the user through an unknown setting [12]. In general, some form of tagging can be applied to the parking slot, provided that tag reading would be easy, automatic, and within cost constraints.

In our proposed system, we opted for QR codes as the main tool for parking slot identification, because of their cheap deployment. The maintenance of QR tags can be ordinarily and easily carried out by ticket inspectors. A QR code is used to store the unique ID of the parking slot, which can then be read through a mobile client device and communicated to the system back-end whenever required throughout a parking transaction.

QR codes placed in the open air may undergo fading, corruption, or scribbling. In these cases, to alleviate the difficulties in detecting the tag contents, the specific code type must be chosen by applying an adequate trade-off between redundancy, compactness, and readability.

Unfortunately, the above two identification methods, taken individually, do not guarantee a robust localization procedure. To cope with this shortcoming, in the proposed system we integrated GPS and QR code information, leveraging also the current parking state information in the target area. In particular, the *Smart Parking Core Service* module primarily considers the QR code information and, if not available, it passes to work on the GPS coordinates provided by the mobile phone. Given the target geographical point and a "tolerance" radius that depends on the GPS localization error, the module can retrieve a list possible parking slots placed around. Considering also the availability status, usage logs, etc. it guesses what the most likely slot is, and updates its status. This approach shows three useful features. First of all, the system can keep working even with damaged QR tags. Secondly, when the GPS procedure is applied, the system can notify the administrators that a QR code is probably damaged. Third, this procedure does not require neither additional information in the database, nor putting in place any sensor.

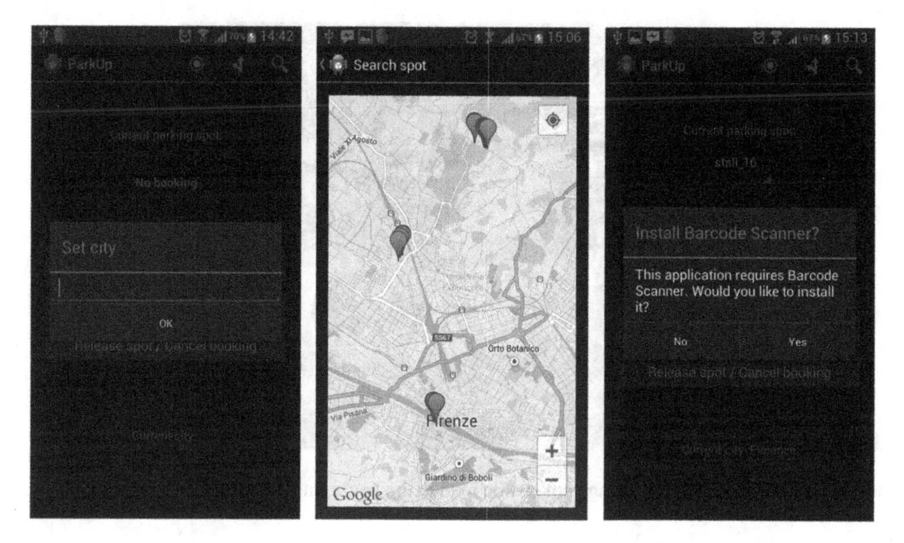

Fig. 5 Automatic searching for an available parking place requires the selection of a destination

5.2 *The Role of the Mobile Application*

A mobile application has been developed to be hosted on end users' devices. This choice shows a lot of benefits with respect to other possible solutions. Nowadays, smart phones and tablets are widely and ordinarily used by many citizens. Smart phones contain various kinds of sensors, such as camera, accelerometer, NFC, etc, allowing people to interact with the environment in several ways. Moreover, through dedicated application stores, users can easily and quickly download and update their software.

The Mobile Application, based on Android, provides drivers with a user-friendly interface for the PGI service, and assists the execution of parking transactions. Information is obtained from the *Mobile End Point* through Wi-Fi or a 3G/LTE cellular network.

In the first place, the end user is required to sign up to the system. The authentication is the first mandatory step for the user. Then, it is possible to search for a convenient parking place within a specific area, to proceed with its occupation, and/or to book a parking spot. The overall procedure takes care of automatically charging the due fare.

At any application access, some previously started transactions may still be ongoing, so the system checks for possible pending reservations or current parking occupancies, which can be dealt with by a user-friendly management console. To initiate a new procedure, the driver must provide some information to look for a convenient parking place within a specified area, such as with the city name and possibly address, and eventually a "tolerance" radius [3]. For example, in the left screenshot in Fig. 5, no pending reservation is detected, and the driver is interested in checking

Fig. 6 The QR tag of the parking spot can be read at the beginning/end of the occupancy

the availability of parking spots in Florence. As the "Ok" button is tapped, the current parking information is retrieved (through a call to a web service) and displayed on a map view. The current status of the whole area (see the central screenshot in Fig. 5) is then graphically displayed by placing status "markers" that possibly show aggregated information instead of per-slot indications, depending on the zoom level. Each single marker can take four different colors: (i) green, for available parking positions; (ii) blue, for available and bookable parking positions; (iii) yellow, for a parking spot already booked by another user; (iv) red, no available place.

At present, the only implemented strategy is PIS, described in Sect. 1, but also the buffered PIS may be reasonably employed. The driver can then decide to get indications to reach the destination place. For this purpose, he/she can take advantage of the satellite navigation system the smartphone is equipped with.

When the user reaches the place and parks the car on the target parking spot, he must notify the system about the start of the occupancy period. A simple scan of the corresponding QR tag is sufficient to perform this action (Fig. 6). Information on the current transactions can be accessed via the management screen, which e.g. reports the elapsed parking time and the amount to be paid so far. An explicit action is required to signal the end of the parking period, and this can be done by a further reading of the QR tag. Calculation and charging of the due fare is performed automatically, improving application usability, and at the same time allowing the system to collect extremely precise information on the occupation history of parking areas. In case a QR code reader is not available on the end user's mobile phone yet, the system assists him/her in downloading it (right screenshot in Fig. 5).

The system also allows for the remote booking of specific, dedicated parking slots. By tapping on a blue marker, a wizard appears to help the user make a reservation. A parking period has to be specified, and the reservation is issued upon the confirmation of the booking details (see the left screenshot in Fig. 7). As already described for the

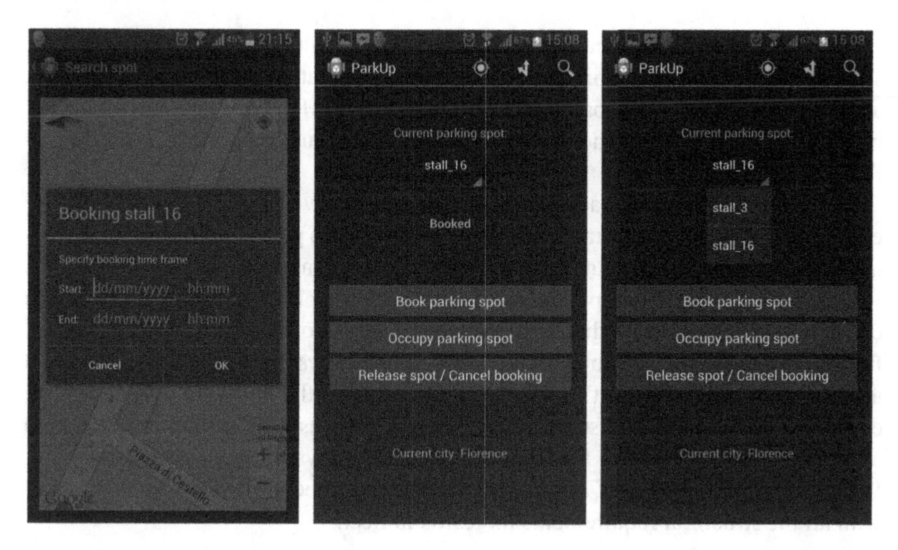

Fig. 7 Booking requires a few actions that can be conveniently carried out by interacting with the mobile application

occupation procedure, the user can browse his own booking information on the management screen (see the central screenshot of Fig. 7). A committed reservation can also be either canceled or extended, according to the user needs. The user can perform multiple bookings for different days and with different parking periods. He can easily switch from one reservation to another through a simple list like a drop-down menu, and perform the desired task, i.e. cancellation or extension (right screenshot in Fig. 7).

5.3 Back-End Components

The core of the whole system is represented by back-end modules that are in charge of managing the parking data repository. These modules, designed as *services*, provide the rest of the system with the basic functionalities. An overall service-oriented organization leads to a loose coupling of the modules, and it directly addresses integration problems, both internal and towards external components eventually present in a Smart City environment.

A natural choice for the implementation of the modules is represented by Web Services. By using language independent XML technologies and HTTP as transport layer, any module can easily communicate with any other. To send and retrieve the necessary data, a client component requires only the WSDL definition of the server counterpart, while the module internals are completely hidden. Services can be easily used by different clients on platforms like Android, iPhone/iPad, and desktop applications, requiring no change on the server side. Additionally, Web Services have shown to be able to accommodate assorted data integration needs in very different application fields [2, 9]. Most of the times expensive efforts to data integration can be replaced by just asking the proper service for the required interface, according to the exposed interface.

Although the communication between the users' devices and the corresponding end-point components may take place over ordinary web protocols, the internal communication support among the back-end Web Services must be carefully chosen [8]. In fact, flexibility and interoperability needs have to be balanced against performance constraints. The system modules have been designed to offer different types of interfaces, and thus we are free to employ at will a suitable system middleware (indicated as "Communication Support" in Fig. 4), with no lack of flexibility. In our system, we equipped the modules with standard SOAP interfaces, as well as with connectors to an ESB (Enterprise Service Bus) that supports fully asynchronous communication. Ultimately, the adoption of a Service Oriented Architecture (SOA) let us comply with all the structural requirements indicated in Sect. 3.

5.4 Implementation Details

The mobile application has been developed in Java through the Android SDK and tested on the Samsung Galaxy S3 with Android 4.2.2. The business logic layer has been implemented as JEE application running on the JBoss Application Server 7.1.1 and integrated in the ESB JBoss Fuse 6.0 for the communication support. The data repository consists in a PostgreSQL DBMS, whose spatial features rely on PostGIS, which is ISO 19125 compliant.

6 Data-Driven Ancillary Services

The large amount of data collected on reservations and occupations is an invaluable source of information on the drivers' behaviors and on how the mobility flows in the city. This information can be extracted from the data by appropriate data mining techniques and used to develop ancillary, yet crucial services. In the following, we will discuss some of these services just as examples of the potentialities inherent in the data collected by the parking management system.

6.1 Dynamic Pricing

The concept of dynamic pricing, combined with parking reservation, is one of the latest trends in the Smart City context. The demand of parking places may vary according to different areas, hours, and week days. For instance, at beach places the request of parking spots is typically higher in week-ends than in working days, and similarly it is more significant in summertime rather than in winter months. Thus, an effective strategy could make the parking fee vary according to the demand, i.e. the higher the demand, the higher the fee. The strategy should take into account not only the different areas (e.g. downtown, residential, commercial, industrial) but also the citizens' needs. The local governments should take care of providing different categories of people (e.g. shoppers, residents, employees) with sufficient parking space, and prevent illegal parking at the same time. The general idea is to apply different parking fees on different periods of the day, at different places, and to different drivers.

The data model described in Sect. 4 can easily accommodate dynamic prices. Fee information is handled on a per-area basis, giving the opportunity to manage parking spots both individually and in aggregates. Furthermore, the system is able to dynamically update the fee of specific parking spots according to the strategies adopted by institutional users or private companies. Strategies should take different parameters into account. For instance, Shoup [19] suggests that fees should be set so that at least a couple of vacant parking slots at each block should be available, whereas Wang et al. [21] proposes a dynamic pricing system based on the occupancy status and the congestion level, analyzing the status of parking spots across the entire city.

Applying different prices according to the real-time parking vacancies may accomplish multiple goals. First of all, institutional users or private companies can maximize their revenue changing their fees. Secondly, the parking price can be used to alleviate the traffic jam problems at real-time. Setting higher parking fees forces the drivers to change their plans in favor of public transport or to re-schedule their activities in a different time/place, reducing the traffic level [20]. Finally, it has been supposed [20] that a uniformly distributed demand over time may let the drivers find a vacant parking space more promptly.

6.2 Short-Term and Long-Term Urban Mobility Planning

The information inferred from data handled by the parking management system can be used by local governments for short-term and long-term urban mobility planning. As regards the short-term planning, the local governments can, for instance, decide to increase the number of bookable parking slots for a limited time period in areas where events are planned at a specific date. The identification of these events can be carried out by exploiting smart services that monitor social networks with the aim of discovering where and when an event (concert, exhibition, party) will take place [1]. Further, when all the parking spots are occupied in areas close to the event place, the traffic ways can be temporally re-organized so as to route the vehicles towards areas with vacant parking places.

As regards long-term mobility planning, the analysis of the spatial and temporal distributions of the parking slot occupations can allow local governments to highlight mostly vacant parking areas close to others that instead are permanently occupied. This different occupation levels often is determined by how the traffic ways are organized to get to a certain point of interest. The local governments could therefore decide to re-organize the ways to achieve a more uniform exploitation of the parking areas. Upon the actual implementation of this long-term planning decision, its impact can be monitored and validated by using the parking management system. Also, the analysis of the parking slot reservations could suggest that the number of bookable slots is too low with respect to the requests. Thus, the local government could decide to increase the number of this type of slots. On the other hand, this analysis could highlight that the number of reservations is quite low with respect to the number of bookable slots, probably because making a reservation is considered too expensive. The local government could therefore decide to reduce the fee for encouraging users to reserve slots. Indeed, reservations represent a valuable help for local governments, because this service allows to monitor the parking occupation state *in advance* and therefore it permit to plan counteractions for possible congestion to come.

7 Conclusion

In this chapter, we have presented how the Internet of Things paradigm can enable the efficient and low-effort management of urban parking, thus providing a small, yet relevant contribution to the implementation of an effective Smart City. We have described the design and the implementation of a system that allows identifying and handling single parking spots, making their reservation, occupation, and dynamic price management possible.

We have first determined the system requirements and discussed a model for the parking data. Then, we have described the architecture of the system and its components. In particular, we have analyzed in detail the on-field identification of the parking slot and the client-side application on the mobile device, which enables drivers to reserve, occupy, and release parking slots. Finally, we have discussed some aspects of the back-end components.

The data collected by the platform allow inferring crucial information on the occupation state of parking slots in specific areas at precise time intervals. This information can be used by local governments for, e.g., defining appropriate pricing schemes so as to promote parking areas not particularly occupied at specific hours or re-arranging the ways of the road so as to force drivers to pass along mostly vacant parking areas.

The application is going to be integrated into an ICT platform that will feature tools and services for innovative and sustainable mobility. The platform has been developed in the framework of the SMARTY (SMARt Transport for sustainable citY) project funded by the Tuscany region (Italy).

Acknowledgments The presented application comes from the refactoring and integration processes carried out in the framework of the SMARTY project, funded by "Programma Operativo Regionale (POR) 2007-2013" - objective "Competitività regionale e occupazione" of the Tuscany Region. The authors would also like to acknowledge Roberto Martucci and Pietro Lorefice for their contributions to early prototypes of the application.

References

1. Anastasi, G., Antonelli, M., Bechini, A., Brienza, S., D'Andrea, E., De Guglielmo, D., Ducange, P., Lazzerini, B., Marcelloni, F., Segatori, A.: Urban and social sensing for sustainable mobility in smart cities. In: Proceedings of The Third IFIP Conference on Sustainable Internet and ICT for Sustainability, IEEE Computer Society (2013)
2. Bechini, A., Cimino, M., Lazzerini, B., Marcelloni, F., Tomasi, A.: A general framework for food traceability. In: International Symposium on Applications and the Internet Workshops (SAINTW'06), pp. 366–369. IEEE Computer Society, Los Alamitos (2005). doi:10.1109/SAINTW.2005.10
3. Bechini, A., Marcelloni, F., Segatori, A.: A mobile application leveraging QR-codes to support efficient urban parking. In: Proceedings of The Third IFIP Conference on Sustainable Internet and ICT for Sustainability, IEEE Computer Society (2013)
4. Bohli, J.M., Sorge, C., Westhoff, D.: Initial observations on economics, pricing, and penetration of the internet of things market. SIGCOMM Comput. Commun. Rev. **39**(2), 50–55 (2009). doi:10.1145/1517480.1517491
5. Chaouchi, H. (ed.): The Internet of Things: Connecting Objects. No. 420 in ISTE. Wiley-ISTE, London (2010)
6. Chou, S.Y., Lin, S.W., Li, C.C.: Dynamic parking negotiation and guidance using an agent-based platform. Expert Syst. Appl. **35**(3), 805–817 (2008). doi:10.1016/j.eswa.2007.07.042
7. Feki, M., Kawsar, F., Boussard, M., Trappeniers, L.: The internet of things: the next technological revolution. Computer **46**(2), 24–25 (2013). doi:10.1109/MC.2013.63
8. Gama, K., Touseau, L., Donsez, D.: Combining heterogeneous service technologies for building an Internet of Things middleware. Comput. Commun. **35**(4), 405–417 (2012). doi:10.1016/j.comcom.2011.11.003
9. Gazzè, D., Bechini, A., Avvenuti, M., Tesconi, M., Marchetti, A.: Integration of external data in document workflows via web services. In: Proceedings of IC3K-KMIS 2012, pp. 346–351. SciTePress (2012). doi:10.5220/0004143203460351
10. Geng, Y., Cassandras, C.G.: A new "smart parking" system infrastructure and implementation. In: Proceedings of EWGT2012–15th Meeting of the EURO Working Group on Transportation, September 2012, vol. 54, pp. 1278–1287. Elsevier, Paris (2012). doi:10.1016/j.sbspro.2012.09.842
11. Giuffrè, T., Siniscalchi, S.M., Tesoriere, G.: A novel architecture of parking management for smart cities. In: Proceedings SIIV-5th International Congress–Sustainability of Road Infrastructures 2012, vol. 53, pp. 16–28. Elsevier (2012). doi:10.1016/j.sbspro.2012.09.856
12. Hammadi, O.A., Hebsi, A.A., Zemerly, M.J., Ng, J.W.P.: Indoor localization and guidance using portable smartphones. In: Proceedings of the 2012 IEEE/WIC/ACM Int' Joint Conferences on Web Intelligence and Intelligent Agent Technology WI-IAT '12, vol. 3, pp. 337–341. IEEE Computer Society, Washington (2012). doi:10.1109/WI-IAT.2012.262
13. Han, D., Zhang, J., Zhang, Y., Gu, W.: Convergence of sensor networks/internet of things and power grid information network at aggregation layer. In: 2010 International Conference on Power System Technology (POWERCON), pp. 1–6 (2010). doi:10.1109/POWERCON.2010.5666553
14. Kopecký, J., Domingue, J.: ParkJam: Crowdsourcing parking availability information with linked data (demo). In: 9th Extended Semantic Web Conference (2012)

15. Lu, R., Lin, X., Zhu, H., Shen, X.: SPARK: A new VANET-based smart parking scheme for large parking lots. In: IEEE INFOCOM 2009, pp. 1413–1421 (2009). doi:10.1109/INFCOM.2009.5062057
16. Mathew Samuel, S., Atif, Y., Sheng, Q., Maamar, Z.: Building sustainable parking lots with the Web of Things. Pers. Ubiquit. Comput. **1**, 1–13 (2013). doi:10.1007/s00779-013-0694-7
17. Polycarpou, E., Lambrinos, L., Protopapadakis, E.: Smart parking solutions for urban areas. In: 2013 IEEE 14th International Symposium and Workshops on a World of Wireless, Mobile and Multimedia Networks (WoWMoM), pp. 1–6 (2013). doi:10.1109/WoWMoM.2013.6583499
18. Rodier, C.J., Shaheen, S.A.: Transit-based smart parking: an evaluation of the San Francisco Bay area field test. Transp. Res. Part C: Emerg. Technol. **18**(2), 225–233 (2010). doi:10.1016/j.trc.2009.07.002
19. Shoup, D.C.: Cruising for parking. Transp. Policy **13**(6), 479–486 (2006)
20. Teodorović, D., Lučić, O.: Intelligent parking systems. Eur. J. Oper. Res. **175**(3), 1666–1681 (2006)
21. Wang, H., He, W.: A reservation-based smart parking system. In: 2011 IEEE Conference on Computer Communications Workshops (INFOCOM WKSHPS), pp. 690–695 (2011)
22. Whiteing, T., Browne, M., Allen, J.: City logistics: the continuing search for sustainable solutions. In: Waters, D. (ed.) Global Logistics and Distribution Planning, pp. 308–320. Kogan Page Business Books (2003)
23. Yang, J., Portilla, J., Riesgo, T.: Smart parking service based on wireless sensor networks. In: IECON 2012–38th Annual Conference on IEEE Industrial Electronics Society, pp. 6029–6034 (2012). doi:10.1109/IECON.2012.6389096

An Integrated System for Advanced Multi-risk Management Based on Cloud for IoT

Maria Fazio, Antonio Celesti, Antonio Puliafito and Massimo Villari

Abstract This chapter presents the Cloud computing technology as strategic solution for the deployment of IoT application and solutions. Cloud computing is a new ICT paradigm able to offer products and solutions as services. Thus, it allows the delivery of on-demand virtual resources (e.g., computational resources, storage systems, applications, data centers,...) over the Internet on a pay-for-use basis. Also, the distributed nature of Cloud computing guarantees high availability of resources dynamically adapting their allocation to specific requirements of the system. The research community together with big business companies are focusing their efforts in the adoption of Cloud for a massive interaction with the physical environment. To show a such trend, we present an on-going project, called SIGMA, which exploits Cloud technologies to acquire, integrate and compute heterogeneous data from several sensor networks for controlling and monitoring both environmental and industrial production systems. Specifically, we describe a new Cloud framework at the basis of the whole SIGMA architecture, in order to show benefits in the adoption of a such technology. The framework is compliant with the Sensor Web Enablement standard specifications and makes use of a plug-in platform to integrate heterogeneous sensing infrastructures. It allows to build abstract objects for accessing sensing devices and observations pandering to the Internet of Things needs.

M. Fazio (✉) · A. Celesti · A. Puliafito · M. Villari
DICIEAMA, University of Messina, Contrada di Dio, S. Agata, 98166 Messina, Italy
e-mail: mfazio@unime.it

A. Celesti
e-mail: acelesti@unime.it

A. Puliafito
e-mail: apuliafito@unime.it

M. Villari
e-mail: mvillari@unime.it

S. Gaglio and G. Lo Re (eds.), *Advances onto the Internet of Things*,
Advances in Intelligent Systems and Computing 260, DOI: 10.1007/978-3-319-03992-3_18,

1 Introduction

According to a recent Gartner report [1], there will be 30 billion devices connected to the Internet by 2020. In this way, we assume a reference scenario where a plethora of heterogeneous devices and Sensor Networks (SNs) are interconnected and share data and information. It is therefore natural to think about possible ways and solutions to face an all-encompassing challenge, where such an ecosystem of geographically distributed sensors and actuators may be discovered, selected according to the functionalities they provide, interacted with, and may even cooperate for pursuing a specific goal.

Cloud computing provides a very flexible technology, which is able to integrate monitoring devices, storage devices, analytics tools, virtual infrastructures and client delivery systems. It offers theoretically unlimited computing and storage capabilities, and efficient communication services for transferring terabyte flows between data centers. This means a hard support in the management of enormous amounts of data generated for IoT purposes, which have to be stored, processed and presented in a seamless, efficient and easily interpretable form.

Both the IoT and Cloud technologies address two important goals for distributed system: high scalability and high availability. All these features make the Cloud Computing a promising choice for supporting IoT services. IoT can appear as a natural extension of Cloud Computing implementations, where the Cloud allows to access IoT based resources and capabilities, to process and manage IoT environments and to deliver on-demand utility IoT services such as sensing/actuation as a service.

This chapter aims to show new opportunities arising from the exploitation of Cloud computing for IoT purposes. Thanks to the high level of virtualization of the Cloud, sensing devices equipped with communication technologies (e.g., RFID reader, GPS receiver, weather stations, biosensors, cameras, smartphones, medical devices, surface physics, motion sensors,...) can be virtualized as Cloud resources, characterized by specific sensing properties and location. The virtualization of a device or a set of devices organized in a Sensor Network (SN) allows to abstract each sensing framework and provide a common interface toward the physical environment. Such abstract components become objects in the IoT, able to interact each other or with other objects.

Sensed information is generally acquired by independent administrations deploying their own monitoring infrastructure and software architecture. Sharing such information is strategic. The idea of such a massive scale data sharing is leading towards the concept of system of systems, which aims to achieve task-oriented integration of different ecosystems provided by independent public and private organizations into a “federated Cloud”. In a Cloud Federation, several independent, heterogeneous, private/hybrid Clouds collaborate each others for increasing their productivity and efficiency. This approach guarantees an effective integration among services offered by different sensing systems. Moreover, traditional Cloud services, such as storage and computing, can improve efficiency and scalability in IoT application deployment, since processing and making correlations on a lot of data can be very complex. Therefore, computing, storage and sensing become complementary aspects in a big picture, and a comprehensive approach from the IoT perspective is needed to optimally coordinate their interactions, thus creating a pervasive infrastructure interacting with the surrounding environment.

This scenario has been envisaged from many different perspectives, along several research activities and projects, on which institutions and governments are spending many efforts. This is also in line with the technological trend that identifies personal and mobile Clouds as the hottest Cloud topics of 2012. To discuss such a trend, we present the SIGMA (sensor Integrated System in cloud environment for the Advanced Multi-risk Management) project. SIGMA proposes a layered architecture whose function is to acquire, integrate and compute heterogeneous data from several sensor networks (weather, seismic, volcanic, water, rain, car and marine traffic, environmental, etc.), in order to strengthen control and monitoring environmental and industrial production systems. Collected data are useful for the prevention and management of risk situations through services provided to citizens and businesses, both public and private.

The Cloud framewerk designed in SIGMA is based on CLEVER, a flexible framework for secure inter-Cloud communication and event notification developed at the University of Messina [2]. CLEVER has been extended with three Cloud utilities: (1) C-COMPUTING, which dynamically creates and executes Virtual Machines (VMs) on CLEVER hosts; (2) C-STORAGE, which stores data within a database deployed in a distributed fashion; (3) C-SENSOR, able to virtualize different types of sensing infrastructures into the Cloud, adding new capabilities for data processing.

The remaining of the chapter is organized as follows. In Sect. 2, we present the Cloud computing paradigm. An overview of the state of the art on the integration of sensing technologies into the Cloud is discussed in Sect. 3. In Sect. 4 we give some hints on risk evaluation issues, since it is the main target of the SIGMA project. The SIGMA project and its reference architecture is introduced in Sect. 5. Then, in Sect. 6, we describe the Cloud framework designed for the SIGMA project to monitor energy consumption in industrial sites. Section 7 concludes this work with lights to future works.

2 Briefly on Clouds

Ian Foster [3] describes Cloud Computing as a large-scale distributed computing paradigm that is driven by economies of scale, in which a pool of abstracted, virtualized, dynamically-scalable, managed computing power, storage, platforms, and services are delivered on demand to external customers over the Internet. Until now, we have assisted to the steady rising of hundreds of independent, heterogeneous Cloud providers managed by private subjects yielding various services to their clients. There is not limit to the possible scenarios that can adopt the Cloud computing paradigm. For example, recently, ICT experts are looking at Cloud computing for improving the efficiency of their systems and, also for reducing the power consumptions and

CO_2 emissions of their datacenters through virtual machine migration and software consolidation techniques.

In order to provide a flexible use of resources, Cloud computing makes wide use of virtualization techniques to treat traditional hardware resources like a pool of virtual ones. In addition, virtualization enables resources migration, regardless of the underlying physical infrastructure. Using virtualization and aggregation techniques, Cloud computing offers its available resource *as a Service* rather than as physical product. In particular, according to the NIST formalization [4], it provides services at three different levels:

- *Infrastructure as a Service* (IaaS): it provides consumers with computation, storage, networks data transfer and other computing resources. Consumers are able to deploy and run arbitrary software, which can include operating systems and applications. For example, this type of service is offered by Amazon EC2, Rackspace, Salesforce.
- *Platform as a Service* (PaaS): it makes consumers able to deploy their own applications onto the Cloud infrastructure using programming languages and tools supported by PaaS Cloud providers. Typical examples of PaaS services are given by Social platforms as well Facebook, Twitter, LinkedIn, Google Apps.
- *Software as a Service* (SaaS): it represents the capability given to consumers of accessing provider's applications running on a Cloud infrastructure. Many Cloud services are already available at this level, such as Amazon Storage, DropBox Storage, Google Map, Google Docs and Microsoft Office Online.

Cloud computing exploits whatever virtual technologies for making an abstraction on data, processing, and storage [5]. Virtual Machines (VMs) represent the typical example of how virtualization technology can be used in Cloud. Cloud costumers are able to preconfigure VMs and to deploy them on the Cloud infrastructure, without any further configuration. VMs may collect data, execute their elaboration, migrate the data if necessary, expose APIs to be used from other VMs being executed in different Clouds and so on. To provide the effective integration of sensing technologies into the Cloud, specific virtualization techniques for monitoring systems need to be exploited [6]. In Sect. 3, we present current solutions in literature to integrate sensing resources and Cloud computing.

3 Related Work and Background

The exploitation of Cloud technologies for providing a scalable management of sensing data is a hot topic being investigated in literature. In this direction, in [7], the authors describe a context-oriented data acquisition and integration platform for Internet of Things over a Cloud computing environment. The work well discusses the representation of information based on an XLM-oriented approach, but it is not clear enough how they leverage the Cloud resources.

As stated in [8], ubiquitous sensor networks may enable sensing data collection at many points and places in real-time. Usually, collected data are saved on data storage systems, such as PCs, database servers or Web servers. The authors proposed the utilization of a free Cloud data service, such as Twitter, exploited to share observed data. They introduced the concept of ambient sensor Cloud system by using both Cloud computing and the Arduino-based Open Field Server platform. People may benefit of these data accessing them by Twitter as Followers.

Another example of Cloud-based Social Platforms for sensing is reported in [9]. The initiative is aimed to investigate the *mood* of users. Here the authors extrapolate mood information from social feeds (e.g., Twitter) and associate to them sensing data (e.g., weather). The mathematical model applied to a *mood space* tries to characterize users, their day-lives and tweets. Their assessment is conducted in off-line computation mode, and doesn't deal with massive computation constrains.

Authors in [10] have designed a Sensor-Cloud infrastructure enabling end-users to dynamically create virtual sensor groups. The introduction of *virtual items* (e.g., sensors, sensor groups, servers, etc.) is interesting indeed, but how they impact on the management of Cloud resources is not very clear. They use conventional SQL-based database to collect data, but this choice may represents a bottleneck in the architecture in terms of performance.

The Cloud data management service reported in [11] aims at massive sensing data management in the Cloud and is based on the Hadoop framework. The authors assert that traditional relational database management systems are a bottleneck in scenarios dealing with ultra-large-scale data sets. Thus, their Cloud data center adopts the Hadoop FileSystem (HDFS) for organizing clusters aimed at the Cloud.

A new architecture in which Server Webs and Server Sensors interact each others using a redirectable stream-oriented channels is described in [12].

4 Risk Evaluation

Risk is the potential that a chosen action or activity leads to a loss. This concept implies that each choice has influence on the outcome. Almost any human intent carries some risk, but some are much more relevant than others. We face risks in many human activities characterized by an *hazard* as well economics investments, building infrastructures and transport of people and/or goods. With the term of *hazard* we mean every risky condition, which can potentially cause harm, like human injury or death, damage to the environment, damage to physical assets and loss of production.

The Risk Investigation is a way for evaluating the hazard of an activity and Fig. 1 shows how it is possible to perform it. The first step is characterized by two independent tasks:

(1) Data Collection, for gathering all the necessary information useful to characterize the context. The specific data types depend on the environment and/or activities. For example, in an industrial production control, data types include air pollution, temperature, power consumption, hydraulic charge pressure, air compression, motion,

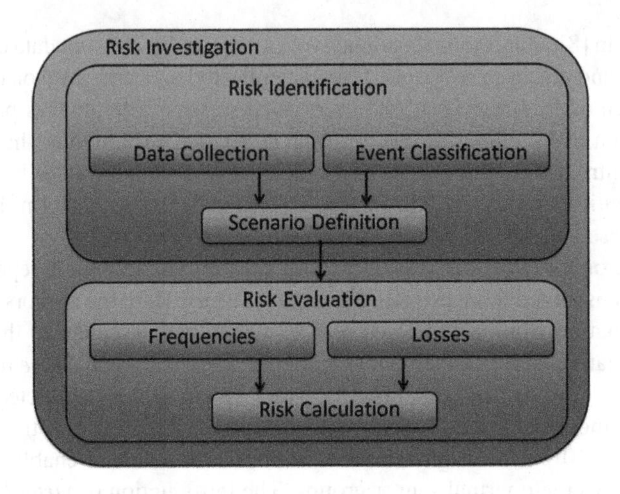

Fig. 1 Risk investigation activities

optical positioning, weight, acceleration, chemical composition, gases, liquid flow and so on.

(2) Event Classification, for specifying which events cause loss and the impact of the loss. In our scenario, damaging events include on-the-job injury, accidents, explosions, environmental damages.

The second step of the Risk Evaluation aims to merge information on data and events in order to clearly define the scenario where the activity will be carried out. The formal description of the scenario allows to perform the evaluation of the risk. Figure 1 shows the main functionalities for Risk Investigation in any human activities falling into the topic of this work.

In the Risk Evaluation assessment, it is possible to identify three typologies of risk:

- Individual Risk (IR): as defined by AIChE/CCPS [13], it represents the probability that an average unprotected person, permanently present at a certain location, is killed in a period of one year due to an accident resulting from a hazardous activity.
- Societal Risk (SR): it is a measure of risk for a group of people. It is most expressed in terms of the frequency distribution of multiple casualty events (F/N curve, described later).
- Environmental Risk (ER): it represents the potential threat of adverse effects on living organisms and environment by effluents, emissions, wastes, resource depletion, etc., arising out of an organization's activities.

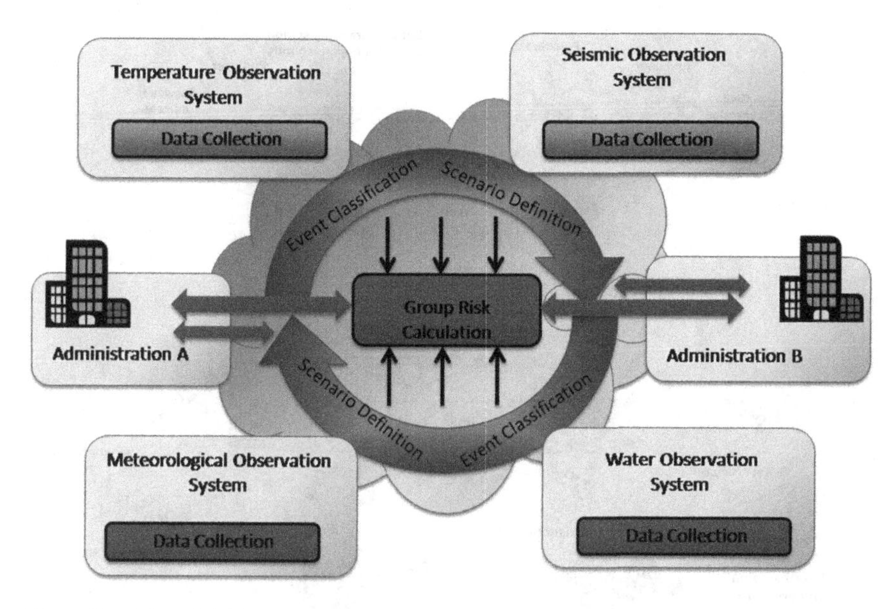

Fig. 2 Cloud-based risk investigation

4.1 Risk Investigation in the Cloud

A leader organization responsible for risk investigation coordinates efforts over each task in order to provide the risk calculation. The cooperation of different entities involved in an industry for the monitoring of risks is based on a rigid hierarchical organization, where higher elements are responsible for the lower. In a Cloud environment, thanks to its distributed paradigm, the architectural organization of work can be redesigned. All the sites, buildings and sectors involved in risk investigation are organized in a federation. Each of them has a specific role and responsibilities and gives its contribution to the federation, according to a flat organization model of tasks. This approach improves the flexibility of the whole system, where competences and activities are independent from each other.

As shown in Fig. 2, Temperature, Water, Seismic and Meteorological Observation Systems are responsible to provide Data Collection through their monitoring infrastructures to the Cloud. At the same time, other sites provide services for the Scenario Definition. However, each entity specifies properties of the scenario according to different parameters (geographical area, legislation, HS policies,...). Risk Calculation is performed through the elaboration of risks models through Cloud virtual resources. All these contributions are orchestrated within the Cloud in order to offer seamless services.

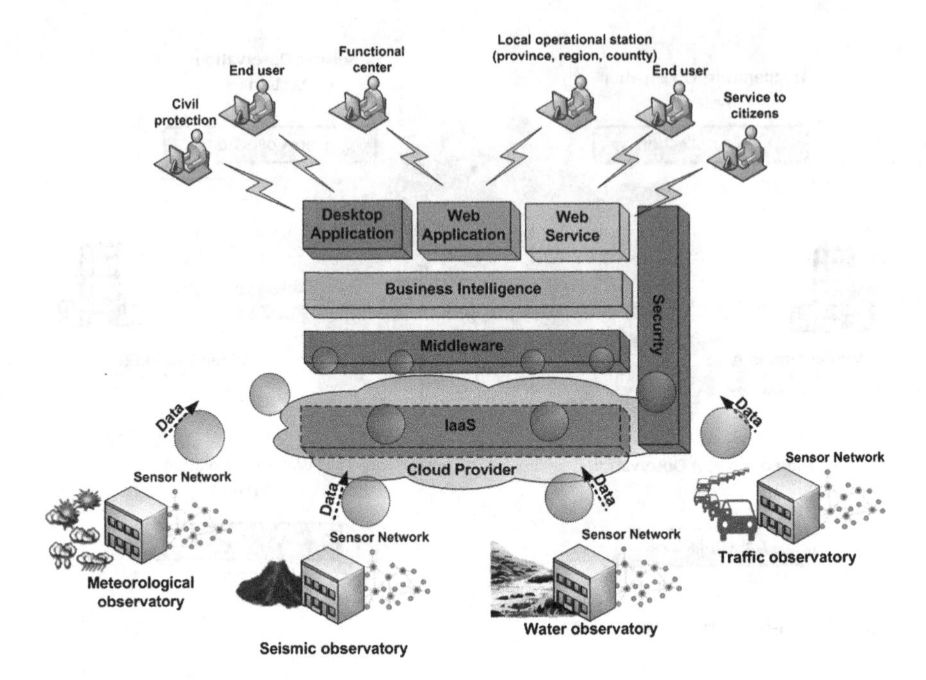

Fig. 3 Overall SIGMA architecture

5 The SIGMA Project

The objective of the SIGMA project is to manage the risk situation in both the industrial production process and in the territory in order to provide support for the preparation of response plans. For example in the industry field, analyzing data coming from different ICT equipments and the surrounding environment may be possible to control the production, instead considering the territory, analyzing data coming from car traffic sensor network of a given area may be possible to provide useful information to the population and relevant authorities in case of particular traffic jam due to rare events such as a big social event or a natural disaster.

The project will also include the functionality to simulate emergency situations in order to allow the training of personnel responsible for emergency management. How depicted in Fig. 3, the layers that make up the architecture are six. At the lowest layer there are different sensor networks. Some of them are already installed on territory, as the SIAS network that consists of a series of weather stations to support the agriculture industry, the water observatory that consists of a series of hydrometric stations and rainfall to support the design of water projects and the INGV networks for monitoring seismic and volcanic activities in Sicily, Italy. In addition to the existing sensor network, SIGMA also foresees the makeup of dense network in order to integrate the existing networks, for multiparameter monitoring of sensitive areas and increased hydrological, hydro geological, geological, seismic, volcanic land risk, and integration with other networks such as that for car and naval traffic monitoring with GPS and GSM systems, in order to provide a more diverse and accurate data. The architecture also provides for the use of new types of sensors called "smart sensors", capable of an initial high layer processing of collected information [14]. Many industrial sites also have their own private network monitoring systems that can be integrated for early identification of risk situations for employees and citizens in general. At this layer different transmission networks of information are also placed, from sensor networks to higher levels of architecture. Recent events have demonstrated, in tactical scenarios and in areas affected by natural or anthropogenic disasters, the need for robust integrated telecommunications systems in which are available high speed voice, data and video connections [15]. In addition, the industrial production has to be properly controlled in case of rare natural disaster. At upper level the system is based on virtualized and distributed resources provided by a Cloud computing framework. This layer is based on CLEVER, a flexible framework for inter-Cloud communications and event notification [2]. It includes specific component for virtual infrastructure set up and management, sensing environment integration and data retrieval and storage. The advantages of the framework come from the fact that it will provide computation and flexible storage capacity with enhanced performance, thus facilitating the integration of unstructured networks that make available large amounts of data to be stored and processed. The infrastructure is in fact able to expand and downsize itself in a flexible and automatic way. At higher level is the Middleware layer, an intermediate software layer that, through a series of interfaces, gathers data from various heterogeneous networks, standardizing them and making it available at Business Intelligence. The Business Intelligence level is responsible for process data, implementing the actual business logic of the architecture. At this level, through a series of algorithms, many complex problems are solved and the results are supporting the industrial plant or territory monitoring and management activities. The highest level of architecture is finally represented by the Application layer that takes care to create interfaces for user interaction with the system (e.g. Functional Centers, Operating Rooms, etc ...). The architecture provides three types of high-level interfaces: Desktop Application, Web Application and Web Services. The first two are pure graphical interfaces, made respectively in Desktop and Web area, with which users can interact with the system. The Web Service interface instead is conceived as a means of communication for interaction with other independent systems. Given the high importance of data that are processed by the system, security is a fundamental requirement and needs special attention in the design phase. To ensure a high degree of reliability, the layer that handles security is designed to embrace all levels of architecture, so that making the system easy to administer in terms of safety, and secure in its entirety. The project includes significant field-testing activities with particular emphasis on integrated management of environmental and industrial risk situations. The objective is to demonstrate, with the help of some use cases oriented to territory multi-risk management, that the "systematization" of infrastructure that are not interoperable today, produces add value both for those providing the resources (e.g., sensors, data storage and processing) and for the users, making it possible to offer services with high added value. In particular two areas of test bed will be identified, on which experiment the technology developed in the project. A first case study refers to the integrated volcanic phenomena and territory management and will use the data from the integrated sensor networks form INGV, SIAS and the water observatory. The second case study will instead be a more industry specific and will focus on critical sites such as industrial plant site of ST. The case studies will use the potential of the integrated telecommunications network for transporting data from remote sites (sensor network) to the centers of information processing. In particular, we will use a test bed that uses innovative types of satellite terminals with antennas with planar phased-array technology, that allows the connection of moving vehicles with the sensor network (local connection), and with Network Control Center (a satellite connection). This test bed allows the transmission of data collected locally, even in the absence of communication infrastructures in emergency situations in which is indispensable to ensure a secure and reliable connection between teams called to intervene on the ground and decision-making and coordination centers.

6 The Cloud Framework in SIGMA

In this section we focus our attention on the Cloud framework designed to support functionalities for monitoring and risk control in the SIGMA project. It is based on CLoud-Enabled Virtual EnviRonment (CLEVER), that is a Cloud framework developed at the University of Messina [2], which aims at the setup and management of an overlay network for the interaction of many Clouds spread over the Internet. It organizes nodes of the Cloud infrastructures according to a cluster-based approach, as shown in Fig. 4.

CLEVER nodes contains a host level management module, called Host Manager (HM). A single node may also include a cluster level management module, called Cluster Manager (CM). The CM contains the intelligence for treating and analyzing all incoming data, whereas the HM has lower level functionalities. Indeed it represents the remote agent of the CM used for the execution of specific tasks. Thus, in the cluster, we have the CM at a higher layer and, at a lower layer, many HMs depending on it. At least one active CM has to be deployed on each cluster but, in order to ensure higher fault tolerance, some redundant CMs remain in a listening state to detect if the active CM turns off. A CM acts as an interface between Cloud clients (software entities, which can exploit the Cloud) and the software running on the HMs.

The CM receives commands from the clients, gives instructions to the HMs, elaborates information and finally sends results to the clients. It also performs the management of resources (e.g., uploading, discovering, etc.) and the monitoring of the overall state of the cluster (e.g., workload, availability, etc.). A HM perform operative tasks for running Cloud utilities and it can be seen as a gateway towards the physical infrastructure. For example, it instantiates VMs and run them on the physical hosts, or gathers sensed data from the SNs forwarding them to the CM.

Both CMs and HMs are composed by several sub-components, called *agents*, which are designed to perform specific tasks. To improve the readability of the

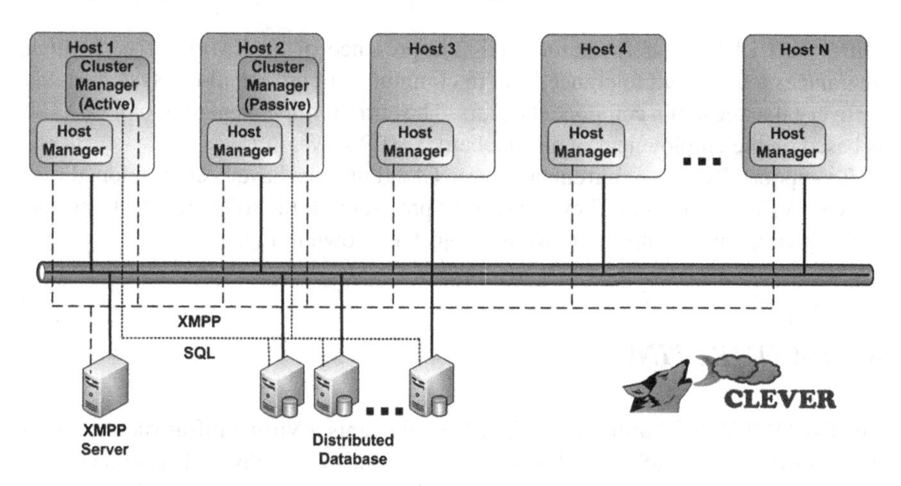

Fig. 4 Cluster-based organization of CLEVER nodes

chapter, we name Cluster Agent (CA) an Agent in the CM and Host Agent (HA) an Agent in the HM, but we emphasize that, from a technical point of view, there is not any difference in the design of CAs and HAs.

To enable the interaction among the Cloud components, CLEVER supports three types of communication: Internal Remote Method Invoker (IRMI), External Remote Method Invoker (ERMI), Notification. The IRMI communication refers to the message exchanging protocol among agents within the same manager (both in CMs and HMs). ERMI communications allow a CA to exchange messages with HAs. In fact, each CA has knowledge of the agents in the HMs that depend on the CM itself. On the contrary, HAs do not have knowledge of the CAs at the upper-layer of the hierarchy. Even though this design choice seems to limit the interoperability among the agents, it allows to reduce the complexity of inter-module communications and to increase the scalability, fault-tolerance and availability of the system. To allows communications from an HM to the CM, CLEVER uses the Notifications, which are sent from an HA to the CM without specifying the CA interested in the communication. For example, in the provisioning of sensed data, an HA gathers data from a SN and forward them towards the cloud independently from the particular services that will manipulate these pieces of information at the upper-layer.

To implement ERMI and Notifications, the XMPP protocol [16] has been used. In fact, XMPP is able to offer:

1. decentralization (i.e., no central master server should exist: such capability in native on the XMPP) in a way similar to a p2p communication system for granting fault-tolerance and scalability when new hosts are added in the infrastructure;
2. flexibility to maintain system interoperability;
3. native security features based on the use of channel encryption and/or XML encryption.

Thus, the CLEVER architecture needs the presence of an XMPP Server, which guarantees a high fault tolerance level in communications and allows system status recovery if a crash of a component occurs. The current implementation of CLEVER is based on the employment of an Ejabberd XMPP server [17].

To support SIGMA requirements, three Cloud utilities have been developed into the CLEVER framework. They are shortly presented in the following sections and their functionalities support the whole project as shown in Fig. 5.

6.1 C-COMPUTING

The C-COMPUTING utility of CLEVER implements a Virtual Infrastructure Manager (VIM) at the IaaS layer for the management of the physical resources of a datacenter (i.e., a cluster of machines). The VIM acts as a dynamic orchestrator of Virtual Machines (VMs). The main pupose of such an utility is to setup VMs (e.g., preparing disk images, setting up networking, executing VMs on the CLEVER hosts according to their workload, data location and other parameters, and so on), regardless of the underlying used Hypervisor (currently the C-COMPUTING utility supports QEMU/KVM, Virtualbox, and WMware). The basic operations of C-COMPUTING are:

- managing the VMs images, by means of images discovery, transfer and uploading;
- managing the VMs, providing functions to start, stop, suspend, shut-down, destroy, and migrate VMs;
- monitoring the VMs behavior and performance, in terms of CPU, memory and storage usage.

The C-COMPUTING utility allows to put in practice the elastic computing mechanism typical of the Cloud computing paradigm by means of VM migration mechanisms. In fact, the elastic computing allows to scale up/down Cloud infrastructures according to particular workloads. In simple works, the Cloud infrastructure can scale up instantiating new VMs when required, and it can scale down, e.g, suspending, shutting-down, or destroying VMs. The Business Intelligence algorithms developed into the SIGMA project will be executed on several VMs into the C-COMPUTE system and the number of necessary VMs will be dynamically adapted to the requirements of the Business Intelligence layer, through the middleware that works as interface between the Cloud and the Business Intelligence layer, as shown in Fig. 5. The dynamic allocation of VMs is a very important feature for informative systems that need variable computation capabilities, such as the SIGMA one, with peaks in the processing resources demand in particular time period or at the occurrence of specific events.

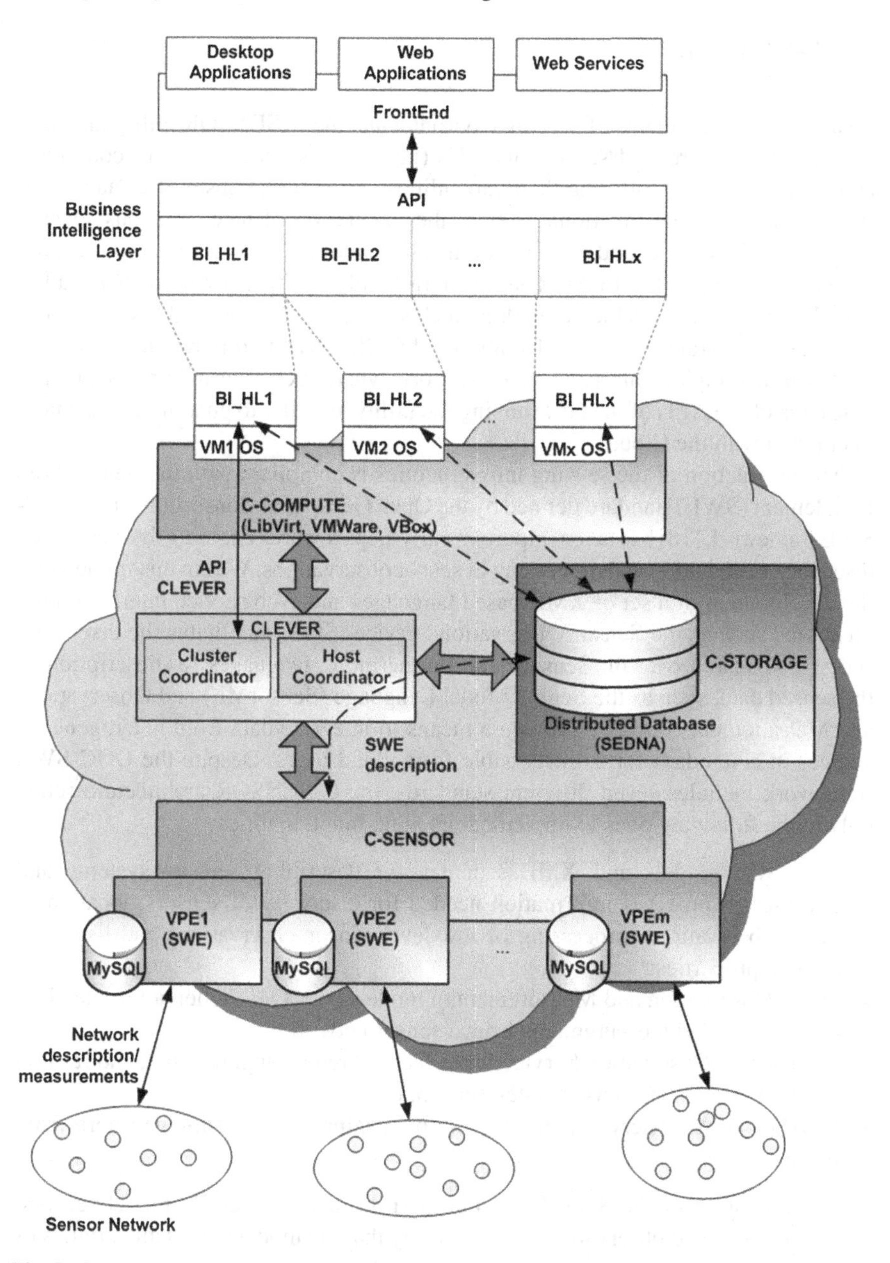

Fig. 5 CLEVER components in the SIGMA projects

6.2 C-SENSOR

To meet the requirements of sensing environments, the C-SENSOR utility manages the physical resources of sensors and SNs (i.e., sensors nodes, network communication equipment,...), offering their capabilities and observations at the PaaS layer. It implements all the functionalities for the integration of heterogeneous sensing infrastructures in the Cloud hiding underneath technologies and for the provisioning of sensed data. Thus, C-SENSOR has to interact with a sensing environment to gather data. Due to many available networking technologies for sensors and SNs, it is very important to guarantee a great adaptability of C-SENSOR to different technologies. To this aim, it implements a plug-in framework, where each plug-in implements specific calls of the APIs of the HM running the utility, in order to guarantee a seamless interaction with the Cloud.

The abstraction of the sensing infrastructures is compliant with the Sensor Web Enablement (SWE) standard defined by the Open Geospatial Consortium. The OGC-SWE framework [18] has taken important early steps towards enabling the web-based discovery, exchanging and processing of sensor observations. Within this framework, the development of a set of XML-based languages and Web service interface specifications, such as the Sensor Observation Service (SOS), facilitates the discovery, access and search over the sensor data. Furthermore, languages for description of the sensed data, such as the Sensor Model Language (SensorML) and Observations and Measurements (O&M), provide a means to integrate data from heterogeneous sources in a standard format accessible from cloud users. Despite the OGC-SWE framework includes seven different standards, the C-SENSOR architecture refers only to the following ones to implement its main functionalities:

- SensorML: models and XML schemas for describing sensors systems and processes; it provides information needed for discovery of sensors, location of sensor observations, processing of low-level sensor observations and listing of taskable properties;
- O&M (Observation and Measurements): models and XML Schema for encoding observations and measurements from a sensor network;
- SOS (Sensor Observation Service): interface for requesting, filtering, and retrieving observations and sensor system information;
- SAS (Sensor Alert Service): interface for publishing and subscribing to alerts from sensors.

The execution of the SOS Agent into a HM aims to carefully model sensors, sensor systems, and observations in such a way that the model covers all varieties of sensors and supports requirements of all users of sensor data.

6.3 C-STORAGE

The set of data necessary to enable Cloud services is stored within a specific database deployed in a distributed fashion and managed through the C-STORAGE utility. C-STORAGE contains the overall set of information related to the framework (e.g. the current state of the VMs, sensing data, state of the sensing infrastructure,...). Since the database could represent a centralized point of failure, it has to be developed according to a well structured approach, for enabling fault tolerance features. The best way to achieve such features consists of using a Distributed Database. Also, C-STORAGE has been designed according to plug-in approach, in order to leave high flexibility in terms of storage requirements. The plug-in devised for sensed data is able to interact with the Sedna native XML database [19]. In fact, Sedna natively supports the XML data schema (as well as SWE standards). Even, it is opensource and allows to create incremental hot backup copies of a database and supports ACID transactions. Sedna dynamically generates the descriptive schema from any type of data and maintains them by using an incremental approach. This solution makes the storage of information on sensors and observations very flexible. Thanks to the XPath and XQuery capabilities of Sedna, the system easily retrieves information from the database formatting them according to the SWE specifications.

To support huge amount of data fot IoT application, a new plug-in for the Hadoop file system (HDFS) can be developed. HDFS is highly fault-tolerant file system and it is designed to be deployed on low-cost hardware. It provides high throughput access to application data and is suitable for applications that have large data sets. Moreover, the Map-Reduce functionality of HDFS can be employed for processing scalable elaborations.

6.4 The Security Layer

How depicted in Fig. 3, the security layer transversely involves all the layers of the SIGMA architecture. This is due to the fact that each layer presents different security requirements that need to be properly addressed. With reference to the Cloud framework, the communication system of CLEVER, that is based on the XMPP, has to be secured in order to guarantee data confidentiality, integrity, and non-repudiation. In order to achieve totally secure communications, each exchanged message has to be signed and encrypted by each CLEVER component (i.e., CM or HM). Since the XMPP does not not natively provides such a security features, in the SIGMA project has been needed to extend the XMPP security mechanisms in order support the following functionalities:

- **Digital identity management**. Each component during the in-band registration (i.e., an automatic enrollment of a client on the XMPP server) with the XMPP server requires a digital certificate to a trusted Certification Authority (CA) through the Simple Certificate Enrollment Protocol (SCEP).

- **Signed message exchange**. Each component should be able to sign a message sent to another one.
- **Encrypted message exchange**. Each component should be able to perform a total or partial encryption of a message.
- **Private chat rooms**. The communication system should allow the management of private chat room with restricted access to authorized components.
- **Encrypted chat rooms.** The communication system should allow the management of private and encrypted chat rooms. The key exchange between the communicating components should take place according to a PKI schema. The component that play the role of "moderator" instantiate a new chat room associating a session key. When a new component wants to join the communication, the "moderator" component sends the session key encrypted with the public key of the new component itself.

The XMPP security extension has been performed according to the *XEP 0027* [20] specification that describes the use of Jabber with the Open Pretty Good Privacy (OpenPGP—RFC 4880—[21]). OpenPGP is an interoperable specification that provides cryptographic privacy and authentication for data communications. As highlighted by the *internet draft*, the XEP 0027 does not represent a standard, although it could be in the future, but it describes a possible solution for authentication and data encryption in end-to-end XMPP communication [22]. In addition, the access to data coming from the different sensor networks has to be restricted to authorized software component of the above layer.

7 Conclusions and Future Works

The main aim of this chapter is to motivate the adoption of Cloud technologies for IoT purposes. Cloud offers several services as distributed, efficient, scalable and federated solutions, which can support the abstraction, storage, management of thighs. It guarantees the integration of heterogeneous physical devices and systems spread across a world wide area able to capture environmental measurements. The huge amount of data gathered from such sensing infrastructures are posted to the Cloud and then abstracted, stored, analyzed, processed and manipulated according to the IoT application requirements.

The chapter presents the on-going SIGMA project, which is focused on risk detection in industrial production. In particular, we have explained how the Cloud framework has been designed to support a high level abstraction of sensing infrastructures and observations according to the SWE specifications. The framework aslo provides storage and computing utilities for a complete support of IoT requirements.

We are currently working to optimize the sensor activity and in particular we are testing aggregation algorithms able to reduce network traffic and increase the overall efficiency of the network.

References

1. Magrassi, P.: A world of smart objects: the role of auto identification technologies. In: Strategic Analysis Report, Gartner. Stamford (CT), USA (2002)
2. Celesti, A., Tusa, F., Villari, M.: Puliafito., A.: Integration of CLEVER clouds with third party software systems through a REST web service interface. In: 17th IEEE Symposium on Computers and Communication (ISCC'12), pp. 827–832. Cappadocia, Turkey (2012)
3. Foster, I., Zhao, Y., Raicu, I., Lu, S.: Cloud computing and grid computing 360-degree compared. In: Grid Computing Environments Workshop, 2008. GCE '08, pp. 1–10 (2008)
4. NIST: Cloud Computing Reference Architecture http://www.nist.gov/customcf/get_pdf.cfm?pub_id=909505 Dec 2011
5. Vernik, G., Shulman-Peleg, A., Dippl, S., Formisano, C., Jaeger, M., Kolodner, E., Villari, M.: Data on-boarding in federated storage clouds. In: IEEE CLOUD 2013 IEEE 6th International Conference on Cloud Computing June 27–July 2, 2013, Santa Clara Marriott, CA, USA (Center of Silicon Valley) (2013)
6. Fazio, M., Paone, M., Puliafito, A., Villari, M.: Huge amount of heterogeneous sensed data needs the cloud. In: International Multi-Conference on Systems, Signals and Devices (SSD 2012). Chemnitz, Germany (2012)
7. Chen, Y.S., Chen, Y.R.: Context-oriented data acquisition and integration platform for internet of things. In: Conference on Technologies and Applications of Artificial Intelligence (TAAI 2012), pp. 103–108 (2012)
8. Hirafuji, M., Yoichi, H., Kiura, T., Matsumoto, K., Fukatsu, T., Tanaka, K., Shibuya, Y., Itoh, A., Nesumi, H., Hoshi, N., Ninomiya, S., Adinarayana, J., Sudharsan, D., Saito, Y., Kobayashi, K., Suzuki, T.: Creating high-performance/low-cost ambient sensor cloud system using openfs (open field server) for high-throughput phenotyping. In: Proceedings of SICE Annual Conference (SICE 2011), pp. 2090–2092 (2011)
9. Yerva, S., Jeung, H., Aberer, K.: Cloud based social and sensor data fusion. In: 15th International Conference on Information Fusion (FUSION 2012), pp. 2494–2501 (2012)
10. Yuriyama, M., Kushida, T., Itakura, M.: A new model of accelerating service innovation with sensor-cloud infrastructure. In: Annual SRII Global Conference (SRII 2011), pp. 308–314 (2011)
11. Bao, Y., Ren, L., Zhang, L., Zhang, X., Luo, Y.: Massive sensor data management framework in Cloud manufacturing based on Hadoop. In: 10th IEEE International Conference on Industrial Informatics (INDIN 2012), pp. 397–401 (2012)
12. Melchor, J., Fukuda, M.: A design of flexible data channels for sensor-cloud integration. In: 21st International Conference on Systems Engineering (ICSEng 2011), pp. 251–256 (2011)
13. AIChE/CCPS: Guideline for Chemical Process Quantitative Risk Analysis, 2nd edn. New York (2000)
14. Fazio, M., Merlino, G., Bruneo, D., Puliafito, A.: An architecture for runtime customization of smart devices. In: 2013 12th IEEE International Symposium on Network Computing and Applications (NCA), pp. 157–164 (2013)
15. Fazio, M., Paone, M., Puliafito, A., Villari, M.: HSCloud: Cloud architecture for supporting homeland security. Int. J. Smart Sens. Intell. Syst. **5**(1), 246–276 (March 2012)
16. The Extensible Messaging and Presence Protocol (XMPP) protocol: http://tools.ietf.org/html/rfc3920
17. Ejabberd, the Erlang Jabber/XMPP daemon, http://www.ejabberd.im/ (2012)
18. Reed, C., Botts, M., Davidson, J., Percivall, G.: OGC sensor web enablement: overview and high level architecture. IEEE Autotestcon, pp. 372–380 (2007)
19. Sedna, Native XML Database System: http://modis.ispras.ru/sedna/ December 2011
20. XEP-0027: Current Jabber OpenPGP Usage, http://xmpp.org/extensions/xep-0027.html
21. OpenPGP Message Format, http://www.rfc-editor.org/info/rfc4880
22. Celesti, A., Fazio, M., Villari, M., Puliafito, A.: SE CLEVER: a secure message oriented middleware for cloud federation. In: IEEE Symposium on Computers and Communications (ISCC 2013). Split, Croatia (2013)

Towards Internet Intelligent Services Based on Cloud Computing and Multi-Agents

Domenico Talia

Abstract Cloud computing systems provide large-scale infrastructures and services for Internet computing applications. They are "elastic" therefore are able to adapt to user and application needs. Clouds are used through a service-oriented interface that implements the *-as-a-service paradigm to offer Cloud services on demand. This chapter discusses Cloud computing models and architectures, their use in distributed applications, and examines analogies, differences and potential synergies between Cloud computing and multi-agent systems. Since the combination of Clouds and agents can benefit from semantics, we also discuss use of ontologies in this context. Our analysis is lead having in mind the goal of implementing Internet-based complex systems and intelligent applications by using of Cloud systems, software agents, and ontologies. In particular, the convergence of interests between multi-agent systems that need reliable distributed infrastructures and Cloud computing systems that need intelligent software with dynamic, flexible, and autonomous behavior can result in new systems and applications.

1 Introduction

European Commission Vice President Neelie Kroes, responsible for the European Digital Agenda, recently presented a vision of what the Internet could look like in the coming years. This vision is based on millions of active (intelligent) things talking to other active things, and a greater degree of security and privacy for people online. "Tomorrow's Internet landscape could look very different", she said. "New smart systems, available on the go. New social media replacing the old. Cloud computing that is scalable, flexible, everywhere. Enormous data sets, used to benefit science,

D. Talia (✉)
ICAR-CNR and University of Calabria, Rende, Italy
e-mail: talia@deis.unical.it

S. Gaglio and G. Lo Re (eds.), *Advances onto the Internet of Things*, Advances in Intelligent Systems and Computing 260, DOI: 10.1007/978-3-319-03992-3_19,

healthcare, our economy, and our everyday lives. And a more secure and a more private Internet".

In this scenario, Cloud computing provide elastic services, high performance and scalable data storage to a large and everyday increasing number of users [1]. Cloud computing enlarged the arena of distributed computing systems by providing advanced Cloud computing enlarged the arena of distributed computing systems by providing advanced Internet services that complement and complete functionalities of distributed computing provided by the Web, Grid computing and peer-to-peer networks. In fact, Cloud computing systems provide large-scale infrastructures for high-performance computing that are dynamically adapt to user and application needs.

Today Clouds are mainly used for handling highly intensive computing workloads and for providing very large data storage facilities. Both these goals are combined with the third goal of potentially reducing management and use costs. At the same time, multi-agent systems (MAS) represent a distributed computing paradigm based on multiple interacting agents that are capable of intelligent behavior. Multi-agent systems are often used to solve complex problems by using a decentralized approach where several agents contribute to the solution by cooperating one each other. One key feature of software agents is the intelligence that can be embodied into them according to some collective artificial intelligence approach that needs cooperation among several agents that can run on a parallel or distributed computer to achieve the needed high performance for solving large complex problems keeping execution time low.

Cloud computing and multi-agent systems can benefit from the use of semantics technologies both in the implementation of basic levels and middleware and in the development of large-scale applications that include software agents running on Cloud platform. Ontologies can be exploited to improve the task of agents in annotating, searching, and classifying resources (things) and facilitating the task of Clouds services in supporting scalable Internet of Things applications.

Although several differences exist between Cloud computing and multi-agent systems, they are two distributed computing models, therefore several common problems can be identified and several benefits can be obtained by the integrated use of Cloud computing systems and multi-agents. The research activities in the area of Cloud computing are mainly focused on the efficient use of the computing infrastructure, service delivery, data storage, scalable virtualization techniques, and energy efficiency. In summary, we can say that in Cloud computing the main focus of research is on the efficient use of the infrastructure at reduced costs. On the contrary, research activities in the area of agents are more focused on the intelligent aspects of agents and on their use for developing complex applications. Here the main problems are related to issues such as complex system simulation, adaptive systems, software-intensive applications, ontology management, distributed computational intelligence, and collective learning.

Despite these differences, Cloud computing and multi-agent systems share several common issues and research topics in both areas have several overlaps that need to be investigated. In particular, Cloud computing can offer a very powerful, reliable, predictable and scalable computing infrastructure for the execution of multi-agent systems implementing complex agent-based applications such when modeling and simulation of complex systems must be provided. On the other side, software agents can be used as basic components for implementing intelligence in Cloud computing systems and, in the more general scenario of Future Internet, making systems and applications more adaptive, flexible, and autonomic. Issues such as resource management, service provisioning, smart behavior can be efficiently supported by MAS in designing and running large-scale Internet applications.

For these reasons and for others that we discuss later, this chapter investigates research work in the two areas and point out potential synergies that deserve to be analyzed. The chapter discusses Cloud computing models and architectures, their use in parallel and distributed applications, and examines analogies, differences and potential synergies between Cloud computing and multi-agent systems. Analysis is led having in mind the goal of implementing high-performance complex systems and intelligent applications by using both Cloud computing systems and software agents. Section 2 introduces Cloud computing concepts and reviews some research activities. Section 3 presents multi-agent systems and research topics that are related to Cloud computing. Section 4 presents some ideas on using intelligent software agents to improve the performance and functionality of Clouds. Section 5 illustrates how Cloud computing platforms can be used for the efficient execution of MAS. As a case study, in this section the use of semantic web techniques and ontologies is discussed. Section 6 concludes the chapter.

2 Cloud Concepts and Models

Since the Cloud computing paradigm has been conceived several definitions have been given. Some of them focus on on-demand dynamic provisioning of processing and storage resources, others emphasize the service-oriented interface and the exploitation of virtualization techniques. The National Institute of Standards and Technology (NIST) have given a complete reference definition [7]. NIST defined Clouds as follows: *"Cloud computing is a pay-per-use model for enabling available, convenient, on-demand network access to a shared pool of configurable computing resources (e.g., networks, servers, storage, applications, services) that can be rapidly provisioned and released with minimal management effort or service provider interaction"*. Moreover, according to NIST: *"Cloud model promotes availability and is comprised of five key characteristics, three delivery models, and four deployment models"*.

The key features of Clouds are: On-demand self-service, ubiquitous network access, location independent resource pooling, rapid elasticity, and pay per use. Figure 1 summarizes the main characteristics of Cloud computing systems both from the technical side and the business side [4].

The delivery models of Clouds are very important because they define three different types of Cloud computing systems:

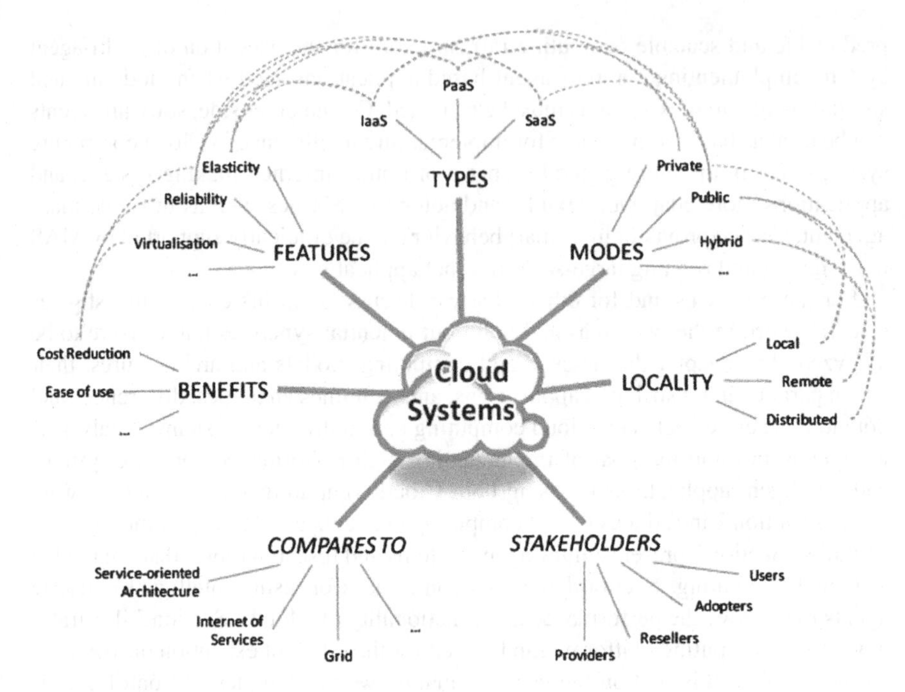

Fig. 1 Main characteristics of Clouds (from [4])

- *Infrastructure as a Service (IaaS).* The capability provided to the user is to rent computing, storage, networks, and other computing resources where the user is able to deploy and run software, which can include operating systems and/or applications. The user does not manage or control the hardware Cloud infrastructure but has control over operating environments, storage, deployed applications, and possibly select networking components. Examples for commercial Cloud infrastructures are Amazon EC2 and Rackspace.
- *Platform as a Service (PaaS).* The functionality provided to the user is to deploy onto the Cloud infrastructure consumer-created applications using programming languages, compilers and toolkits supported by the provider (e.g., Java, .Net). The consumer does not manage or control the underlying cloud infrastructure, network, servers, operating systems, or storage, but the consumer can control the deployed applications and possibly the application hosting environment configurations.
- *Software as a Service (SaaS).* The capability provided to the consumer is to use the provider's applications running on a Cloud infrastructure and accessible from various client devices through a thin client interface such as a Web browser (e.g., web-based email). The consumer does not manage or control the underlying cloud infrastructure, network, servers, operating systems, storage, or even individual application capabilities, with the possible exception of limited user-specific application configuration settings.

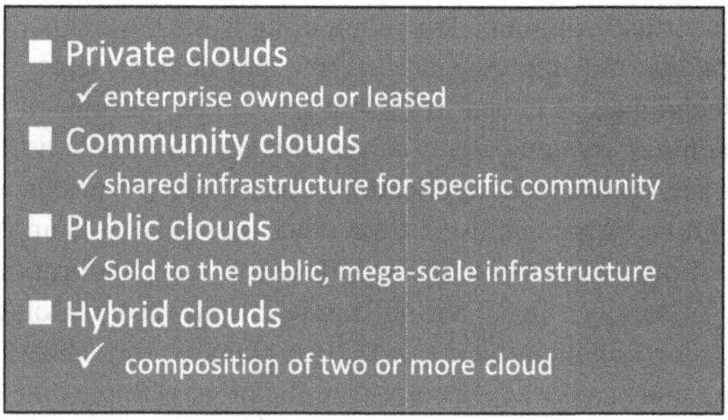

■ Private clouds
✓ enterprise owned or leased
■ Community clouds
✓ shared infrastructure for specific community
■ Public clouds
✓ Sold to the public, mega-scale infrastructure
■ Hybrid clouds
✓ composition of two or more cloud

Fig. 2 Deployment models for Clouds

2.1 Cloud Deployment Models

About five years ago, when the first Cloud infrastructure has been deployed by Amazon, the online bookseller company that took the decision to start a new business selling computing resources to companies and private users, the only deployment model was the *Public Cloud* one. It is a pay-per-use IaaS Cloud infrastructure that is owned by an organization selling Cloud services to the general public or to enterprises. Thus, it is public because it can be rent by anyone for developing and/or running any kind of applications. To use Amazon services, users must provide a credit card account and can spend from few cents to thousands or millions of dollars depending on the number of used resources and the usage time.

After this early Cloud version, other deployment models different from *Public Clouds* have been designed and implemented (see Fig. 2):

- *Private Cloud.* The Cloud infrastructure is owned or leased by a single organization and is operated only for that organization. No public access to it is permitted. This model can be used in case of strict data privacy and/or security requirements.
- *Community Cloud.* The Cloud infrastructure is shared by a limited number of organizations and supports a specific community that has shared concerns (e.g., goals, security requirements, policy, and compliance issues).
- *Hybrid Cloud.* This fourth class of Cloud infrastructure is a composition of two or more Clouds (private, community, or public) that although they are unique entities, are combined together by standardized or proprietary technology that enables data and application portability (e.g., Cloud federation).

Cloud computing is the most recent result of the advancement of several computer technologies both from the hardware side, such as virtualization and multi-core architectures, and from the software side like cluster computing, Grid computing, Web services, service-oriented architectures, autonomic computing, and large-scale data storage. Indeed, today Clouds extend the Internet panorama, by providing a new computing paradigm that facilitates the use of Internet as a worldwide computing platform.

In particular, virtualization in Cloud computing is the key element that separates system functionality and implementation from physical resources. By exploiting virtualization techniques, a Cloud infrastructure can be partitioned in several parallel virtual machines, dynamically configured according to the user requirements and devoted to run independent applications concurrently. Virtualization separates applications from hardware and users from other users giving them the feeling that a large-scale computing infrastructure is devoted to their applications by meeting a given quality of service (QoS). Virtualization is also used to isolate applications avoiding that if one fails that other can fail too. Finally, virtualization is a way to improve security and privacy of concurrent applications running on the same Cloud.

2.2 Example of Cloud Systems

As we mentioned, the Cloud computing paradigm represents an advancement of the existing computing services available over the Internet. In particular, Cloud infrastructures adopted the Web services paradigm for delivering new capabilities beyond the traditional Web capability.

Several companies set up large Cloud facilities and built programming environments where developers can program applications as Cloud software services. Just to mention some example, Amazon on his EC2 and S3 Cloud platforms implemented Elastic BeanStalk, Microsoft implemented .Net technology on Azure, Google provides the AppEngine, and VMware has Cloud Foundry.

On the other side, the research community developed open source software that can be deployed and configured on servers, computer farms or data centers for implementing private, public, community or hybrid Cloud infrastructures or for inter-Cloud computing facilities. Examples of these systems are OpenNebula, Eucaliptus, OpenQRM, Puppet, and OpenStack. These open source software projects are also working to systems and services that allow Cloud-to-Cloud interoperability and federation.

3 Multi-Agent Systems

An agent is a computational entity that acts on behalf of another entity (or entities) to perform a task or achieve a given goal. Agent systems are self-contained software programs embodying domain knowledge and having ability to behave with a specific degree of independence to carry out actions needed to achieve specified goals. They are designed to operate in a dynamically changing environment.

Agents typically include a set of features. The main features of agents include the following:

- *Autonomy:* the capacity to act autonomously to some degree on behalf of users or other programs also by modifying the way in which they achieve their objectives.

- *Pro-activity:* the capacity to pursue their own individual set goals, including by making decisions as result of internal decisions.
- *Re-activity:* the capacity to react to external events and stimuli and consequently adapt their behavior and make decisions to carry out their tasks.
- *Communication and Cooperation:* the capacity to interact and communicate with other agents (in multiple agent systems), to exchange information, receive instructions and give responses and cooperate to fulfill their own goals.
- *Negotiation:* the capability to carry out organized conversations to achieve a degree of cooperation with other agents.
- *Learning:* the ability to improve performance and decision making over time when interacting with the external environment.

Although a single agent can act and run to perform a given task, the agent paradigm was conceived as a distributed computing model where a set of agents interact one another by exchanging information and cooperating to perform complex tasks where interaction, intelligence, adaptation and dynamicity are key issues to be handled.

This means that even if we can define an agent in isolation, the agent paradigm can find its complete exploitation if we consider agents as entities acting in a collection of agents, therefore implementing the so called multi-agent system paradigm. In fact, it is rather difficult to imagine that an agent will exist and operate only as a stand-alone entity and will never interact with other agents (real or artificial) in its environment. Also information agents, or personal agents, which are mainly supposed to work as stand alone entities in solving problems, will certainly improve their behavior and achieved results if cooperate with other agents to receive information, to delegate task execution or to exchange knowledge that improve the agent role and contribution. According to these considerations, the social dimension of agents is one of its essential features.

As stated by Sycara [11], the characteristics of MASs are that

- each agent has incomplete information or capabilities for solving a problem and, thus, has a limited viewpoint on the global task to be done;
- there is no system global control;
- data are decentralized; and
- computation is asynchronous.

These features are typical of decentralized computing paradigms. In fact, being a distributed computing paradigm, multi-agent systems share several characteristics with other distributed paradigms like actors (which we can consider as their progenitors), concurrent objects, peer-to-peer networks, Grid computing, sensor networks, autonomic computing, and Cloud computing. At the same time, it is worth to notice that agents possess some properties, as discussed before, that differentiate them from other distributed computing models.

Commonalities and differences among these distributed computing models can be exploited for the integrated use of some of the technologies that are based on them. For example, decentralized applications based on multi-agent systems can be developed on Grid systems or on peer-to-peer networks. At the same time, applications based on sensor networks can use distributed intelligence techniques by means of a multi-agent system with learning and pro-activity features.

In the past years several agent programming environments supporting specific agent architectures and providing libraries of interaction protocols like Jason, 3APL, JACK, Claim, SyMPA, JADE, Cougaar, Jadex, and ZEUS have been developed. Moreover, software engineering methodologies like Gaia, Tropos, and AUML have been designed to analyze and design agent-based systems. Efforts have been done to standardize some features or facilities of agent systems, such as has been done with FIPA and KQML for inter-agent communication. These environments, toolkits and methodologies are enabling technologies for implementing MAS applications on traditional computing systems. However they can be more interesting if they will be available of distributed computing infrastructures like Grid, Cloud or P2P networks for supporting the development of large-scale MAS applications achieving high performance and scalability.

However, despite the potential common space where agent technology and Cloud computing infrastructures can be effectively used to produce innovative models, techniques, systems and applications, till today only a few research activities that make use of both these technologies are performed. In the literature a very limited number of chapters can be found on agents and Cloud integration [2, 3, 5, 10].

In the next two sections, we discuss two main approaches for the integrated use of agents and Cloud systems. The first one is based on the principle that agent flexibility, intelligence, pro-activity, and autonomy can be used in Cloud computing platforms to produce new advanced Cloud solutions and services that offer new functionalities and intelligent services that todays are not yet available in current Cloud computing infrastructures.

The second one is centered on the idea that Cloud infrastructures can offer an ideal platform where run MAS-based systems, simulations and applications because of its large amount of processing and storage resources that can be dynamically configured to run large-scale MAS-based software at unprecedented scale. In this context, we discuss the use of ontologies in resources search and classification.

4 Clouds Using Agents

Cloud computing is a novel technology that has been designed and implemented in the past five years, mainly due to industry that was looking to a large-scale scalable computing infrastructure for implementing and selling service-oriented commercial solutions.

Whereas much of the current effort on Cloud computing was devoted to the production of Cloud infrastructures and technologies for supporting virtualization and data centers, little attention has been devoted to introduce innovative methods for users and developers to discover, request, assemble and use Cloud computing resources. Autonomous and flexible agents and MASs are suitable tools for negotiating user access, automating the resource and service discovery, and composition, trading, and harnessing of Cloud resources.

A new discipline, called *agent-based Cloud computing* must be set for providing agent-based solutions founded on the design and development of software agents for improving Cloud resources and service management and discovery, SLA negotiation, and service composition. Autonomous agents can make Clouds smarter in the interaction with users and more efficient in allocating processing and storage to applications.

In large-scale data centers, agents can search, filter, query and update the massive volumes of data that are stored. We can envision a scenario where *Cloud agents* working on our and operating systems behalf, to provide intelligent data access services, monitoring services, processor-to-application assignment strategies, and energy-efficient use of Cloud computing infrastructures.

Research activities must be carried out to implement effective agent-based solutions for Cloud computing. This work should be done towards the three different *-as-a-Service delivery classes. In IaaS infrastructures, agents can be used to help the intelligent provisioning of basic resources to user applications, whereas in Paas infrastructures, agent can play a role in the efficient deployment and execution of programming environments that developers use for application implementation. Finally, in SaaS Cloud infrastructures, agents can be programmed to optimize the use of applications provided as services and the management of the underlying hardware/software infrastructure taking care of its efficient utilization and, at the same time, for maintaining the declared QoS.

In Clouds, there also is the need to design and implement techniques and methodologies that adapt to the dynamic behaviors of Cloud computing environments. Autonomic techniques may help providers and users to reach this goal. Multi-agent systems that are able to handle with changing configurations, heterogeneity, and volatility, can provide a promising approach for addressing this requirement. Last but not least, security and trust are two very critical issues in Cloud computing as data and software are stored, accessed and run on machines that are not owned or directly managed by owners of data and software. Agent-based models and algorithms for trust and security in Cloud infrastructures could be very useful.

In summary, if agent-based solutions will be introduced in the software infrastructure of Clouds we will have:

- Intelligent and flexible Cloud services,
- Autonomous and pro-active services,
- Autonomic Clouds.

5 Agents Using Clouds

Complex agent-based applications or large-scale simulations based on MASs often require high-performance computing systems and large data storage devices. Therefore, Cloud infrastructures can offer an ideal platform where to run MAS-based systems simulations and applications because of its large amount of processing and memory resources that can be dynamically configured for executing large agent-based software at unprecedented scale.

Agent-based applications can rely on Cloud computing infrastructures to access and use vast amounts of processors and data. So this approach would allow offloading the compute-intensive agents to the appropriate subsets of processes and storage elements in a Cloud. The entire MAS application can run on a Cloud infrastructure or only the most compute-intensive part of it can be hosted in the Cloud, whereas the light part can run on a local server or simply on the client PC. In this way agents can become more efficient and, at the same time, lighter and smarter. This can be obtained because, by using powerful Cloud facilities, agents can improve their intelligence and accurateness by running more sophisticated algorithms. In fact, the amount of storage and processing power of a Cloud-enabled MAS is larger than in other computing environments, making it more powerful.

Cloud-enabled agents can couple agents and large-scale dynamic distributed computing platforms bringing big new opportunities to the agent computing area and expanding agent's knowledge beyond the possibilities offered by traditional computing platforms.

Virtualization mechanisms offered by Cloud computing can be exploited for efficient composition of parallel machines where to execute large scale concurrent agents with real-time constraints or needing high performance for achieving results in reasonable time.

Agents implemented in Cloud systems can adapt to available virtual machines by using the basic properties of agents such as autonomy, pro-activity, negotiation and learning. Since Clouds are elastic, they can expand and shrink based on demand of users or applications. This property is very useful for the scalable execution of MAS applications and simulation that are able to adapt to the available resources.

In summary, agents can find in Cloud computing infrastructures the appropriate platform where to run and access large data. This opportunity must be exploited for implementing efficient MASs and, from a more general point of view, for advancing the way to design and implement a new generation of large-scale software agents.

5.1 Semantic Search on Clouds

Crucial to the implementation of complex middleware and applications on Cloud infrastructures is the automation of the management tasks, which can be implemented as a set of software agents, which support the different phases of a resource or a service life cycle, including publication and discovery. Existing solutions to distributed resources and service (in other words, Internet things) discovery mostly rely on centralized architectures and perform only syntactic matching between service requests and descriptions [6].

As resource and service discovery in distributed infrastructures is about finding relevant services, the overall quality of a discovery service is determined not only by usual QoS measures such as performance, reliability and availability, but also by its precision, which measures how many of the discovered objects are relevant, and how relevant they are. Accurate service discovery should be able to find best approximate matches usable for the service requester. Service discovery supported by Cloud infrastructures has to deal with a large number of services described using different approaches and languages, and managed by distinct service providers. In such heterogeneous environments, syntactic keyword and taxonomy-based matching can be insufficient to achieve high-precision service discovery [8]. Furthermore, centralized architectures go against the view of Cloud infrastructures that are inherently decentralized and are unlikely to go through the growing up rate of incoming requests. Thus, solving the problems of both service publishing/discovering and centralization is of great importance to enable effective information delivery in massive-scale Cloud and distributed infrastructures.

A novel approach can be based on designing a highly-decentralized framework providing semantic-based service publishing/discovery for automatic information/service delivery in massive-scale Cloud infrastructures. This approach can exploit Semantic Web technologies to provide Cloud services with meaningful and machine-understandable descriptions that enable to semantically characterize service definitions in order to improve the precision of discovery. From an architectural point of view, a fully decentralized approach based on multi-agent systems could be followed as they are more autonomous, reliable and scalable in a distributed setting more than in centralized and hierarchical architectures. Furthermore, as compared to centralized approaches, this approach fits better in solving the discovery problem as in Cloud and distributed infrastructures it is more realistic to think about a set of autonomous service providers (i.e., agents), rather than a centralized service repository.

Just to show how such an approach can be implemented, a realistic hypothesis to implement semantic discovery in large computing infrastructures is to combine structured networks (i.e., DHTs) with semantic overlay networks (SONs) [9]. The principle behind SONs is that node connections are influenced by content, so that nodes having a given resource are connected to nodes with similar resources (that is belonging to the same class). Thus, "semantically" related nodes form a SON. Queries are routed to the appropriate SONs, increasing the chances that matching resources will be found quickly, and reducing the search load on nodes that have unrelated content. DHTs and SONs can benefit from each other in the sense that a SON can be constructed by exploiting the DHT mechanisms and hence the former can help to light the way to the semantics-free content publishing and retrieval approach of the latter. Since distributed infrastructures have to deal with a high level of heterogeneity, this hypothesis needs to take into account the semantic interoperability problem. To address the semantic interoperability issue techniques based on semantic mapping or matching in decentralized systems can be exploited.

A practical implementation of this solution includes the use of ontologies and SONs combined with a distributed software layer implemented by a large number of distributed software agents runningmulti-Cloud infrastructures that realize the distributed backbone for a massive number of Internet intelligent things that fill find in the semantic- and agent-enhanced Cloud infrastructure the needed support for connection, communication, interaction and cooperation.

6 Final Remarks

The Internet of Things is as a dynamic global network infrastructure connecting intelligent "things" with self-configuring capabilities based on standard and interoperable communication protocols. In this future scenario, physical and virtual "things" have identities, physical attributes, and virtual personalities and use intelligent interfaces, and are seamlessly integrated into the information network. To implement this infrastructure several technologies will be necessary. Here we presented a few of them.

We discussed in the chapter how Clouds and multi-agent systems can be used for implementing new intelligent Internet "things", services and applications. Scientific areas and issues involved in carrying out research work that will produce intelligent Cloud services and high-performance multi-agent systems on Clouds are mentioned and sketched [12]. We also discussed, by an example, the use of ontologies in this context for resource search and classification.

Intelligent objects ("things") can benefit from a virtual representation of them implemented as a software agent. Such virtual "things" can embed every physical characteristic of the object, but also improve it with intelligence, adaptive communication, ambient awareness, negotiation capabilities, and pro-activeness. When objects are many, run complex algorithms, and need efficiency, they can find in Cloud platforms the opportune storing and computing infrastructure.

The convergence of interests between multi-agent systems that need reliable distributed infrastructures and Cloud computing systems that need intelligent software with dynamic, flexible, and autonomous behavior will result in new Internet systems, services, and applications. Both research communities must be aware of this opportunity and should put in place the joint research activities needed to reach that goal. In particular, the Internet of Things paradigm can benefit from those scientific areas that can provide scalable hardware and software platforms united to intelligent, mobile and semantic-aware software agents.

References

1. Armbrust, M., Fox, A., Griffith, R., Joseph, A.D., Katz, R., Konwinski, A., Lee, G., Patterson, D., Rabkin, A., Stoica, I., et al.: A view of cloud computing. Commun. ACM **53**(4), 50–58 (2010)
2. Aversa, R., Di Martino, B., Rak, M., Venticinque, S.: Cloud agency: a mobile agent based cloud system. In: 2010 International Conference on Complex, Intelligent and Software Intensive Systems (CISIS), pp. 132–137. IEEE (2010)

3. Cao, B.Q., Li, B., Xia, Q.M.: A service-oriented qos-assured and multi-agent cloud computing architecture. In: Cloud Computing, pp. 644–649. Springer, Heidelberg (2009)
4. Group, C.C.E.: The future of cloud computing. Tech. rep, European Commission (2010)
5. Lopez-Rodriguez, I., Hernandez-Tejera, M.: Software agents as cloud computing services. In: Advances on Practical Applications of Agents and Multiagent Systems, pp. 271–276. Springer (2011)
6. Mastroianni, C., Talia, D., Verta, O.: Designing an information system for grids: Comparing hierarchical, decentralized p2p and super-peer models. Parallel Comput. **34**(10), 593–611 (2008)
7. National Institute of Standards and Technology (NIST): Cloud computing (2011). www.nist.gov/itl/cloud/. Website consulted on September 12, 2013
8. Pirró, G., Talia, D.: Ufome: An ontology mapping system with strategy prediction capabilities. Data Knowl. Eng. **69**(5), 444–471 (2010)
9. Pirrò, G., Talia, D., Trunfio, P.: A dht-based semantic overlay network for service discovery. Future Gener. Comput. Syst. **28**(4), 689–707 (2012)
10. Sim, K.M.: Towards complex negotiation for cloud economy. In: Advances in Grid and Pervasive Computing, pp. 395–406. Springer (2010)
11. Sycara, K.P.: Multiagent systems. AI magazine **19**(2), 79 (1998)
12. Talia, D.: Clouds meet agents: Toward intelligent cloud services. Internet Computing, IEEE **16**(2), 78–81 (2012)

Chatbots as Interface to Ontologies

Agnese Augello, Giovanni Pilato, Giorgio Vassallo and Salvatore Gaglio

Abstract Chatbots are simple conversational agents using "pattern matching rules" to carry out the dialogue with the user and various expedients to improve their credibility. However, the rules on which they are based on are too restrictive and their language understanding capability is very limited. Nevertheless chatbots are widespread in several applications, especially to provide information to users in a new and enjoyable way. In this chapter we describe different chatbot architectures, exploiting the use of ontologies in order to create clever information suppliers overcoming the main limits of chatbots: the knowledge base building and the rigidness of the dialogue mechanism.

1 Introduction

Human computer interaction is a challenging research area having the aim of building more and more usable human-computer interfaces. The use of natural language as an interface allows a fulfilling interaction and a greater accessibility by expert and inexpert users to the system. For these reasons in last years there has been a growing interest toward the use of conversational agents. Research on this kind of systems involves natural language understanding and the analysis and management

A. Augello (✉) · G. Pilato
Viale delle Scienze, ICAR-CNR, ed. 11, Palermo, Italy
e-mail: augello@pa.icar.cnr.it

G. Pilato
e-mail: pilato@pa.icar.cnr.it

G. Vassallo · S. Gaglio
Viale delle Scienze, DICGIM-UNIPA, ed. 6, Palermo, Italy
e-mail: giorgio.vassallo@unipa.it

S. Gaglio
e-mail: salvatore.gaglio@unipa.it

S. Gaglio and G. Lo Re (eds.), *Advances onto the Internet of Things*,
Advances in Intelligent Systems and Computing 260, DOI: 10.1007/978-3-319-03992-3_20,

of conversational practices. However, natural language is characterized by many ambiguities that adult human beings can resolve through their own cultural baggage built with everyday experiences. These difficulties led to the development of simple dialogue systems, called chatbot as an alternative to advanced dialogue systems. The main limits of chatbot technology regard their knowledge representation as well as their information retrieval and dialogue capabilities. Even if the simple technology allows to easily implement a dialogue system, the obtained conversation is limited by pattern matching rules on which chatbots are based. The traditional chatbot dialogue capability is too rigid, it can answer to the user only if there is a pattern which matches the question in its knowledge base. Chatbot are lacking in the intuitive capability of human beings to see meaning, relationships and possibilities beyond the reach of senses. Besides, the chatbot knowledge base is expensive and boring to create and, as highlighted in [1], all possible user questions have to be considered at design time. In our opinion chatbots can be improved by: (a) simplifying the knowledge base design process, extending the chatbot knowledge base with other information repositories, generalizing as much as possible the chatbot pattern-template modules (as an example it could be possible to write generic question answers modules, which the chatbot can adapt and properly complete exploiting other information repositories to answer to a specific user question), using unsupervised methodologies to train automatically the chatbot knowledge base exploting information available in big textual corpora, (b) enhancing and making more clever their answering mechanism providing chatbots of inferential capabilities about the conversation topics and semantic analysis capabilities in order to handle the different ways of user expression analyzing similarity relations among words.

In last years we have proposed hybrid chatbot architectures combining symbolic and sub-symbolic methodologies for knowledge representation and reasoning, simulating rational and intuitive capabilities of human brain involved in dialogue understanding. In particular the rational component of the chatbot brain is implemented by means of ontologies. In the next sections, after a brief discussion on the state of the art, the main modules of the proposed architectures are described and some related applications will be discussed.

2 State of the Art

Chatbots are conversational agents using "pattern matching rules" to carry out a simple conversation by means of the detection of rules and keywords into the user sentences. During the conversation the chatbot looks for a lexical matching between the user query and a set of question-answer modules stored in its knowledge base. The answers are given using a set of predefined responses. They also use various techniques to improve their credibility, for example they can store user information and preferences, keep trace about the history and the current topic conversation.

A.L.I.C.E. (Artificial Linguistic Internet Computer Entity) is a chatbot developed in the Lehigh University (Pennsylvania) and distributed with GNU license.

The knowledge of Alice chatbot is composed of question-answer modules, called categories and described by Aiml (Artificial Intelligence Mark-up Language) language [2]. Aiml is a *mark-up* language in which specific tags are defined to properly interpret the meaningful elements of the sentence written by the user. In particular the tag *aiml* encloses the categories (question-answers modules) belonging to the Aiml file, each category is described by the tag *category*, composed of a tag *pattern* enclosing a sentence that will be compared to the user question, and a *template* tag which encloses the rules to generate the chatbot answer. Other optional tags are also present. The presence of the special symbols "_" and "*", called "wildcards" in the pattern allows the chat-bot to answer also when in its knowledge base there is not a pattern exactly equal to the user question, the value associated to a wildcard can be read using the *star* tag. The template can contain other Aiml tags; in this case the answer is dynamically composed by properly analyzing them.

Special functions are carried out by the *topic*, *that* and *srai* tags. The first tag can be used before a group of categories to set its subject, or can be used inside the template to set or get the conversation topic. The second tag is located between the pattern and template tags and allows the chatbot to remember its last answer and keep trace of the conversation. The *srai* tag allows to recursively call other categories, activating again the pattern matching algorithm which manages the dialogue with the user. Different systems are aimed to the improving of the chatbots technology. In [3] the knowledge base of a community of chatbots, each one expert in a specific topic is coded in a semantic vector space allowing them to estimate their own competence about questions asked by the user. In the CyN project [4] the pattern matching interpreter of Alice, called Program N is linked to OpenCyc [5], the Open source version of Cyc, the largest common sense knowledge base available today. This allows the chatbot to exploit the great quantity of knowledge available in Cyc and to generalize the Aiml categories using OpenCyc information. Reasoning and inferential mechanisms are also included in chatbot systems proposed in [6–8]. In [9] it was proposed the creation of a lexicon extended with thesaurus for simulate a sort of semantic analysis. The implemented chatbot prototype can learn many new lexemes and syntagma structures during the interaction with the user. A deeper attention to conversational features is given into iAiml system [10], where rules to treat intentional information have been added to Aiml KB, exploiting the Conversational Analysis Theory (CAT) as linguistic base for consider intentionality in adjacent pairs in dialogue.

3 Ontology Based Architectures for Conversational Information Suppliers

In this section we discuss some architectures we have introduced in order to the enhance both knowledge representation and dialogue capabilities of chatbots. In particular during our research activities we have integrated symbolic and sub-symbolic knowledge representation approaches to provide chatbots of both rational and

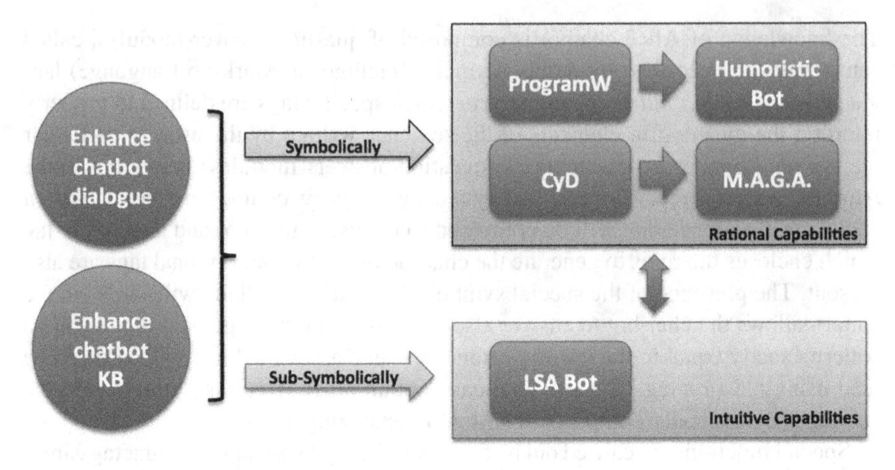

Fig. 1 Systems implemented to enhance chatbot dialogue and KB

intuitive dialogue capabilities. Three main systems are described: the first two implementing the rational and the intuitive components of the chatbot KB respectively, and the last the hybrid rational-intuitive KB architecture.

3.1 Integration of ProgramD and ResearchCyc: Cyd

The idea of extending the chatbot knowledge base with the information stored in a wide common sense repository such as Cyc, has been developed at the Daxtron Laboratories with the implementation of the CyN project [4]. The aim of the project was to provide a natural language interface to the Cyc ontology, making its content available to a wide number of users. in particular CyN is the integration of the Aiml interpreter written in C++ Program N to OpenCyc.

The project is open and easily extensible; therefore we have considered adapting it to the Aiml interpreter written in Java Program D. This choice allows for the exploiting of several advantages deriving from the use of Java language and using the Java API available to access the Cyc knowledge base. At present the technology, called CyD (Cyc + ProgramD) for consistency with CyN, integrates the version of Cyc available for research purpose ResearchCyc with ProgramD. The aim of CyD is to enhance the chatbot KB with the information available in ResearchCyc and provide these conversational systems with common sense reasoning and inferential capabilities. To this end the Aiml language was enriched with new tags to allow the chatbot to query the Cyc ontology directly from its KB rules. The chatbot template can contain these new "ad hoc" AIML tags which transform it into a "meta-answer" that must be processed by the OpenCyc inference engine to produce the chatbot answer most appropriate to the user question. The main tags for communication with Cyc are illustrated in what follows [4]:

The Cyc responses are embedded in a natural language sentence according to the rules of the template. As an example the following category that allows the chatbot to verify if a concept belongs to a specific collection.

```
<category>
       <pattern>IS * A *</pattern>
       <template>
          <cycsystem>
             (cyc-query '
              (#$genls
                 <cycterm><star index="1"/></cycterm>
                 <cycterm><star index="2"/></cycterm>
               )#$EverythingPSC)
           </cycsystem>!!
         </template>
</category>
```

A possible dialogue could be:

User: *Is Palermo a City?*
Chatbot: *Yes!!*

The integration with Cyc has several advantages among which being the benefit of making categories writing easier; it is possible to think of different ways to express general topics creating generic, default categories. The string matching the wildcard symbol belonging to the category pattern will be searched into the Cyc KB; if a corresponding Cyc constant is found, the chatbot can analyze all the information associated to it exploiting the different Cyc predicates.

For example, it can extract a definition of the constant by means of the *#$comment* predicate, verify if the constant is a collection or a collection instance and analyze the other related concepts by means of the *#$isa* or *#$genls* predicates. Or in specific contexts it can analyze "ad hoc" predicates.

Is possible to arrange all these information together, creating dynamically exhaustive answers, not present in the traditional AIML knowledge base. The possibility of exploiting a wide, common-sense ontology, like Cyc, makes the dialogue more fluent and reduces the number of default answers needed in traditional chatbots to fill up their " cultural" gaps.

Additionally, the possibility of interacting with the Cyc knowledge base by means of natural language allows inexpert users to create or extend ontologies with new concepts, facts and relations, without having to learn CycL statements.

3.1.1 A Cyd Application to User Mobile Conversational Assistants: MAGA

The CyD technology has been used into a more complex system to implement multimodal natural language user assistants. This project advanced with an analysis of

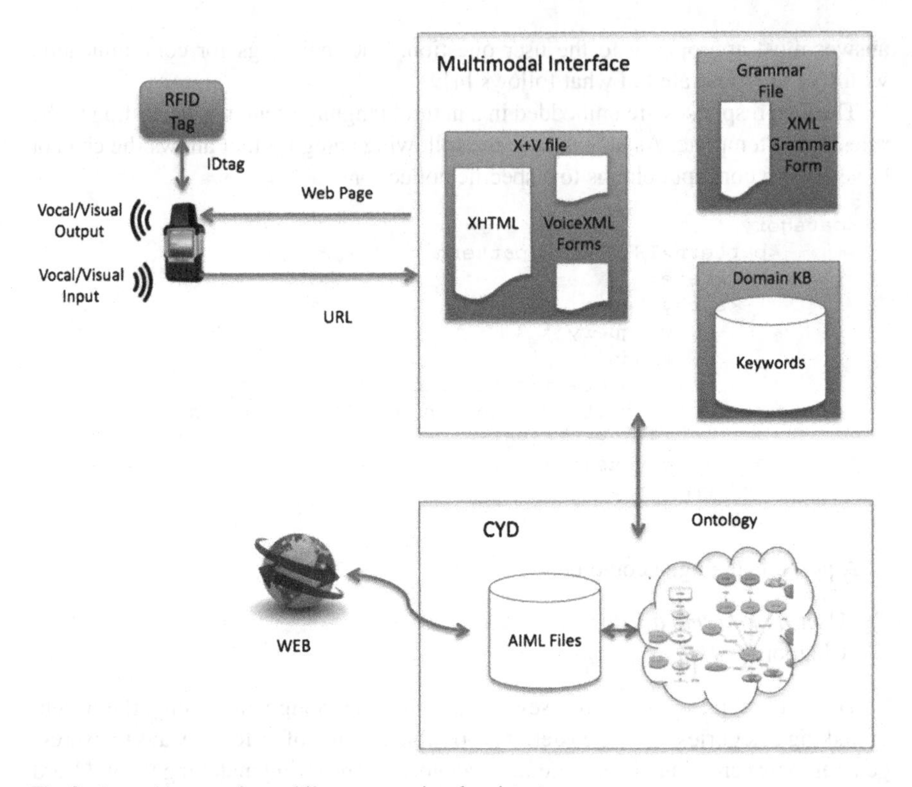

Fig. 2 An architecture for mobile, conversational assistants

the potentialities offered by ubiquitous computing technology and by personal mobile devices, equipped with third-generation wireless communication technologies, which can be exploited with the aim of providing people with useful information in relation to the environment. The aim is to offer an assistance service to users for information retrieval, accessible from mobile devices like PDAs and smartphones, in order to satisfy mobility needs of the user. We applied this architecture to the cultural heritage field by proposing a multimodal guide created to provide information during cultural heritage sites tours, and named it M.A.G.A. (Mobile Archaeological Guide at Agrigento) [11].

An overview of the M.A.G.A. architecture, based on a client-server paradigm, is shown in Fig. 2.

The main feature of the system is the integration of different technologies. In fact, the application is accessible using a PDA equipped with an RFID-based, autolocalization module, while the information retrieval service is provided by means of a spoken, natural language interaction with a conversational agent, integrated with reasoning capabilities based on CyD technology.

In particular, the RFID module allows the system to sense changes in user's environment, and to automatically adapt itself. The chatbot is provided with inferential capabilities thanks to the use of CyD technology. It is possible define an "ad hoc" microtheory for the specific context, properly defining the collections of concepts and facts regarding the analyzed domain. It is possible to exploit the information already present in Cyc, by hooking up the created microtheory to some of the existing ones.

The dialogue between the chatbot and the user is verbally carried on,by using a spoken, natural language in addition to traditional visual and keyboard (or stylus)-based commands. To this aim we used XHTML + Voice [12], X + V for short, while the multimodal interface has been developed with the IBM Multimodal Tools 4.1.2.2 [13] for WebSphere Studio V5.1.2. The speech recognition process is carried out through an "ad-hoc" built-in speech grammar, including a set of rules which specifies utterances that a user may say. The system is accessible through a web page in a multimodal browser from the handheld device. The interaction occurs through the loading of X + V pages, which can be triggered by user vocal and visual command or RFID detection; in every page the user can have a vocal dialogue with the chatbot. The chatbot searches for the best match rule in its knowledge base. The AIML rule can directly produce an answer, or it might be necessary to query Cyc in order to construct a more suitable answer to the user's request. The chatbot answer could also be the result of a query to standard search engines which will search for local or remote documents related to the user query. The interaction between the application running on the PDA and the system is also started by the detection of a RFID tag, which is used to estimate the PDA position within the environment. According to this feature, people can go on asking questions about the current object to the chatbot with vocal queries, or they can discard the information and continue their tour.

The system is easily adaptable to application domain changes: the grammar is easy-fitting and can be improved with minimal effort, the choice of using Cyc allows the system developers to exploit the large amount of data already organized and described in this ontology. This makes the system more adaptable to domain changes, as it is not necessary to write the entire set of knowledge every time, but only the most specific. For this reason the system was applied for other purposes: to provide services in a university campus [14], to assist users in shopping activities [15].

3.2 Integration of a ChatBot and WordNet: ProgramW

The chatbot knowledge base has been also extended with the information available in the WordNet lexical database. The idea has been made concrete with the setting up of an open source project, called ProgramW, available at the sourceforge web site [16]. Program W is an AIML interpreter written in Java, which extends Program D technology in order to enable the interaction between the chatbot and WordNet. In particular we created new Aiml tags which allow chatbots to query WordNet directly from the rules belonging to its knowledge base. Chatbots can exploit information about lexical terms, for example their lemmas or the corresponding glosses (*wordnetlemma* and *wordnetgloss* tags) and evaluate existing relations between words defined in

Wordnet such as synonymy (*wordnetsynset* tag), antonymy (*wordnetantonym* tag) hyponymy-hypernymy (*wordnethyponym* and *wordnethyperny* tags), meronymy-holonymy (*wordnetmeronym* and *wordnetholonym* tags). It is also possible to find the relationship joining two words by means of *wordnetrelation* tag.

This integration can be used to improve the language analysis capability of the chatbot agent.

3.2.1 An Application of ProgramW: EHeBby, the Humoristic Bot

In recent years there was an interest in enhancing the realness of interaction with the conversational agents, providing these systems with the capability of change their behaviour according to the conversation content like in conversation between human beings. An important feature in social human interactions is represented by the capability of generating and understanding humour, therefore it is auspicious to reproduce this ability also in conversational agents [17]. Therefore a humoristic chatbot was proposed; it is capable of generating humorous expressions, proposing riddles to the user, telling jokes and ironically answering the user. Besides this, the chatbot is capable of detecting, during the conversation with the user, the presence of humorous expressions, listening and judging jokes, and reacting by changing the visual expression of the avatar, according to the perceived level of humour. The generation of humorous text is well suited for conversational agents. As a matter of fact, it is possible to define inside the chatbot knowledge base, composed of question-answer modules, the funniest answers most fitting to the user query. We focused our attention to the recognition of humour rather than to the generation.

As a consequence we have analysed the literature techniques in computational humour aimed to the recognition of humour in very short sentences ("one-liners") for the design of a humorous conversational agent. A humour recognition methodology has been implemented in a chatbot, through the research, inside the sentences introduced by the user, of the main features that characterize the text as humoristic, in particular the features which can be computationally detected, such as alliteration, antinomy and adult slang, according to what was suggested by the authors of [18] The core of the system consists of the chatbot knowledge base, composed of three kinds of AIML categories:

- the set of standard Alice categories, which allow the chatbot to hold a general conversation with the user;
- a set of categories aimed at the humorous sentences generation. These categories allow the chatbot to answer the user in a humoristic way, by means of jokes.
- a set of categories which allow the chatbot to recognize an humoristic intent in the user sentences. This feature is obtained by connecting the chatbot knowledge base to external resources, such as the lexical dictionary WordNet and the CMU pronouncing dictionary [19], in order to detect the presence of humorous linguistic features in the sentence.

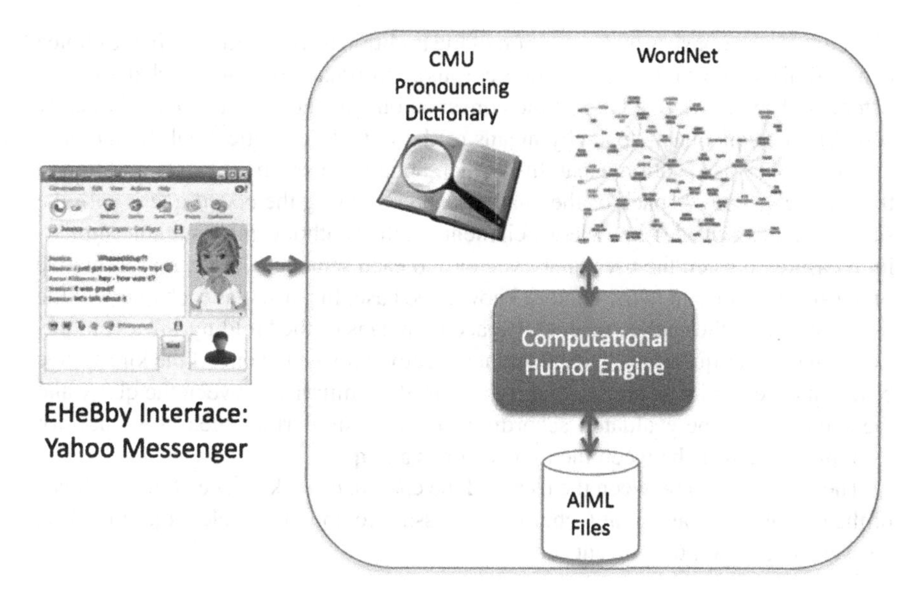

Fig. 3 Humoristic Bot architecture

The humour recognition has been implemented by means of the recognition, into the user sentences, of the peculiar humoristic texts features selected by Mihalcea and Strapparava [18].

3.3 Hybrid Architectures for Chatbots

3.3.1 Introduction of an "Intuitive" Component in the Chatbot's Brain: LSAbot

The aim of this system is to add some sort of intuitive reasoning ability to chatbots, attempting to overcome the rigid pattern matching rules, proposing a sort of "semantic" matching". We believe that this intuitive-associative capability can be obtained using the LSA (Latent Semantic Analysis) methodology [20]. The representation of information in a LSA based semantic, "conceptual" space and a consequent sub-symbolic geometric representation of the chatbot knowledge space, can help to better design a human-like conversational interface provided with intuitive, associative capabilities.

The proposed approach involves the following steps:

- Chatbot creation and training;
- Generation of a semantic space where to code the chatbot competences;
- Choice of the dialogue management technique necessary to obtain the selection of the best answer to deal with the current subject.

The first step consists in the creation of the traditional AIML KB and in the choice of a natural language micro-documents corpus appropriate to extend the chatbot competences. The knowledge base of the conversational agent is then sub-symbolically coded in the conceptual space by means of the Latent Semantic Analysis methodology [20] The association capability is then obtained mapping the user question in the same conceptual space of the chatbot and comparing the coded query with the sub-symbolically coded knowledge elements of the chatbot using a suitable similarity measure between the user query vector and each sentence vector, representative of the answers present in the chatbot knowledge base. In particular each query of the user is coded in the same conceptual space by means of the "folding-in" technique. Let q be the user query and $\mathbf{q}$ its associated vector, let s be the one of the knowledge base sentences and $\mathbf{s}$ its corresponding vector; the similarity between the query and the sentence can be evaluated according to vector similarities measures, like, for example, the cosine between the two vectors $\mathbf{s}$ and $\mathbf{q}$.

The conversation between the user and the chatbot can take place after the choice of the dialogue management technique, necessary to obtain the selection of the best answer to deal with the current subject.

4 Integration of CyD and LSAbot: A Rational/Intuitive Chatbot Architecture

In this architecture common sense and intuitive reasoning capabilities have been integrated in a chatbot. The main idea is to allow the chatbot to exploit information stored in both a common sense ontology and in a semantic space, but the peculiar feature relates the interconnection of these two different knowledge bases. In fact we propose the induction of a sub-symbolic layer in an ontology by means of a mapping of the concepts into a semantic space. Given a specific Cyc microtheory its mapping is performed building a LSA semantic space where the corpus training is composed of the comments associated to the ontology concepts and other micro-documents, each associated to a single concept. To each concept are associated one or more documents, which are represented by vectors into the space. Therefore each concept is projected in the space, and a layer of sub-symbolic, semantic relationships between concepts, given by their mutual geometric distance in the space, is automatically created. In this way the ontology concepts can be related to each other by means of their sub-symbolic relationships as well as by means of the relationships defined into the ontology. The conversational agents can exploit this sub-symbolic layer to retrieve semantic relations between ontological concepts already stored in the KB which are not easily reachable by means of the traditional ontology rules but that are more easily reachable through associative sub-symbolic paths.

Therefore the chatbot brain is comprised of two different but interconnected areas as shown in Fig. 4.

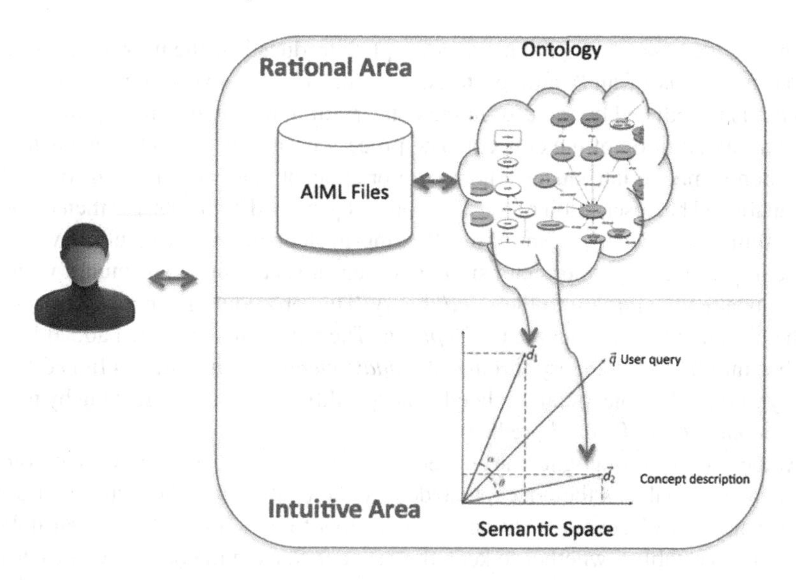

Fig. 4 Rational/Intuitive chatbot architecture

The former is a rational area, made of the Cyc ontology and the standard AIML knowledge base of the chatbot. The latter is an intuitive area, made of a semantic space in which both Cyc concepts, AIML categories and user queries are mapped.

Each ontology concept is encoded as a point in the multi-dimensional semantic space using its Cyc definition and its related documents and is, as a result, identified by a set of vectors. The geometric similarity measure, establishes a correlation between the vectors and therefore a semantic sub-symbolic link among the concepts.

In particular given a concept c_k, and its associated vector $\mathbf{c}_k$, the set CR of concepts sub-symbolically conceptually related to the concept c_k can be obtained determining the n concepts associated to the set of vectors whose similarity measure with c_k is higher than an experimentally fixed threshold T:

$$CR = \{\mathbf{c}_i, \;\; with \;\; i = 1, \dots, n \;\mid\; sim(\mathbf{c}_k, \mathbf{c}_i) \geq T\} \tag{1}$$

where $sim(\mathbf{c}_k, \mathbf{c}_i)$ is a properly defined geometric distance, related to the cosine between the vectors $\mathbf{c}_k$ and $\mathbf{c}_i$.

The chatbot can exploit the semantic layer through new specific Aiml tags introduced for this interaction. In particular, the *relatedConcept* tag allows the chatbot to retrieve the concepts conceptually related to a specific ontology concept, while *sentenceConcept* tag allows the chatbot to retrieve the concepts related to a sentence. In summary, the chatbot can dialogue with the user, exploiting its standard knowledge base or properly query the ontology by means of "ad-hoc" defined Aiml tags, but it can also exploit the intuitive, sub-symbolic layer through other Aiml tags, created for trigger an intuitive behaviour of the chatbot. Furthermore, the chatbot can improve its knowledge base adding unknown concepts introduced by the user in the conversation, by means of an "ontology targeting" mechanism. Every time an unknown concept is introduced by the user in the conversation, the chatbot invites the user to give him a definition of the concept. The process is similar to what happens in real life when someone introduces a new term or concept, and we ask them for further explanation. The description of the new concept provided by the user is then mapped in the semantic space by means of the "folding-in" technique. The resulting vector is then compared through the vector similarity measure according to formula 1 with the concepts already stored into the microtheory. This procedure is obtained by means of the aforementioned *sentenceConcept* tag. The new concept is then added by the chatbot into the microtheory through the *addConcept* tag and then is linked to the sub-symbolically conceptually related concepts through the *cyc-assert* tag by means of a *#$conceptuallyRelated* predicate.

At the end of the dialogue, the conversation files can be analyzed by the ontology expert which analyzes the concepts added by the chatbot together with their associated values of closeness with the Cyc concepts to which they have been linked. He then can establish whether to keep the new concepts in the ontology or to delete it. The proposed architecture was evaluated in [21] mapping the concepts of two different microtheories of Cyc in a semantic space, a smaller one in order to validate the proposed technique on the entire set of microtheory concepts and a greater one in order to carry out a test on one practical scene. The chosen application domains are the american academic structure, described by *AcademicOrganizationMt* microtheory, and the *BiologyMt*. The following dialogue is a small example of a possible interaction with the chatbot.

User: *I need information about departments related to the study of plants.*
Chatbot: *Do you want information about the BiologyDepartment?*
User: *Can you indicate me other related Departments?*
Chatbot: *I have found a relation with the Agriculture Department.*
User: *Can you give me information about it?*
Chatbot: *It is a specialization of AcademicDepartment, instances of which conduct teaching and research in Agriculture.*
User: *Can you give me information about some Private University?*
Chatbot: *Sorry, I don't have any information about it. Can you help me in finding a definition of Private University?*
User: *The definition is a university that is run without the control of any government entity.*
Chatbot: *Ok, I can add this information In my KB. It is right to relate this concept to the University concept?*
User: *Yes, it is.*

The following category allows the chatbot to understand the user information requests about a generic argument matched in the pattern by the wildcard *. Its value is recovered in the template through the *star* tag ,stored in the variable userSentence and hidden to the user by means of the *think* tag.

```
<category>
   <pattern> I NEED INFORMATION ABOUT * </pattern>
   <template>
         <think>
           <set name='userSentence'></star>
          </set>
         </think>
         <srai>RELATED CONCEPT</srai>
   </template>
</category>
```

The following category is recursively called by the previous one, by means of the *srai* tag. In its template the Cyc concept most related to the user request is searched through the *sentenceConcept* tag. If a concept is detected, it is stored in the *cycConcept* variable, and the chatbot asks to the user if the concept detected is that for which he was querying.

```
<category>
    <pattern> RELATED CONCEPT </pattern>
    <template> Do you want information  about
                    <set name='cycConcept'>
                       </sentenceConcept >
                     </set>?
    </template>
</category>
```

If the user asks for an information related to an unknown concept, the chatbot asks him for a definition.

```
<category>
    <pattern>
      CAN YOU GIVE ME INFORMATION   ABOUT *
     </pattern>
      <template>
    <think><set name='cycConcept'>
              <cycterm></star></cycterm>
           </set></think>
      <condition>
        <li name=' cycConcept' value='NULL'>
        Sorry, I don't have any information about it.
        Can you help me in finding a    definition of </star>?
        </li>
        <li>... </li>
      </condition>
      </template>
</category>
```

The user gives to the chatbot the concept definition, and the chatbot searches for conceptually related concepts to which the new one can be linked.

```
<category>
    <pattern> THE DEFITION IS * </pattern>
    <template>
```

```
                <think> <set name='userSentence'></star></set>
                </think>
                Ok, I can add this information in my KB.
                It is right to relate this concept to the
                <set    name='conceptuallyRelated'>
                    </  sentenceConcept >
                </set> concept?
            </template>
        </category>
```

5 Conclusion

In this chapter we summarizes several works proposed to design human like, conversational agents overcoming the limitations of traditional rule-based chatbots like Alice. The first step was the formalization of a rational component in the chatbot KB in order to provide chatbots of reasoning and natural language processing capabilities. Afterwards, an intuitive component has been added to the chatbot KB, exploiting the associative properties deriving from natural language elements representation in a semantic space model. Finally the two components have been integrated together with the aim of simulating the main areas of the human brain. The proposed approaches have been implemented and tested in several applications on specific domains. Future works will regard the enhancement of the proposed framework introducing a dynamic analysis of conversation in the semantic, "conceptual" space.

Acknowledgments This work has been partially supported by the PO FESR 2007/2013 grant G63F12000240004 funding the OnSicily project.

References

1. L'Abbate, M., Thiel, U., Kamps, T.: Can proactive behavior turn chatterbots into conversational agents?. In: Proceedings of the IEEE/WIC/ACM International Conference on Intelligent Agent Technology (IAT '05), pp. 173–179. IEEE Computer Society, Washington, DC, USA (2005). doi:10.1109/IAT.2005.49.http://dx.doi.org/10.1109/IAT.2005.49
2. Wallace, R.S.: The anatomy of A.L.I.C.E. In: Epstein, R., Roberts, G., Beber, G. (eds.) Parsing the Turing Test, pp. 181210. Springer Science+Business Media. London. ISBN 978-1-4020-6710-5 (2009)
3. Pilato, G., Vassallo, G., Augello, A., Vasile, M., Gaglio, S.: Expert chat-bots for cultural heritage. Intell. Artif. **2**(2), 25–31 (2005)
4. Kino, H:. Coursey living in CYN: mating AIML and CYC together with program N (2004)
5. Resource. http://www.cyc.com
6. Stock, O.: Language-based interfaces and their application for cultural tourism. AI Mag. **22**(1), 85–98 (2001)
7. Mori, K., Jatowt, A., Ishizuka, M.: Enhancing conversational flexibility in multimodal interactions with embodied lifelike agent. In: Proceedings of the 8th international conference on Intelligent user interfaces (IUI '03), pp. 270–272. ACM, New York (2003). doi:10.1145/604045.604096.http://doi.acm.org/10.1145/604045.604096

8. Koutamanis, A.A.: An intelligent verbal interface for the retrieval of pictorial architectural information. In: Agger, K., Christiansson, P., Howard, R (eds.) Distributing Knowledge in Building Proceedings, vol. 2, pp. 6–13. Aarhus School of Architecture, Aarhus (2002)
9. Zdravkova, K.: Conceptual framework for an intelligent chatterbot. In: Proceedings of the 22nd International Conference on Information Technology Interfaces, 2000. ITI 2000, p. 189, 194, pp. 16–16 (2000)
10. Neves, A.M.M., Barros, F.A., Hodges, C., iAIML: a mechanism to treat intentionality in AIML chatterbots. Tools with artificial intelligence, 2006. In: 18th IEEE International Conference on ICTAI '06, p. 225, 231 (2006). doi:10.1109/ICTAI.2006.64
11. Augello, A., Santangelo, A., Sorce, S., Pilato, G., Gentile, A., Genco, A., Gaglio, S.: MAGA: a mobile archaeological guide at agrigento, workshop giornata nazionale su guide mobili virtuali 2006. ACM-SIGCHI, Torino, 18 Oct 2006 Online Proceedings http://hcilab.uniud.it/sigchi/doc/Virtuality06/index.html
12. Axelsson, J., Cross, C., Ferrans, J., McCobb, G., Raman, T., Wilson, L.: XHTML+Voice profile 1.2, W3C Note. (2004)
13. IBM Multimodal Tools. http://www-306.ibm.com/software/pervasive/multimodal/
14. Sorce, S., Augello, A., Santangelo, A., Pilato, G., Gentile, A., Genco, A., Gaglio, S.: A multimodal guide for the augmented campus. In: Proceedings of the 35th annual ACM SIGUCCS fall conference (SIGUCCS '07), pp. 325–331. ACM, New York (2007). doi:10.1145/1294046.1294123http://doi.acm.org/10.1145/1294046.1294123
15. Antonella, S., Agnese, A., Salvatore, S., Giovanni, P., Antonio, G., Alessandro, G., Salvatore, G.: A virtual shopper customer assistant in pervasive environments. In: Meersman, R., Tari, Z., Herrero, P(eds.) Proceedings of the 2007 OTM Confederated International Conference on the Move to Meaningful Internet Systems—Volume Part I (OTM'07), pp. 447–456. Springer, Berlin (2007)
16. Caronia, A., Pilato, G., Augello, A., Gaglio, S.: Getting started with programW. Resorces http://programw.sourceforge.net
17. Heylen, D.: Talking head says "Cheese!" humour as an impetus for embodied conversational agent research CHI-2003 workShop: humour modeling in the interface
18. Mihalcea, R., Strapparava, C.: Bootstrapping for fun: web-based construction of large data sets for humour recognition. In: Proceedings of the Workshop on Negotiation, Behaviour and Language (FINEXIN 2005), pp. 25–30. Ottawa (2005)
19. CMU Dictionary. http://www.speech.cs.cmu.edu/cgi-bin/cmudict
20. Landauer, T.K., Foltz, P.W., Laham, D.: An introduction to latent semantic analysis. Discourse Process. **25**, 259–284 (1998)
21. Pilato, G., Augello, A., Vassallo, G., Gaglio, S.: Sub-symbolic semantic layer in Cyc for intuitive chat-Bots. ICSC, pp. 121–128. IEEE Computer Society, Los Alamitos (2007)

Body Area Networks and Healthcare

Daniele Peri

Abstract Derived from Wireless Sensor Networks, Body Area Networks, comprise a wide range of typologies with sensor nodes placed on, close to, or implanted in the body that measure physiological signs. The availability of compact mobile computing devices makes it possible to integrate traditional healthcare with new powerful means. New paradigms in public health are arising from these developments, such as e-health and mHealth, and new converging applications can be envisioned. Physiological data acquisition provided by BANs may give care providers a unobtrusive real-time view on patient's health. On the other hand, the patient may be informed, assisted and even given the proper treatment by care providers. In this chapter, recent work on BANs focused on healthcare and mHealth is surveyed.

1 Introduction

The synergistic advancement of mobile computing and wireless technology has vigorously extended the horizon of decades old research fields aimed at providing healthcare with technological aids. Conventional telemedicine applications already see their scopes widened by the availability of more and more powerful and inexpensive iterations of mobile computing concepts. As simple smartphones may in fact provide a capillary access to electronic consultations in underserved areas, even in underdeveloped regions, the deployment of more sophisticated application can be easily envisioned. The important branch of Information Technology in healthcare for which the "electronic-health" (eHealth) term had been previously coined, is thus increasingly changing into "mobile-health" (mHealth) [17], denoting a paradigm shift toward mobile computing devices that are versatile, ubiquitous and wirelessly connected [18]. Wireless Sensor Networks (WSNs), made possible by the integration

D. Peri (✉)
DICGIM, University of Palermo, Viale delle Scienze, Ed. 6, 90128 Palermo, Italy
e-mail: daniele.peri@unipa.it

S. Gaglio and G. Lo Re (eds.), *Advances onto the Internet of Things*,
Advances in Intelligent Systems and Computing 260, DOI: 10.1007/978-3-319-03992-3_21,

of low-power microcontrollers, miniaturized radios, sensors and actuators, supplied other converging approaches to mHealth [1]. In fact, since the early definition of Body Sensor Networks [46] to the emerging and broader characterization as Body Area Networks (BANs), networks with nodes placed or implanted in the human body for healthcare applications have been the infrastructure of choice for a large corpus of research in the field [31, 40]. A survey of recent work on BANs, related technologies, and mHealth applications is provided in the following sections.

2 Body Area Networks

The definition of Body Area Networks (BANs) embraces a wide range of typologies in which sensor nodes capable of wireless communication are placed on, close to, or implanted in the body, and even fitted in textiles [7]. Nodes measure physiological signs for a broad range of applications, not restricted to medical ones [31]. BANs share with Wireless Sensor Networks (WSNs) many challenges, beginning with those related to power consumption and wireless communication. However, the extent of these challenges is in many ways exacerbated by the strict requirements of healthcare. For some of these applications, in fact, monitoring physiological parameter—such as body temperature, blood pressure, heartbeat, glucose levels—at low sample rates may be all that it is needed. Other healthcare applications may instead request high sample rates for prolonged time as in the case, for example, with electroencephalography (EEG) and electrocardiography (ECG) [25].

2.1 Communication

As already stated, one of the primary concern with BANs, as it is with WSNs, is the energy expenditure involved in communication. A consistent interest in literature has been thus placed on both low level—at the physical or Media Access Control layers (MAC)—and routing protocols in the communication stack, in order to reduce power consumption, latency, jitter, and implement Quality of Service and emergency signalling [40], in several cases provided with a critical review of the then ongoing standardization processes [31, 39]. Energy consumption of the communication tasks in WBANs built with conventional WSN-derived sensor nodes, may be reduced by adopting cooperative techniques. In [19] a WBAN integrated with environmental sensors exploiting multi-hop transmission through a cooperative MAC is proposed along with a discussion of similar approaches. A MAC protocol that uses heartbeat for synchronization instead of periodic beacon signals is described in [27]. A battery-aware MAC protocol is proposed in [38]. Along with energy consumption, other specific concerns arise in BANs. The proximity of nodes to the human body poses different threats to reliable and safe radio communication. The interaction of the electromagnetic waves on the body, besides hindering the quality of the radio link, must be prevented to become harmful by evaluating and minimizing the Specific Absorption Rate (SAR) [39]. With the aim to overcome the issues with Radio Frequency (RF) signals based BANs, Intra-body communication (IBC) propose using of the human body as a communication channel, via capacitive of galvanic coupling [34]. Indeed, the recent WBAN IEEE 802.15.6 standard includes non-RF Human-body communication along with RF Narrowband and Ultra Wideband physical layers. A recent survey on IBC can be found in [36]. Other research addresses the issues with radio wave based links by radically changing the transmission technology, for instance resorting to ultrasonic transceivers [8, 13]. Communication links are not only responsible for a great deal of the energy expenditure in BAN nodes, but their data rates represent also limiting factors also for the design of applications.

2.2 Data Compression

Another strategy to limit energy consumption and, at the same time, increasing effective communication data rates, in BANs consists in data compression at either the transmission or the sensing level. The latter may result in reduction of the data to be sent through the radio channel, without the additional computational overhead of data compression algorithms, and it is the path pursued, for instance by Compressed Sensing (CS), methods [10]. ECG and EEG are natural targets for CS methods, as their multichannel signals present relevant interdependence and would normally require high sampling rates [4]. In [48] a fetal electrocardiogram monitoring system is introduced that combines CS algorithms with a Bayesian learning framework to compress multichannel ECG. A similar approach is also applied to multichannel EEG [49]. Distributed data compression schemes, instead, can reduce the communication load when sustained rates of redundant, highly spatial and temporal correlated data, for instance collected by many closely placed sensors in body segment tracking applications, are available [45].

3 BANs, Healthcare and the Internet of Things

The efforts that produced open and interoperable standards for WSNs and BANs and the convergence with mobile computing, and the easy integration with Wide Area Networks, are paving the way for a whole new range of possibilities in healthcare applications. Compared with remote patient monitoring discussed in Sect. 2 that can be considered evolutionary with respect to established Telemedicine practice, the integration of BANs and "Mobile Health" (mHealth) systems poses as a revolution. Besides ubiquitous health monitoring and early detection of abnormal conditions, also active care and healthy lifestyle promotion can be supported by means of convenient, accessible and well accepted devices such as smartphones integrated with implanted or wearable sensors. A great deal of physiological, activity and environmental data can be collected, processed, integrated, stored both locally and remotely, and further processed by care providers. Physiological data includes body temperature, blood glucose level, blood pressure, blood oxygen saturation and heart rate. Physical activity data comprises body posture, level of physical activity and can be easily obtained by motion sensors. Environmental data such as location, temperature, humidity, light, atmospheric pressure, concentrations of air polluting agents, can be collected by either BANs or interconnected environmental monitoring stations. Given the significant processing power provided by mobile devices, anomaly detection [11] and data fusion algorithms [9] can be run locally to extract sensible information from such many heterogeneous sources. For the same reason, applications can depart from centralized paradigms and adopt more scalable and versatile large-scale sharing of information, in the direction of the Internet of Things (IoT). Research on Micro Electro-Mechanical Systems (MEMS) [30] may soon provide implantable actuator nodes allowing BANs to integrate even more electronic and biological systems. It is then possible to envision organs and subsystems of the human body become connected sub-BANs, in turn part of the IoT. A number of devices easily found in homes, such as weight scales, thermometers and blood pressure monitors, could provide preventive medicine applications with physiological data of healthy people for early diagnosis or suggest lifestyle changes. A lot of efforts in designing high-level protocols and frameworks is still required to support this vision. An infrastructure for decentralized development of mHealth applications is proposed in [12]. A relatively recent discussion of BANs and mHealth integration along with applications, challenges and trends can be found in [20]. In the remainder of the section, high-level tasks and issues involved in the design of BAN based mHealth applications are discussed. A few recent examples of the trends in mHealth applications are then presented in Sect. 4.

3.1 Physical Activity Recognition and Classification

Measure of physical activity level in everyday activity may provide additional insightful data not only to monitor patients, but also to support follow-up and rehabilitation. Additionally, healthy people could also benefit from fitness management and preventive healthcare applications monitoring their lifestyle and providing feedback in case, for instance, of inadequate levels of physical activity, bad posture habits, or prolonged exposure to unhealthy environments. A precondition for the development of such mHealth applications is the availability of methods to recognize and classify physical activity. Several approaches are described in literature ranging from tracking of body and limb motion, activity classification and user profiling [3]. Indoor navigation and motion capture methods with conversion from local, on-body, to global coordinate systems, using difference of arrival time and strength of RF signals in heterogeneous BANs are discussed in [16]. Classification of limb movements for kinesiotherapy by means of support vector machines and K-nearest neighbor methods applied to wireless sensors signal strength measures is described in [14]. In [22] a hierarchical architecture based on Artificial Neural Networks is used to classify data from the tri-axial accelerometer of a wireless sensor placed on the chest into states and activities. An incremental learning approach based on probabilistic neural networks and fuzzy clustering for human activity recognition from data provided by wearable sensors is proposed in [43]. In [2] an ear-worn activity recognition device along with wireless ambient sensors is used to detect every-day activities by means of a Bayesian classifier.

3.2 Security and Privacy

In medical applications security is a primary concern. Implantable medical devices (IMDs) have been scrutinized and reverse-engineered leading to the discover of major security flaws as in the case with an implantable cardioverter defibrillator [15], or a glucose monitor and insulin delivery system [26]. Connecting IMDs to BANs poses new security issues. Privacy and safety of patients must be well guarded against health or life threats coming from the connections BANs may sport outside their network [47]. Besides ruling out the possibility of unintended connections, BANs should guarantee that all sensor nodes are placed on the same body [6]. Security measures may exploit properties of physiological signals such as ECG and photoplethysmogram (PPG), to generate identification codes or keys [33]. In [29] ECG data is compressed, then a part of it that is essential for ECG reconstruction is selectively encrypted reducing energy cost without sacrificing too much the data transmission quality. In [41] a symmetric key agreement scheme based on physiological signals that guarantees authenticated communication is proposed for deployment of BAN sensor nodes. Similarly to this approach, in [50] ECG time-variant features are used to generate and share encryption keys. In [37] a framework for secure mHealth application is presented.

3.3 Context and Service-Oriented Architectures

In BANs both users and nodes may move freely, making network topology dynamic, and subject to context changes and connectivity interruptions. Moreover, providing mobile users with services in highly variable environments prompts for the deployment of service discovery, distribution and mapping strategies, as well as context monitoring. A two-layered service discovery and composition architecture based on BAN node clustering by connectivity patterns is proposed in [5]. In [28] a service-oriented framework for context recognitions able to adapt in real-time to both topological and contextual changes is introduced. A mobile monitoring framework for detection of user contexts is presented in [21].

4 Applications

Mobile Health is destined, sooner or later, to assume a central role in healthcare, extending to many of the scopes of telemedicine and Electronic Health, and beyond. A consistent part of the research work in the field is motivated by the underlying idea to support healthcare with tools targeting major health threats but also the challenges posed by population aging, possibly increasing disease prevention, as outlined in Sect. 3. Another recurrent motivation for research in the field is to help reducing healthcare costs by moving as many as possible treatments from hospital to homes. In this section a few examples of research applications representative of the convergent trend between BANs and mHealth are presented.

4.1 Cardiovascular Diseases

Cardiovascular diseases represent the principal cause of death in the more developed countries thus the development of new and better tools for prevention, early diagnosis, treatment and follow-up is constantly sought. For instance, a system for hearth failure prevention is proposed in [42]. The system measures daily heart rate, respiration and activity by means of wireless sensors placed on garments, bed sheets and pillows. A mobile device collects and processes data and encourages the user to perform personalized daily routines, while keeping contact with a remote back-end used by physicians to monitor and guide patients. An application for hypertension management as home care is presented in [24]. A BAN is used to monitor heart rate and blood pressure, while a mobile base unit processes data to asses medication response and detect adverse drug events. The base unit serves as a bidirectional link to the doctor through a server placed in the hospital so that treatment can be personalized in real-time.

4.2 Physical Therapy

Physical therapy is one of the best applicative targets for BANs. Many sensors are in fact required to analyze body posture and movements thus justifying their use even with current technology. In [23] a BAN is part of a gaming application to assist in physical therapy. Body sensor measures are used in the game to perform adaptation required by the patient health status. Real-time personalization of the treatment is claimed to reduce recovery time. It is quite easy to imagine a growing trend of such applications, given that workout video games are common off the shelf products that use comparable technology and that games to support rehabilitation, even if in very narrow scopes, have already been developed [35]. On the other hand, the advancement of wearable sensor nodes may only widen the range of possibilities by reducing costs and easing deployment of BANs, as stated in Sect. 3.1. Tough the main focus of the chapter is already been discussed in Sect. 2.2, in [45] extensive body tracking is used to support physical rehabilitation through Pilates exercises. Analysis of motion data is used to assess whether the patient has performed the exercises correctly, and to provide a feedback and replay of the correct movements. In [32] accelerometer data is classified by a support vector machine to assess severity of symptoms and motor complication of Parkinson's disease.

4.3 Eldery and Passive Care Assistance

Detection of abnormal activities is of great importance to reduce health risks. This is another possible scenario for BAN integrated into the IoT, as recognizing hazardous behaviors may require may different kind of sensors even in confined environments like homes. As an example, in [44] a system based on a wrist-worn wireless sensor node and a base unit to monitor the elderly is presented. The sensor node measures body and ambient temperature, heart rate and accelerations and a classifier is used for lifestyle analysis, and to detect abnormal behaviors potentially leading to health threats. A voice based interface based on simple questions to be answered is adopted for user interaction.

5 Conclusion

Body Area Networks promise to bring connectivity to a new level, making the human body part of the Internet of Things. Applications to improve health, assist during treatments and follow-ups, and even prevent the insurgence of disease would be made possible, given that on-body, wearable and environmental sensor nodes could be integrated by sharing communication and processing protocols. While the enormous potential of BANs cannot be overlooked, they are still at the research stage and it is still necessary that many pitfalls be avoided and issues overcome. However, their deployment is hastened by the raising paradigms of e-Health and m-Health, in a mutual supported development toward intelligent systems that are able to provide healthcare with more accessible and effective means.

Acknowledgments This work has been partially supported by the PON R&C grant MI01_00091 funding the SeNSori project.

References

1. Alemdar, H., Ersoy, C.: Wireless sensor networks for healthcare: a survey. Comput. Netw. **54**(15), 2688–2710 (2010)
2. Atallah, L., Lo, B., Ali, R., King, R., Yang, G.Z.: Real-time activity classification using ambient and wearable sensors. IEEE Trans. Inf. Technol. Biomed. **13**(6), 1031–1039 (2009). doi:10.1109/TITB.2009.2028575
3. Augello, A., Ortolani, M.: Lo Re, G., Gaglio, S.: Sensor mining for user behavior profiling in intelligent environments. Advances in Distributed Agent-Based Retrieval Tools. Studies in Computational Intelligence, vol. 361, pp. 143–158. Springer, Berlin (2011)
4. Balouchestani, M., Raahemifar, K., Krishnan, S.: Wireless body area networks with compressed sensing theory. In: 2012 ICME International Conference on Complex Medical Engineering (CME), pp. 364–369 (2012). doi:10.1109/ICCME.2012.6275663
5. Coloberti, M., Lombriser, C., Roggen, D., Tröster, G., Guarneri, R., Riboni, D.: Service discovery and composition in body area networks. In: Proceedings of the ICST 3rd International Conference on Body Area Networks, BodyNets '08, pp. 7:1–7:4. ICST (Institute for Computer Sciences, Social-Informatics and Telecommunications Engineering), ICST, Brussels, Belgium, Belgium (2008). http://dl.acm.org/citation.cfm?id=1460257.1460267
6. Cornelius, C.T., Kotz, D.F.: Recognizing whether sensors are on the same body. Pervasive Mob. Comput. **8**(6), 822–836 (2012). doi:10.1016/j.pmcj.2012.06.005. http://dx.doi.org/10.1016/j.pmcj.2012.06.005
7. Coyle, S., Lau, K.T., Moyna, N., O'Gorman, D., Diamond, D., Di Francesco, F., Costanzo, D., Salvo, P., Trivella, M., De Rossi, D., Taccini, N., Paradiso, R., Porchet, J.A., Ridolfi, A., Luprano, J., Chuzel, C., Lanier, T., Revol-Cavalier, F., Schoumacker, S., Mourier, V., Chartier, I., Convert, R., De-Moncuit, H., Bini, C.: Biotex-2014; biosensing textiles for personalised healthcare management. IEEE Trans. Inf. Technol. Biomed. **14**(2), 364–370 (2010). doi:10.1109/TITB.2009.2038484
8. Davilis, Y., Kalis, A., Ifantis, A.: On the use of ultrasonic waves as a communications medium in biosensor networks. IEEE Trans. Inf. Technol. Biomed. **14**(3), 650–656 (2010). doi:10.1109/TITB.2009.2039755
9. De Paola, A., Gaglio, S.: Lo Re, G., Ortolani, M.: Multi-sensor fusion through adaptive bayesian networks. AI*IA 2011: Artificial Intelligence Around Man and Beyond. Lecture Notes in Computer Science, vol. 6934, pp. 360–371. Springer, Berlin (2011)
10. Donoho, D.: Compressed sensing. IEEE Trans. Inf. Theory **52**(4), 1289–1306 (2006). doi:10.1109/TIT.2006.871582
11. Farruggia, A.: Lo Re, G., Ortolani, M.: Probabilistic anomaly detection for wireless sensor networks. AI*IA 2011: Artificial Intelligence Around Man and Beyond. Lecture Notes in Computer Science, vol. 6934, pp. 438–444. Springer, Berlin (2011)
12. Forsstrom, S., Kanter, T., Johansson, O.: Real-time distributed sensor-assisted mhealth applications on the internet-of-things. In: 2012 IEEE 11th International Conference on Trust, Security and Privacy in Computing and Communications (TrustCom), pp. 1844–1849 (2012). doi:10.1109/TrustCom.234
13. Galluccio, L., Melodia, T., Palazzo, S., Santagati, G.: Challenges and implications of using ultrasonic communications in intra-body area networks. In: 2012 9th Annual Conference on Wireless On-demand Network Systems and Services (WONS), pp. 182–189 (2012). doi:10.1109/WONS.2012.6152227
14. Guraliuc, A., Barsocchi, P., Potortì, F., Nepa, P.: Limb movements classification using wearable wireless transceivers. IEEE Trans. Inf. Technol. Biomed. **15**(3), 474–480 (2011). doi:10.1109/TITB.2011.2118763
15. Halperin, D., Heydt-Benjamin, T., Ransford, B., Clark, S., Defend, B., Morgan, W., Fu, K., Kohno, T., Maisel, W.: Pacemakers and implantable cardiac defibrillators: Software radio attacks and zero-power defenses. In: IEEE Symposium on Security and Privacy 2008 (SP 2008). pp. 129–142 (2008). doi:10.1109/SP.2008.31

16. Hamie, J., Denis, B., Richard, C.: Joint motion capture and navigation in heterogeneous body area networks with distance estimation over neighborhood graph. In: 2013 10th Workshop on Positioning Navigation and Communication (WPNC), pp. 1–6 (2013). doi:10.1109/WPNC.2013.6533282
17. Istepanian, R., Laxminarayan, S., Pattichis, C.S.: M-health: emerging mobile health systems. In: Istepanian, R., Laxminarayan, S., Pattichis, C.S. (eds.) M-Health: Emerging Mobile Health Systems, XXX, 624, illus. 0–387-26558-9, p. 182. Springer, Berlin (2006)
18. Istepanian, R.S., Pattichis, C.S., Laxminarayan, S.: Ubiquitous m-health systems and the convergence towards 4g mobile technologies. In: M-Health, pp. 3–14. Springer, NY (2006)
19. Ivanov, S., Botvich, D., Balasubramaniam, S.: Cooperative wireless sensor environments supporting body area networks. IEEE Trans. Consum. Electron. **58**(2), 284–292 (2012). doi:10.1109/TCE.2012.6227425
20. Jovanov, E., Milenkovic, A.: Body area networks for ubiquitous healthcare applications: opportunities and challenges. J. Med. Syst. **35**(5), 1245–1254 (2011). doi:10.1007/s10916-011-9661-x. http://dx.doi.org/10.1007/s10916-011-9661-x
21. Kang, S., Lee, J., Jang, H., Lee, Y., Park, S., Song, J.: A scalable and energy-efficient context monitoring framework for mobile personal sensor networks. IEEE Trans. Mob. Comput. **9**(5), 686–702 (2010). doi:10.1109/TMC.2009.154
22. Khan, A., Lee, Y.K., Lee, S., Kim, T.S.: A triaxial accelerometer-based physical-activity recognition via augmented-signal features and a hierarchical recognizer. IEEE Trans. Inf. Technol. Biomed. **14**(5), 1166–1172 (2010). doi:10.1109/TITB.2010.2051955
23. Kifayat, K., Fergus, P., Cooper, S., Merabti, M.: Body area networks for movement analysis in physiotherapy treatments. In: 2010 IEEE 24th International Conference on Advanced Information Networking and Applications Workshops (WAINA), pp. 866–872 (2010). doi:10.1109/WAINA.2010.155
24. Koutkias, V., Chouvarda, I., Triantafyllidis, A., Malousi, A., Giaglis, G., Maglaveras, N.: A personalized framework for medication treatment management in chronic care. IEEE Trans. Inf. Technol. Biomed. **14**(2), 464–472 (2010). doi:10.1109/TITB.2009.2036367
25. Latré, B., Braem, B., Moerman, I., Blondia, C., Demeester, P.: A survey on wireless body area networks. Wireless Netw. **17**(1), 1–18 (2011). doi:10.1007/s11276-010-0252-4. http://dx.doi.org/10.1007/s11276-010-0252-4
26. Li, C., Raghunathan, A., Jha, N.: Hijacking an insulin pump: Security attacks and defenses for a diabetes therapy system. In: 2011 13th IEEE International Conference on e-Health Networking Applications and Services (Healthcom), pp. 150–156 (2011). doi:10.1109/HEALTH.2011.6026732
27. Li, H., Tan, J.: Heartbeat-driven medium-access control for body sensor networks. IEEE Trans. Inf. Technol. Biomed. **14**(1), 44–51 (2010). doi:10.1109/TITB.2009.2028136
28. Lombriser, C., Marin-Perianu, R., Roggen, D., Havinga, P., Troster, G.: Modeling service-oriented context processing in dynamic body area networks. IEEE J. Sel. Areas Commun. **27**(1), 49–57 (2009). doi:10.1109/JSAC.2009.090106
29. Ma, T., Shrestha, P., Hempel, M., Peng, D., Sharif, H., Chen, H.H.: Assurance of energy efficiency and data security for ecg transmission in basns. IEEE Trans. Biomed. Eng. **59**(4), 1041–1048 (2012). doi:10.1109/TBME.2011.2182196
30. Nuxoll, E., Siegel, R.: Biomems devices for drug delivery. IEEE Eng. Med. Biol. Mag. **28**(1), 31–39 (2009)
31. Patel, M., Wang, J.: Applications, challenges, and prospective in emerging body area networking technologies. IEEE Wirel. Commun. **17**(1), 80–88 (2010). doi:10.1109/MWC.2010.5416354
32. Patel, S., Lorincz, K., Hughes, R., Huggins, N., Growdon, J., Standaert, D., Akay, M., Dy, J., Welsh, M., Bonato, P.: Monitoring motor fluctuations in patients with parkinson's disease using wearable sensors. IEEE Trans. Inf. Technol. Biomed. **13**(6), 864–873 (2009). doi:10.1109/TITB.2009.2033471
33. Poon, C.C.Y., Zhang, Y.T., Bao, S.D.: A novel biometrics method to secure wireless body area sensor networks for telemedicine and m-health. IEEE Commun. Mag. **44**(4), 73–81 (2006). doi:10.1109/MCOM.2006.1632652

34. Pun, S.H., Gao, Y.M., Mak, P., Vai, M.I., Du, M.: Quasi-static modeling of human limb for intra-body communications with experiments. IEEE Trans. Inf. Technol. Biomed. **15**(6), 870–876 (2011). doi:10.1109/TITB.2011.2161093
35. Rego, P., Moreira, P., Reis, L.: Serious games for rehabilitation: a survey and a classification towards a taxonomy. In: 2010 5th Iberian Conference on Information Systems and Technologies (CISTI), pp. 1–6 (2010)
36. Seyedi, M., Kibret, B., Lai, D., Faulkner, M.: A survey on intrabody communications for body area network applications. IEEE Trans. Biomed. Eng. **60**(8), 2067–2079 (2013). doi:10.1109/TBME.2013.2254714
37. Sorber, J., Shin, M., Peterson, R., Cornelius, C., Mare, S., Prasad, A., Marois, Z., Smithayer, E., Kotz, D.: An amulet for trustworthy wearable mhealth. In: Proceedings of the Twelfth Workshop on Mobile Computing Systems and Applications, HotMobile '12, pp. 7:1–7:6. ACM, New York, NY, USA (2012). doi:10.1145/2162081.2162092. http://doi.acm.org/10.1145/2162081.2162092
38. Su, H., Zhang, X.: Battery-dynamics driven tdma mac protocols for wireless body-area monitoring networks in healthcare applications. IEEE J. Sel. Areas Commun. **27**(4), 424–434 (2009). doi:10.1109/JSAC.2009.090507
39. Ullah, S., Higgins, H., Braem, B., Latre, B., Blondia, C., Moerman, I., Saleem, S., Rahman, Z., Kwak, K.: A comprehensive survey of wireless body area networks. J. Med. Syst. **36**(3), 1065–1094 (2012). doi:10.1007/s10916-010-9571-3. http://dx.doi.org/10.1007/s10916-010-9571-3
40. Ullah, S., Khan, P., Ullah, N., Saleem, S., Higgins, H., Kwak, K.S.: A review of wireless body area networks for medical applications. IJCNS **2**(8), 797–803 (2009)
41. Venkatasubramanian, K., Banerjee, A., Gupta, S.K.S.: Pska: usable and secure key agreement scheme for body area networks. IEEE Trans. Inf. Technol. Biomed. **14**(1), 60–68 (2010). doi:10.1109/TITB.2009.2037617
42. Villalba, E., Salvi, D., Ottaviano, M., Peinado, I., Arredondo, M.T., Akay, A.: Wearable and mobile system to manage remotely heart failure. IEEE Trans. Inf. Technol. Biomed. **13**(6), 990–996 (2009). doi:10.1109/TITB.2009.2026572
43. Wang, Z., Jiang, M., Hu, Y., Li, H.: An incremental learning method based on probabilistic neural networks and adjustable fuzzy clustering for human activity recognition by using wearable sensors. IEEE Trans. Inf. Technol. Biomed. **16**(4), 691–699 (2012). doi:10.1109/TITB.2012.2196440
44. Winkley, J., Jiang, P., Jiang, W.: Verity: an ambient assisted living platform. IEEE Trans. Consum. Electron. **58**(2), 364–373 (2012). doi:10.1109/TCE.2012.6227435
45. Wu, C.H., Tseng, Y.C.: Data compression by temporal and spatial correlations in a body-area sensor network: a case study in pilates motion recognition. IEEE Trans. Mob. Comput. **10**(10), 1459–1472 (2011). doi:10.1109/TMC.2010.264
46. Yang, G.Z. (ed.): Body Sensor Networks. Springer, London (2006). doi:10.1007/1-84628-484-8. http://dx.doi.org/10.1007/1-84628-484-8
47. Zhang, M., Raghunathan, A., Jha, N.: Towards trustworthy medical devices and body area networks. In: 2013 50th ACM/EDAC/IEEE Design Automation Conference (DAC), pp. 1–6 (2013)
48. Zhang, Z., Jung, T.P., Makeig, S., Rao, B.: Compressed sensing for energy-efficient wireless telemonitoring of noninvasive fetal ecg via block sparse bayesian learning. IEEE Trans. Biomed. Eng. **60**(2), 300–309 (2013). doi:10.1109/TBME.2012.2226175
49. Zhang, Z., Jung, T.P., Makeig, S., Rao, B.: Compressed sensing of eeg for wireless telemonitoring with low energy consumption and inexpensive hardware. IEEE Trans. Biomed. Eng. **60**(1), 221–224 (2013). doi:10.1109/TBME.2012.2217959
50. Zhang, Z., Wang, H., Vasilakos, A., Fang, H.: Ecg-cryptography and authentication in body area networks. IEEE Trans. Inf. Technol. Biomed. **16**(6), 1070–1078 (2012). doi:10.1109/TITB.2012.2206115

Urban Air Quality Monitoring Using Vehicular Sensor Networks

Giuseppe Lo Re, Daniele Peri and Salvatore Davide Vassallo

Abstract The quality of air is a major concern in modern cities as pollutants have been demonstrated to have significant impact on human health. Networks of fixed monitoring stations have been deployed in urban areas to provide authorities with data to define and enforce dynamically policies to reduce pollutants, for instance by issuing traffic regulation measures. However, fixed networks require careful placement of monitoring stations to be effective. Moreover, changes in urban arrangement, activities, or regulations may affect considerably the monitoring model, especially when budget constraints prevent from relocating stations or adding new ones to the network. In this chapter we discuss a different approach to environmental monitoring through mobile monitoring devices implementing a Vehicular Sensor Network (VSN) to be deployed on the public transport bus fleet of Palermo.

1 Introduction

In recent years, the concept of Vehicular Ad-hoc NETwork (VANET) was introduced to refer to a wireless network in which nodes are represented by vehicles, which communicate with each other and with some fixed Access Points [1]. The primary purpose of VANETs is the development of distributed and public road safety-oriented applications, in order to save lives, improve traffic conditions and reduce the environmental impact [2]. Vehicular networks represent, in fact, the heart of the wider project of Intelligent Transportation System (ITS), that is the set of efforts and technologies that add the Information and Communications Technology to transport infrastructure and vehicles [3].

G. Lo Re (✉) · D. Peri · S. D. Vassallo
DICGIM, University of Palermo, Viale delle Scienze, ed. 6, 90128 Palermo, Italy
e-mail: giuseppe.lore@unipa.it

D. Peri
e-mail: daniele.peri@unipa.it

S. Gaglio and G. Lo Re (eds.), *Advances onto the Internet of Things*,
Advances in Intelligent Systems and Computing 260, DOI: 10.1007/978-3-319-03992-3_22,

Taking advantage of the technological progress and results obtained in vehicular networks, it is possible to build systems that detect, process and transmit some environmental features to a central server through radio links or cellular networks. This new network paradigm is called a Vehicular Sensor Network (VSN) [4]. It is a combination of communication networks based on wireless transceivers installed on vehicles and Wireless Sensor Networks (WSNs). WSNs are the sensing infrastructure of distributed systems for the control of environmental monitoring [5], habitat monitoring [6], and ambient intelligence [7], which require the perception of some physical quantities, such as temperature, humidity, ambient lighting, and so on.

A VSN has some properties [4] like:

- no energy constraints comparing to well known WSN applications, because vehicles can provide continuous power;
- high computational capabilities because vehicles can be equipped with sufficient computational resources;
- a vehicle could be equipped with a lot of sensors, so it can produce a large amounts of sensed data;
- mobile and dynamic topology.

The most important advantage of vehicular sensor networks over common static sensor networks is the possibility to carry on measurements on large areas using a small number of sensor nodes [8]. In contrast with traditional wireless sensor networks having their nodes placed in fixed locations, vehicular sensor networks are characterized by dynamic changes in network topology. This attractive feature makes VSNs a potentially cost effective solution to provide monitoring services to a broad class of applications. Vehicles may, for instance, recognize a plate and possibly may send messages to neighboring vehicles enabling the police to track the movements of a specified car [9, 10]. Another interesting application is road surface monitoring [11, 12], or urban pollution detection.

It seems thus the right time to make car an intelligent entity able to reason and cooperate with other vehicles or with environment surrounding it. To achieve this goal, the use of some formalism providing a precise structure to domain knowledge may be desirable. An ontology could fulfill this role as is an "explicit specification of a conceptualization" according to [13], that is a particular abstraction of a set of objects, concepts and relationships, that are to be represented formally for sharing and reuse of such knowledge among entities.

In this chapter, we design the prototype of a mobile system able to collect environmental data like urban air pollution. This vehicular sensor network can be considered as composed by intelligent nodes that monitor and analyze the evolution of environmental data, reporting on the occurrence of some critical issue to a supervisor entity. In order to enable reuse of air quality domain knowledge and to share the structure of such information among people or software agents, sensed data should be organized and structured in an ontology.

The remainder of the chapter is organized as follows. In Sect. 2, motivations about urban air quality and the use of vehicular sensor networks will be discussed.

Section 3 is dedicated to previous urban air quality-monitoring and diagnosis systems. In Sect. 4 we discuss our solution to the problem, describing the architecture and the implementation of the sensor network. In Sect. 5, ontological approach to the problem is described. We finally draw conclusions in Sect. 6.

2 Urban Air Quality Monitoring

In recent years researchers have begun to envision vehicular networks as systems capable of monitoring certain physical features and transmit sensed data via radio links to a server for further analysis. Observing certain environmental data, in different city spots, would in fact allow advanced support systems to detect potential alarm scenarios and suggest appropriate countermeasures. By exploiting the inherent network nodes mobility is possible, therefore, to implement a low cost environmental monitoring system with high spatial coverage.

Among the possible environmental information collectible with a vehicular sensor network, air quality plays no secondary role. This is indeed a major concern in modern cities, because air pollutants have a significant impact on human health and on the environment.

Air quality in urban areas is the result of three components: regional factors, urban term, and hot spot terms [14]. In rural areas, pollution levels depend mainly on the medium-to-long-range transport of pollutants traveling by air masses from other areas. The resulting concentration levels, are generally significantly lower in those areas. In urban areas, air pollution is linked to the set of human activities, such as closed environments heating and lighting, public and private vehicular transportation, or construction activities. Urban pollution varies spatially, as it is reasonable to expect, accordingly to human activities, topography, and local micrometeorology.

The importance of this issue is so high that is regulated by the European Commission in the Directive 96/62/EC, that establishes the basic principles of a common strategy to define and set objectives for ambient air quality in order to avoid, prevent or reduce harmful effects on human health and the environment, assess ambient air quality in the Member States, inform the public, and improve air quality where it is unsatisfactory [15].

Air pollution is usually monitored by highly reliable networks of fixed stations. A monitoring station can accurately measure a wide range of pollutants using conventional analysis tools. However, permanent monitoring stations are frequently placed so as to measure ambient background concentrations or at potential hotspot locations and they are usually several kilometers apart. Moreover, the large cost of acquisition and maintenance limits the number of such facilities, resulting in non-scalability of the system and in an extremely limited spatial resolution of the pollution maps. In fact, the effective range of spatial coverage of a static sensor is quite limited, thus it would take a large number of detectors to monitor a wide area of interest. To overcome these problems, we propose a cost-efficient and sustainable vehicular sensor network. If effectively implemented, a VSN can offer a wider spatial coverage, and a finer granularity of detected characteristics. Rather than deploying static sensors, these detectors can be installed on cars or vehicles of public transport services. Sensors attached on moving vehicles periodically monitor the air quality and transmit gathered information to a central storage system. However, there is a trade-off for this gain in spatial coverage. Temporal coverage of sensed data in a particular position will be lower compared to static sensors readings [8] as a characteristic will be measured in the same position only when the vehicle will cross again that point. This lack of temporal coverage can be handled by increasing network nodes density and so mounting sensors on more vehicles, or distributing sensors on public buses, so that environmental characteristics can be monitored continuously along their routes. Another problem that should not be underestimated concerns corrupted measurements within sensor networks. Sensor nodes may occasionally produce incorrect measurements due to battery depletion, dust on sensors, tampering and other causes. Among the several mathematical methods and algorithms in literature, suitable tools to pre-process such gathered data are Bayesian Networks [16, 17].

In the next section we discuss some existing air pollution monitoring platforms providing a brief overview of the state of the art.

3 Related Work

Monitoring air pollution using low-cost gas sensors has gained high interest in recent years. Previous research has attempted to construct networked air quality-monitoring and diagnosis systems.

CitySense [18], developed by Harvard University researchers in collaboration with BBN Technologies, consists in an approximately 100 wireless devices installed on buildings and streetlights in Cambridge. Similar to CitySense, SensorScope [19] is an example of large-scale distributed wireless environmental monitoring system.

David Hasenfratz et al. propose GasMobile [20], an air pollution smartphone-based monitoring platform. In particular, a simple and scalable system for atmospheric ozone concentration detection has been realized using a low-cost sensor connected via USB to a smartphone. A similar project, developed jointly by UC Berkeley and Intel, is called N-SMARTS [21] which shows the possibility to gather raw air pollution data by attaching sensors to GPS-enabled cell phones.

Several studies propose the use of vehicles to form an extensive air quality monitoring system. For example, OpenSense project [22], proposes to install sensors on buses to form an extensive network of mobile air quality data collection sites. A VSN architecture to measure air quality for microclimate monitoring in city areas, is proposed in [23]. These vehicular nodes roam inside the area of interest and periodically report, through short GSM messages, CO_2 concentration data to a central server for further analysis or data mining. In [24], an air pollution sensing network was tested on the public buses of Sharjah. Mobile Discovery Net (MoDisNet) [25] is a distributed infrastructure based on wireless sensors network and Grid computing technology for air pollution monitoring and mining in London. This system uses a set of vehicles such as buses, trucks, taxis, and vehicles, in which mobile sensors are installed forming a sensing network that cooperates with a grid of static sensors installed on roadside. Mobile Air Quality Monitoring Network (MAQUMON) [26] is a system composed by some car-mounted sensor nodes measuring several atmospheric pollutants. Air pollution is tagged with absolute location and time data by means of an on-board GPS. Periodically, the measurements are uploaded to a server, processed and then published on a portal. In [27], two mobile platforms for fine-grained realtime pollution measurement are presented: a mobile sensing box, deployable on public transportation infrastructure and a personal sensing device (NODE) that can be used to create a social pollution sensing.

In [28], a low power MEMS metal-oxide-sensor array is described in order to detect odor events created within the car cabin. Sensing devices have to classify the air quality level inside the car cabin with a multivariate approach closely related to the perceived AQL by human panelists. As displayed by authors, this system could be used, in combination with the the next generation of heating, ventilation, and air conditioning (HVAC) systems, in order to improve air quality inside the vehicle.

Similarly to these studies, we designed a Vehicular Sensor Network architecture for air quality monitoring to complement the fixed stations monitoring infrastructure of the city of Palermo, Italy. A few vehicles of the public transport bus fleet will be equipped with some wireless air pollution detection sensors and with the necessary components for data transmission to fixed access point acting as gateways to a central server. Vehicles, during their normal route, will sample data in a much more detailed way than is currently possible with static monitoring stations. When vehicles move close to a fixed access point installed on the roadside, for example in traffic light proximity, they may automatically transmit gathered data.

4 The Proposed System

Currently, air quality in Palermo is monitored by a network of ten fixed stations, located in some strategic spots (Fig. 1) spread on the city area. The network started to work on 1st August 1996, and it had several technical adjustments over the years in order to comply with European regulations, until March 2003, when network management activities obtained the UNI EN ISO 9001:2000 certification. All the stations are connected to a central data collection and processing facility via switched telephone lines (ISDN). Each station is equipped with a personal computer that processes data from the analysis equipment and transmits them to the central system. The network also includes four access points to the remote data center and two points for data dissemination to public.

To complement the fixed network, we designed a low-cost monitoring system using a vehicular sensor network. The proposed system will collect, process and distribute data from sensors located on vehicles belonging to the public transportation bus fleet. In the initial stage, a few vehicles will be equipped with some sensors that will measure, on their routes, the concentration of some gases like carbon monoxide,

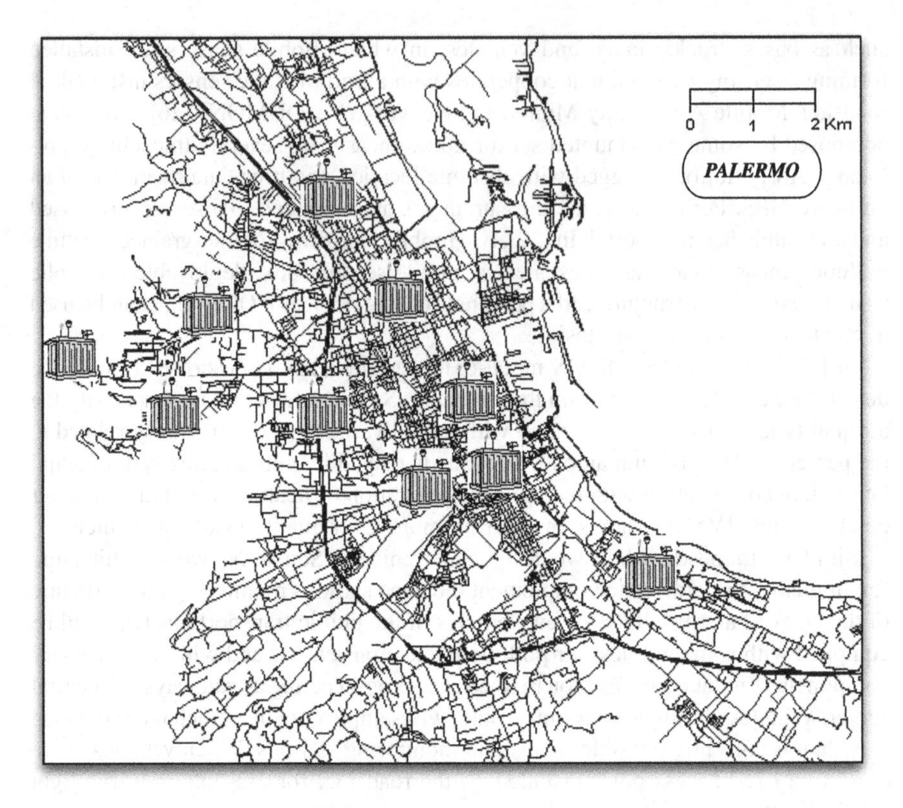

Fig. 1 Geographic positions of the fixed air quality monitoring stations in Palermo. Each station, equipped with a personal computer that processes data from the analysis equipment, is connected to a central data collection and processing facility via switched telephone lines

carbon dioxide, and ozone. Exploiting public buses mobility, even with few sensor nodes, most of the city area can be covered and detailed urban air quality data collected with fine granularity of detected characteristics. Furthermore, temporal coverage problems can also handled and overcome as buses move on fixed and established routes many times during the day.

Figure 2 shows the proposed Vehicular Sensor Network architecture for air quality monitoring. The system is composed by some vehicular sensor nodes, a monitoring server, and some access point installed on roads.

The main components are:

- In-Vechicle nodes, each provided with a microcontroller, communication devices and sensors;
- Gateways, receive data from each node and forwards them to the central server;
- Central server, store gathered data, ensuring integrity, security and availability.

Each vehicular node consists of a central unit and a sensor board. Periodically, the central unit collects the detected air pollution concentration from the sensor board

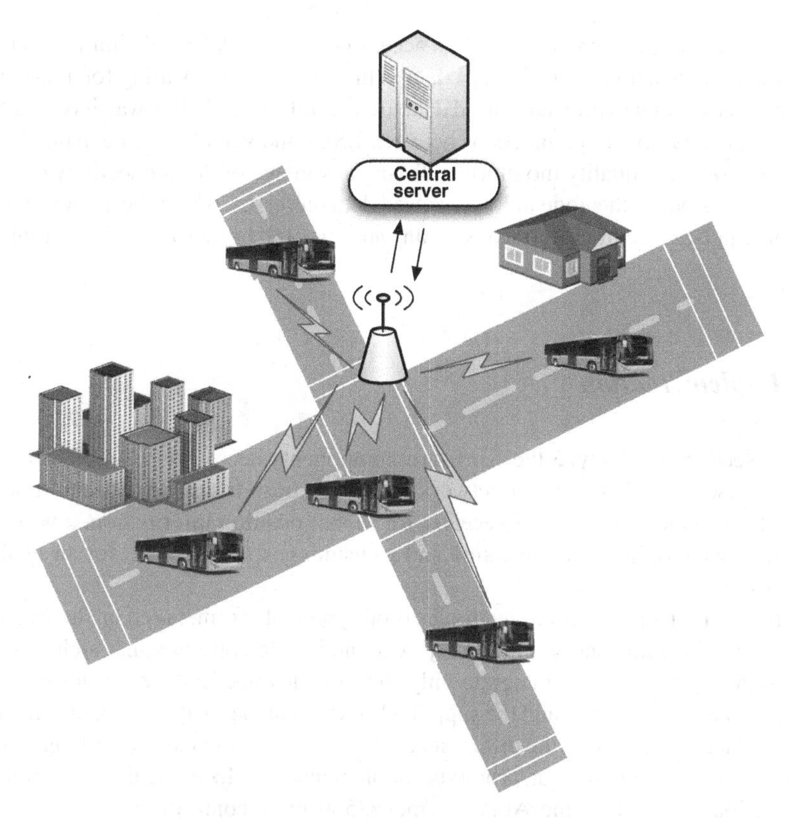

Fig. 2 Vehicular Sensor Network architecture for air quality monitoring. The system is composed by some vehicular sensor nodes that measure different pollutants in the air, some access points installed on the roads, and a monitoring server

and stores the data together with time and the current location given by the Global Position System (GPS) module. This device is so responsible for the aggregation, synchronization and transmission of sensed data to the central server for storage and further processing. Communication is granted through a centralized Vehicle-to-Infrastructure (V2I) architecture. Vehicle-to-Infrastructure concerns the connection between vehicles and fixed access points, also called Road Side Units (RSUs), requiring the placement of such communication devices in external structures, or in convenient places, such as intersections, traffic lights or buildings. A RSU acts as buffer point for the exchange of information between vehicles, so its main task is to extend the network by forwarding data coming from vehicles, to centralized servers or, eventually, to other oncoming vehicles.

The designed application is delay-tolerant and time-based, so it makes no restrictive obligation on sending real-time data to the portal. Instead, each vehicle obeys to a store and forward dissemination policy. It collects sensor data, periodically at regular intervals, and processes them locally before sending them to the central server using opportunistically encountered wireless access points in a delay-tolerant fashion. In particular, every network node keeps data in his own memory waiting for a in-range remote access point to transfer sensed data to a central server. In this way it is possible to reduce the number of connections between RSUs and vehicles, while maintaining high precision and quality monitoring. Taking advantage of the connectivity offered by the access point, the node uploads sensed data on server, where the sensory information can be structured by means of an ontology for further analysis, sharing or reuse.

4.1 Implementation

In this section we analyze the hardware components necessary to implement the system described. To realize a network node we chose to use an Arduino microcontroller, connected to a GPS receiver for global position information, a wireless module for the radio communication and a custom gas sensor board for air quality measuring.

Arduino is an open-source electronics prototyping platform, based on flexible and easy-to-use hardware and software [29]. Arduino is based on a modular architecture, in fact the idea is to easy integrate only the modules needed in each device. The development environment and the supplied libraries can be easily modified using the C/C++ language. Among the different Arduino boards, we chose the Arduino Mega 2560 microcontroller for our prototype implementation. In particular the Arduino Mega 2560 is based on the Atmel ATmega2560 microcontroller which offers 54 digital I/O pins, 16 analog inputs, 4 UART serial port, a 16 Mhz clock, a USB connection, and an ICSP header. The large number of I/O pins facilitates the inclusion of gas sensors and other communication components like the Xbee Pro shield for data transmission.

The core of the monitoring system is the gas sensor board provided with several semiconductor sensors. These sensors exploit the change of the conductivity caused by the absorption of gaseous pollutants on a semiconducting surface [30]. Their innards comprise a support in various materials such as aluminum, silicon, and ceramics, a heating resistor, and a sensing layer that is composed of a metal-oxide material such as tin dioxide (SnO_2) or zinc oxide (ZnO). At working temperature, a set of electrochemical reactions between the atmospheric oxygen and oxide granules are established. These reactions modulate and regulate the electronic flow between the grains of the sensing element, changing its resistivity, and so giving information about a precise gas concentration. These devices are particularly sensitive to temperature changes, therefore it is necessary to control the air flow directed towards the sensor head when mounted on a moving vehicle. The behavior of the sensor with respect to temperature changes is not definable by a mathematical equation, and the temperature drop of the heating element induces an error in gas concentration measures. This effect is negligible for low gas concentrations, but becomes significant for higher levels of pollution rate.

In order to validate the proposed design, in a preliminary phase, the VSN will be deployed on a single vehicle belonging to the public transportation bus fleet of Palermo and a few access points. This platform installed on the bus will monitor the following parameters:

- Temperature
- Relative humidity
- Nitrogen dioxide (NO_2)
- Carbon dioxide (CO_2)
- Carbon monoxide (CO)
- Ozone (O_3).

For data gathering, some Xbee-based Access Points will be installed on streets, and will communicate with a central server that can be implemented on a low consumption single-board computer like the Raspberry Pi board [31]. The Raspberry Pi, designed to run Linux kernel-based operating systems, is based on an ARM1176JZF-S 700 MHz processor, and includes 512 MB of RAM, two USB ports, audio and video output, and uses a Secure Digital card as boot and long-term storage. On the Pi board Lighttpd, a lightweight web server application, allow the Pi to serve dynamic HTML pages backed by a SQLite database, in which gathered information is organized in such a way as to ensure integrity, security and availability.

In the next section we propose an ontological approach to structure the domain knowledge concerning the air monitoring problem in our design.

5 Ontological Approach

Urban air pollution management requires advanced modeling and information processing techniques. Artificial intelligence provides several techniques and technologies that can solve efficiently different environmental problems. AI techniques present advantages over more traditional numeric modeling approaches, which require heavy computational resources and need as input complex data, often not easily available [32]. It is very important to make decisions in environmental protection management, so a multi-agent system (MAS) approach could be applied as a valid and robust solution. A multi-agent system is a network of software agents that interact to solve problems that are beyond the individual capacities or knowledge of each problem solver [33]. An agent can be a physical or virtual entity that can act autonomously and has skills to achieve its goals and tendencies. According to [34], the technology of intelligent agents and multi-agent systems provides the software infrastructure for the implementation of distributed environmental systems that can monitor and control the environmental quality.

In our context, the vehicular sensor network can be modeled as a multi-agent system composed by a set of intelligent entities. Each vehicle has some running software agents, which monitor and analyze the evolution of environmental data like air quality, and report to a supervisor agent if some critical problem can occur. According to Chomsky theories, information data, gathered by sensors installed on vehicles, have to be represented in a formal way so that intelligent decision agents can speak and reason on it. A way to formalize information is the definition of a correct ontology, that allow us to define messages with meaning and without ambiguity. One of the more common goals in developing ontologies is sharing common understanding of the structure of information among people or software agents [35]. The interaction between agents depends mainly on the adoption of a conceptualization, that is, a formal representation of the reality of a specific situation, so ontologic step is fundamental in our system, in order to define a common vocabulary for researchers who need to share information in this domain, but also for knowledge-based programs defining which queries and assertions are exchanged among agents.

The use of sensing devices and wireless sensor networks is raising, so an increasing volume of heterogeneous data, data formats, and measurement procedures is generated. The ontological model provides a way to manage the sensors and the accompanying volume of generated data. It can also be used to validate sensor readings and to sort out faulty sensor information as described in [36], in which the W3C Semantic Sensor Network Incubator group (SSNXG) defined an OWL 2 ontology to describe the capabilities and properties of sensors, the act of sensing and the resulting observations.

In order to formalize air quality concepts, an ontology may be used to structure air pollution domain concepts. Because one of the most important ontology features is its reusability, we considered to structure our knowledge domain according to AIR_POLLUTION_Onto. This ontology is dedicated to air pollution analysis and control, and has been actually used in MAS_AirPollution [37].

Urban air pollution structured data will be understandable and processable by agents whose goal is to monitor and mitigate pollution effects through different types of applications. For instance, combining 3D urban models, atmospheric factors and air pollution information, it is possible to estimate the quality of air in some urban areas [38]. Another interesting application could be road traffic management. Using the information provided by environmental and street ontology, it would be possible to develop some intelligent software agents that can set a recommended maximum speed for a vehicle, or manipulate traffic lights in order to manage the number of vehicles on the roads of a particular city area. This actions may contribute to disperse atmospheric pollutants thus reducing their concentration in building bordered areas.

6 Conclusions

Urban atmospheric pollution is a crucial issue for many urban areas, making it necessary to monitor and control gas pollutants concentration. Vehicular Sensor Networks are one interesting recent development in wireless and mobile networking. They are extremely multifunctional and may be useful for different applications, such as environmental monitoring. In this chapter we proposed a VSN architecture for urban air quality monitoring. The main advantage of our approach is the economy and the simplicity of the system. Using just a few vehicles belonging to a public transportation fleet, a fine-grained monitoring system able to cover all the city areas can be deployed. Moreover we propose the use of ontology as the most suitable tool to enable efficient machine-to-machine cooperation.

Acknowledgments This work has been partially supported by the PON R&C grant MI01_00091 funding the SeNSori project.

References

1. Hartenstein, H., Laberteaux, K.: VANET: Vehicular Applications and Inter-networking Technologies, vol. 1, Wiley (2010) (Online Library)
2. Da-Jie, L., Dow, C.R.: Introduction to ITS and NTCIP. Telematics Communication Technologies and Vehicular Networks: Wireless Architectures and Applications. In: Huang, C.M., Chen, Y.S. (eds.) Hershey: IGI Global, 32–57 (2010)
3. Dimitrakopoulos, G., Demestichas, P.: Intelligent transportation systems. IEEE Veh. Technol. Mag. **5**(1), 77–84 (2010)
4. Gavrilovska, L.: Application and Multidisciplinary Aspects of Wireless Sensor Networks. Springer, London (2011)
5. Anastasi, G., Lo Re, G., Ortolani, M.: Wsns for structural health monitoring of historical buildings. In: Proceedings of the 2nd Conference on Human System Interactions, HSI '09, pp. 574–579, (2009)
6. Anastasi, G., Farruggia, O., Lo Re, G., Ortolani, M.: Monitoring high-quality wine production using wireless sensor networks. In: 42nd Hawaii International Conference on, System Sciences HICSS '09, pp. 1–7, (2009)
7. De Paola, A., Gaglio, S., Lo Re, G.: Sensor9k: A testbed for designing and experimenting with WSN-based ambient intelligence applications. Pervasive Mobile Comput. **8**(3), 448–466 (2012)
8. Wong, K.J., Chua, C.C., Li, Q.: Environmental monitoring using wireless vehicular sensor networks. In: Proceedings of the 5th International Conference on Wireless Communications, Networking and Mobile Computing WiCom 09, pp. 1–4, (2009)
9. Lee, U., Zhou, B., Gerla, M., Magistretti, E., Bellavista, P., Corradi, A.: Mobeyes: smart mobs for urban monitoring with a vehicular sensor network. IEEE Wireless Commun. **13**(5), 52–57 (2006)
10. Song, H., Zhu, S., Cao, G.: Svats: A sensor-network-based vehicle anti-theft system. In: The 27th IEEE Conference on Computer Communications INFOCOM 2008, pp. 2128–2136, (2008)
11. Eriksson, J., Girod, L., Hull, B., Newton, R., Madden, S., Balakrishnan, H.: The pothole patrol: using a mobile sensor network for road surface monitoring. In: Proceedings of the 6th International Conference on Mobile Systems, Applications, and Services ACM, pp. 29–39, (2008)
12. Mednis, A., Strazdins, G., Liepins, M., Gordjusins, A., Selavo, L.: Roadmic: Road surface monitoring using vehicular sensor networks with microphones. Networked Digital Technologies, pp. 417–429. Springer, Heidelberg (2010)
13. Gruber, T.R.: Toward principles for the design of ontologies used for knowledge sharing? Int. J. Hum. Comput. Stud. **43**(5), 907–928 (1995)
14. Fenger, J.: Urban air quality. Atmos. Environ. **33**(29), 4877–4900 (1999)
15. Management and quality of ambient air. http://europa.eu/legislation_summaries/other/l28031a_en.htm. (2005). Web site consulted on 26 July 2013

16. De Paola, A., Gaglio, S.: Lo Re, G., Ortolani, M.: Multi-Sensor Fusion Through Adaptive Bayesian Networks. AI*IA 2011: Artificial Intelligence Around Man and Beyond. Lecture Notes in Computer Science, vol. 6934, pp. 360–371. Springer, Heidelberg (2011)
17. Lo Re, G., Milazzo, F., Ortolani, M.: A distributed bayesian approach to fault detection in sensor networks. In: Proceedings of the IEEE Global Communications Conference GLOBECOM, 2012, pp. 634–639, (2012)
18. Murty, R.N., Mainland, G., Rose, I., Chowdhury, A.R., Gosain, A., Bers, J., Welsh, M.: Citysense: An urban-scale wireless sensor network and testbed. In: Proceedings of the IEEE Conference on Technologies Homeland, Security, pp. 583–588, (2008)
19. Barrenetxea, G., Ingelrest, F., Schaefer, G., Vetterli, M.: Wireless sensor networks for environmental monitoring: the sensorscope experience. In: Proceedings of the IEEE International Zurich Seminar on, Communications, pp. 98–101, (2008)
20. Hasenfratz, D., Saukh, O., Sturzenegger, S., Thiele, L.: Participatory air pollution monitoring using smartphones. In: Proceedings of the 1st International Workshop on Mobile Sensing: From Smartphones and Wearables to Big Data (2012)
21. Honicky, R., Brewer, E.A., Paulos, E., White, R.: N-smarts: networked suite of mobile atmospheric real-time sensors. In: Proceedings of the Second ACM SIGCOMM Workshop on Networked Systems for Developing Regions ACM, pp. 25–30, (2008)
22. Aberer, K., Sathe, S., Chakraborty, D., Martinoli, A., Barrenetxea, G., Faltings, B., Thiele, L.: Opensense: open community driven sensing of environment. In: Proceedings of the ACM SIGSPATIAL International Workshop on GeoStreaming, pp. 39–42, (2010)
23. Hu, S.C., Wang, Y.C., Huang, C.Y., Tseng, Y.C.: Measuring air quality in city areas by vehicular wireless sensor networks. J. Syst. Softw. **84**(11), 2005–2012 (2011)
24. Al-Ali, A., Zualkernan, I., Aloul, F.: A mobile gprs-sensors array for air pollution monitoring. IEEE Sens. J. **10**(10), 1666–1671 (2010)
25. Ma, Y., Richards, M., Ghanem, M., Guo, Y., Hassard, J.: Air pollution monitoring and mining based on sensor grid in london. Sensors **8**(6), 3601–3623 (2008)
26. Völgyesi, P., Nádas, A., Koutsoukos, X., Lédeczi, Á.: Air quality monitoring with sensormap. In: Proceedings of the 7th International Conference on Information Processing in Sensor, Networks, pp. 529–530. (2008) (IEEE Computer Society)
27. Devarakonda, S., Sevusu, P., Liu, H., Liu, R., Iftode, L., Nath, B.: Real-time air quality monitoring through mobile sensing in metropolitan areas. In: Proceedings of the 12th IEEE International Conference on Peer-to-Peer, Computing (2013)
28. Blaschke, M., Tille, T., Robertson, P., Mair, S., Weimar, U., Ulmer, H.: Mems gas-sensor array for monitoring the perceived car-cabin air quality. IEEE Sens. J. **6**(5), 1298–1308 (2006)
29. Arduino web site. http://www.arduino.cc/ (2013). Web site consulted on 12 Sept 2013
30. Yamazoe, N., Sakai, G., Shimanoe, K.: Oxide semiconductor gas sensors. Catal. Surv. Asia **7**(1), 63–75 (2003)
31. Raspberry pi web site. http://www.raspberrypi.org (2013). Web site consulted on 18 Sept 2013
32. van Aalst, R.: Guidance report on preliminary assessment under EC air quality directives. European Environment Agency (1998)
33. Ferber, J.: Multi-agent systems: An Introduction to Distributed Artificial Intelligence, vol. 1. Addison-Wesley, New York (1999)
34. Weiss, G.: Multiagent Systems: A Modern Approach to Distributed Artificial Intelligence. The MIT press, Cambridge (1999)
35. Noy, N.F., McGuinness, D.L., et al.: Ontology development 101: A guide to Creating your first Ontology. Stanford University, Stanford (2001)
36. Compton, M., Barnaghi, P., Bermudez, L., García-Castro, R., Corcho, O., Cox, S., Graybeal, J., Hauswirth, M., Henson, C., Herzog, A., et al.: The ssn ontology of the w3c semantic sensor network incubator group. Web Semant. Sci. Serv. Agents. World Wide Web **17**, 25–32 (2012)

37. Oprea, M.M.: Air_pollution_onto: an ontology for air pollution analysis and control. In: Proceedings of the Artificial Intelligence Applications and Innovations III, pp. 135–143. Springer (2009)
38. Metral, C., Falquet, G., Karatzas, K.: Ontologies for the Integration of Air Quality Models and 3D City Models. In: Teller, J., Tweed, C., Rabino, G. (eds.) Conceptual Models for Practitioners, Società Editrice Esculapio, Bologna (2008)

Concentrated Solar Power: Ontologies for Solar Radiation Modeling and Forecasting

Antonino Piazza and Giuseppe Faso

Abstract This chapter considers the possibility of formally representing implicit and explicit knowledge of solar radiation modeling and forecasting by means of ontologies, with particular reference to the implications of concentrated solar power. The various applications discussed in the literature include various methods, such as spectral, parametric and empirical models, artificial neural networks and fuzzy logic approaches, as well as satellite and ground based imaging techniques. We want to use the principles of semantic technologies and formal ontologies, to represent knowledge in solar radiation models. The purpose is to capture the semantics of information and realize a system for sharing and re-use of knowledge in this domain.

1 Introduction

The growing demand for electricity in recent years has prompted research into the exploitation of renewable energy sources. Among the various technologies available, concentrated solar power (CSP) is probably the most promising for large-scale systems [11]. Due to its dependence on the source, the availability of information on solar beam radiation and solar radiation modeling and forecasting have become critical (Sect. 4) [19]. Most of the studies carried out in this area have therefore focused on our understanding of processes related to solar radiation fluxes and their interaction with the atmosphere. There has been a proliferation of various radiation models that calculate and predict the irradiance variability on different spatial and temporal scales [22]. The complexity and the increasing number of radiation models therefore requires organization and integration of knowledge. Semantic technologies,

A. Piazza (✉) · G. Faso
DICGIM, University of Palermo, Viale delle Scienze ed. 8, 90128 Palermo, Italy
e-mail: antoninopiazza@hotmail.it

G. Faso
e-mail: giuseppe.faso@unipa.it

S. Gaglio and G. Lo Re (eds.), *Advances onto the Internet of Things*,
Advances in Intelligent Systems and Computing 260, DOI: 10.1007/978-3-319-03992-3_23,

such as ontologies, promise to considerably improve the representation and sharing of a models' knowledge, and thus support decision processes, given the automation of design procedures [3, 15].

This chapter investigates how semantic ontologies can be used to represent information and knowledge about solar radiation modeling and forecasting. A formal representation and informatic systems can reduce data uncertainty and improve the model selection process as a function of the constraints imposed by different situations.

To facilitate understanding, a brief overview of CSP plant technologies and solar radiation concepts have been provided in Sects. 2 and 3. In Sect. 4, we explain the importance of solar resources data in the various stages of development and management of a CSP plant. In Sect. 5, we discuss solar radiation models and issues relating to model selection processes and data retrieval. Then, in Sect. 6, we describe the reasons for an ontological conceptualization in the solar radiation modeling and forecasting domain. The basic structure of a possible solar radiation modeling and forecasting ontology is described in Sect. 7, and the major features of the ontologies are examined. Finally, we conclude with general considerations on how these ontologies can be related to each other to create a more complete ontology that includes not only solar radiation modeling and forecasting, but also other aspects of solar energy systems.

2 CSP Plant Technologies

Concentrated Solar Power (CSP) systems use mirrors to reflect and concentrate sunlight onto receivers to collect solar energy and raise the temperature of a working fluid (carrier). This high-temperature fluid is used to support industrial processes or generate electricity using conventional methods such as steam rankine cycles. There are a variety of mirror shapes, sun-tracking methods and ways to collect and concentrate sunlight [22].

The four main types of commercial CSP concentrating technologies, shown in Fig. 1, are:

- parabolic dishes;
- central receivers (also called solar towers);
- parabolic troughs;
- linear fresnel systems.

A parabolic dish-shaped reflector concentrates sunlight onto a receiver located at the focal point of the dish. The concentrated beam radiation is absorbed into a receiver to heat a fluid or gas (air) used to generate electricity in a small piston or Stirling engine, or a micro turbine, attached to the receiver [22].

A central receiver or solar tower consists of a series of large mirrors, called heliostats, placed around a tower. Each of them has a separate tracking motion system which makes it possible to position the mirrors so that the reflected sunlight is focused

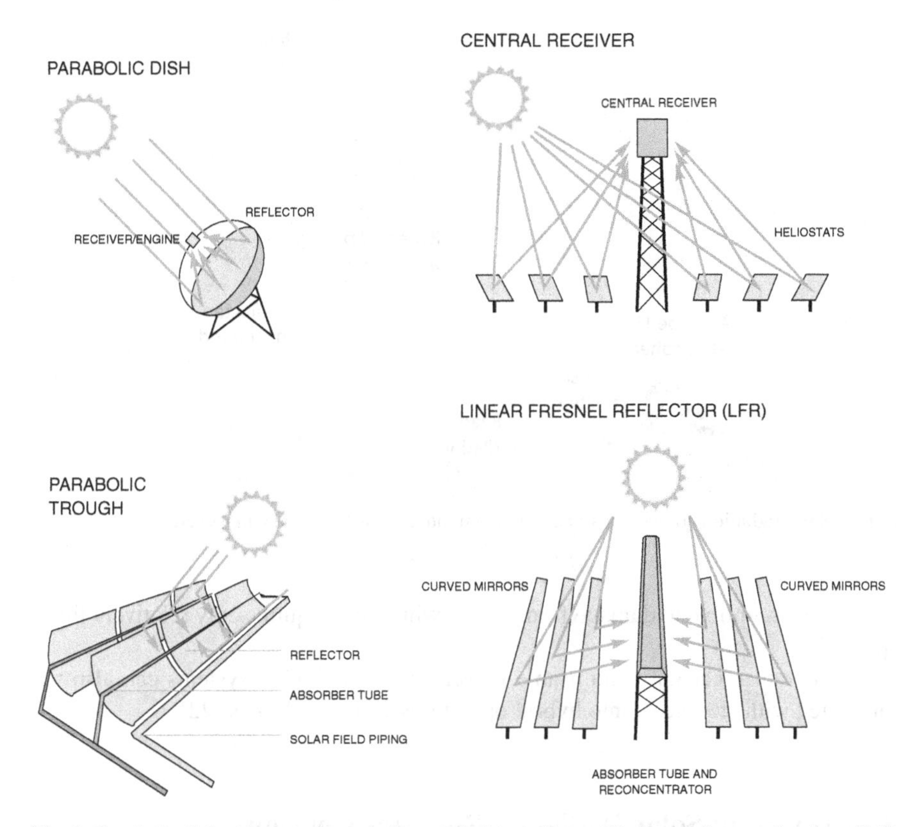

Fig. 1 Parabolic dish, solar tower, parabolic trough, Linear Fresnel Reflector. Image taken from [16]

onto a central receiver mounted at the top of a tower. A heat-transfer medium in this central receiver absorbs the highly concentrated radiation reflected by the heliostats and converts it into thermal energy [22].

Parabolic trough technology uses trough-shaped mirrors to concentrate sunlight onto thermally efficient receiver tubes running the length of the trough, and placed in the trough's focal line. The trough is parabolic along one axis and linear in the orthogonal axis, and is usually designed to track the sun along one axis. A thermal transfer fluid runs through the receiver tubes to absorb the concentrated sunlight [22].

A linear Fresnel Reflector consists of a series of nearly-flat or slightly curved reflectors which concentrate solar radiation onto absorber tubes placed several meters above the mirrors. Each absorber tube is equipped with a secondary reflector, which is open on its lower side, and whose purpose is to ensure that the totality of the reflected rays is focused on the receiver, directly or after an internal reflection. This system is line-concentrating, similar to a parabolic trough, with the advantages of low costs for structural support and reflectors. Due to its design, the Fresnel collector

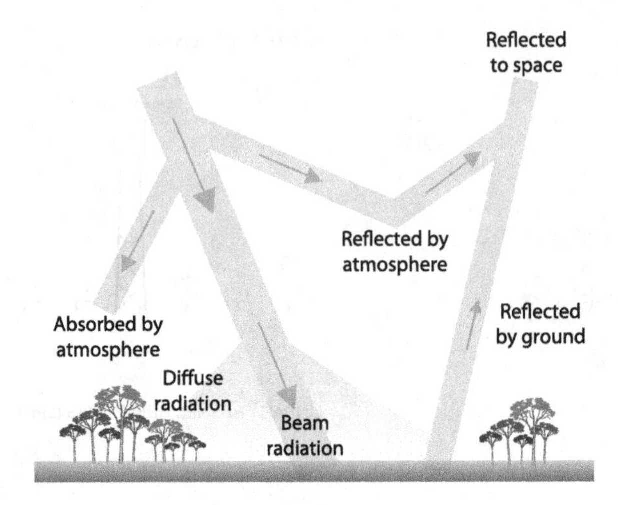

Fig. 2 Solar radiation components resulting from interactions with the atmosphere

is also unsusceptible to damage from strong winds and requires only relatively little space [22].

To offer firm capacity and usable power on demand, CSP systems can also be integrated with storage or into hybrid operations with fossil fuels [22].

3 Overview of Solar Radiation Resource Concepts

Naturally, the fuel source for all concentrated solar power is the sun and in particular, the electromagnetic energy emitted by the continuous nuclear reactions in its interior, which propagates in space and comes to earth. As a result of its passage through the atmosphere, the solar radiation is separated into the following components [18]:

- solar beam radiation, that part of solar radiation which directly reaches the earth's surface in parallel beams, undeviated by clouds, fumes or dust in the atmosphere;
- a diffuse component, generated by the scattering of the solar radiation in the atmosphere;
- that part of solar radiation which is reflected by the ground and the atmosphere.

These components are shown in Fig. 2.

To measure these components, the following quantities are associated with solar radiation [19]:

- Direct Normal Irradiance (DNI);
- Direct Horizontal Irradiance;
- Diffuse Horizontal Irradiance (DHI);
- Global Horizontal Irradiance (GHI).

These quantities are used to assess the solar radiation resource and to characterize a particular location. DNI is the amount of solar radiation received directly from the sun, falling onto a plane perpendicular to the direction of the sun. Direct Horizontal Irradiance differs from DNI in that it is measured on a flat horizontal plane. DHI represents the energy flux density of the solar radiation incoming from the entire sky dome on a horizontal surface, excluding the direct beam coming from the sun's disk. GHI is defined as the total energy from sunlight, both direct and diffuse, that reaches the unit area horizontal to the surface of the earth [19].

Surface based measurements of DNI and GHI are best measured using well calibrated pyrheliometers and pyranometers, but such measurements can only be made on a sparse network, given the operation and maintenance costs involved [19].

4 Why Solar Resource Data are of Importance to Concentrated Solar Power

Concentrated solar power plants need solar beam radiation to operate. Since the radiation is almost directly proportional to the amount of power generated by a power plant, knowledge of the local area's meteorological information and solar radiation data is essential in developing and planning concentrated solar power plants [22].

The availability of high quality information on solar radiation makes a positive impact on all of the steps in a CSP project [19]:

1. site selection
2. predicted plant output
3. temporal performance and operating strategy.

Site selection requires a large amount of solar data, such as annual and seasonal average solar energy, and climate changes in the different locations. However, direct measures of solar radiation are not always available, due to the limited spatial representativeness of actinometric stations and their limited global distribution. Also, data are often provided in a raw format, so additional processing is required. It is therefore necessary to derive the information of interest on solar beam radiation using appropriate models that process the data available [4, 8, 19].

After identifying areas favorable to the development of CSP technology, site selection for the installation of a solar power plant proceeds with a more detailed analysis and using more stringent exclusion criteria in narrowing down the area to be investigated. In addition to the simple average annual solar radiation, it is necessary at this stage to extrapolate more detailed information to predicted plant output and conduct a feasibility study. Of course, the prospects of application must also take account of other aspects unrelated to solar radiation, such as the commercial value of the area, the conformation of the ground, its distance from urban centers (fine dust and pollution can reduce the fraction of direct radiation reaching the ground) and other technical, commercial and environmental constraints [19].

After the realization and startup of the plant, measurements and estimates of solar radiation continue to be needed, as they make it possible to check the performance of the system and define the operating strategy. Verification is performed by comparing the value of the plant's energy production with that calculated using solar measurements and estimates. This allows us to characterize the models for the specific concentrated solar power plant and highlights design or accomplishment errors, solar radiation or plant modeling errors, and also highlights possible malfunctions. Then, the predicted solar radiation is indispensable for the evaluation of the power fed into the grid, for the management of the plant and for the definition of energy costs [17, 19].

This advance knowledge makes it possible, for example, to program maintenance for days on which radiation is low, and to secure the plant components in case of sudden changes in cloudiness, but above all, it enables system managers to compensate for the variable nature of the solar source which causes discontinuities in energy production, through appropriate management of thermal energy storage or complementary fossil fuel production systems [19].

5 Solar Radiation Models

The solar radiation reaching the earth's upper atmosphere is a quantity rather constant in time. It is dependent on slight variations, over short and long periods, in the radiation emitted by the sun and it is mostly dependent on the earth's elliptical orbit. For modeling purposes, the power radiated by the sun is set to a constant value, the solar constant $G_{SC} = 1366.1\ W/m^2$ and the variations in extraterrestrial solar radiation (ETR) are determined only by the earth's elliptical orbit. The earth's orbital eccentricity causes a variation in the earth-sun distance of $\pm 1.7\,\%$ with a consequent ETR variation of $\pm 3.4\,\%$ from G_{SC} during the year. To compute extraterrestrial solar irradiance on a horizontal surface, it is sufficient to apply the cosine law: $G_{0,ext} = G_{ext} \cos \theta_z$, where G_{ext} is the ETR and zenith angle θ_z may be expressed as a function of geographical latitude, sun declination angle, and hour angle [2].

Crossing the atmosphere, solar radiation is partially scattered and absorbed at different wavelengths by a number of factors, such as gases, clouds and dust (see Fig. 2 in Sect. 3). Thus, the solar radiation at ground level is much more variable than that observed at the top of the atmosphere. Obtaining reliable radiation data at ground level therefore requires systematic measurement. However, in most countries, the spatial density of solar radiation measuring stations is inadequate, due to high costs and maintenance requirements [2]. Various models have therefore been explored by many researchers to estimate, with reasonable accuracy, the solar radiation from other available meteorological data.

Among these models, there are those based exclusively on physical considerations and others based on statistical approaches [2].

The physical models, such as Leckner's spectral model [9], study radiant energy exchanges within the earth-atmosphere system. Parameters used as inputs in the physical models include:

- astronomical factors (solar constant, earth-sun distance, solar declination and hour angle);
- geographical factors (latitude, longitude and altitude);
- geometrical factors (surface azimuth, surface tilt angle, solar altitude, solar azimuth);
- physical factors (albedo, scattering of air molecules, water vapor content, scattering of dust and other atmospheric constituents);
- meteorological factors (atmospheric pressure, cloudiness, etc).

However, spectral models seem to be too difficult to use in engineering applications, because they require accurate meteorological data and an exact knowledge of the atmosphere composition. Using these, simpler parametric models for solar irradiance were developed [12], such as Yang's hybrid model [20, 21], Paulescu and Schlett's model [13], Gueymard's Parametric Solar Irradiance Model (PSIM) [6], and so on.

On the other hand, the models based on statistical and empirical approaches [2]:

- describe the quantities of interest statistically;
- investigate the statistical relationships among the main solar radiation components and the spatial correlation between simultaneous solar data at different places;
- seek a statistical interrelationship between the main components of solar irradiation and other meteorological parameters which may be available, such as sunshine duration, cloudiness, temperature, etc.

The models, derived from the Angström-Prescott Relation [14] for example, have been widely applied to estimate global solar radiation from sunshine duration [12]. However these models, by relying on empirical relationships, are not spatially independent, and their performance is only valid if the context of application is similar to that of development.

More modern methods are able to provide solar radiation information derived from satellite images. Satellites observe the earth-atmosphere system and provide continuous information for very large areas at a temporal resolution of up to 15 minutes and a spatial resolution of up to 1 km. Several methods use this information to transform the radiance (the physical magnitude actually measured by the satellite sensor) into the cloud index, which is a relative measure of cloud cover. The solar irradiation at the earth's surface is then derived from the cloud index [12].

Furthermore, as explained in Sect. 4, the need for high accuracy forecasts on multiple time horizons is becoming increasingly important for the solar energy industry. A large number of solar forecasting computation models have therefore been developed. The most popular forecasting methods used in the solar energy domain include ARIMA, Markov chains, Bayesian inference, and several other approaches developed in the field of artificial intelligence, such as genetic algorithms, expert systems, artificial neural networks, fuzzy systems and some hybrid systems that combine multiple techniques [7].

This proliferation of models implies a certain difficulty in model selection and in the accuracy of the information obtained. The applicability of a model is also influenced by the availability of measures and input data required. There are several data repositories and services that provide measurements and estimates of various quantities on continental and global scale, including:

- NASA SSE (an archive of over 200 satellite-derived meteorology and solar energy parameters, globally available at a resolution of 1x1 degree);
- NREL/USA (the US dynamic solar atlas; on 40 km^2 grid cells it provides monthly averages of daily total solar resources available to flat plate photovoltaic modules and concentrators);
- HelioClim-1 (Solar radiation calculated from Meteosat images by the Heliosat 2 method. Coverage: field of view of Meteosat satellite, Europe and north Africa);
- World Radiation Data Centre (Worldwide solar radiation database).

However, the various data repositories have different levels of precision and detail, as well as covering different geographical areas. In addition, data representation is not standardized, and the data are not always in agreement. Furthermore, the use of different terminology often makes common understanding impossible, thereby impeding automatic information processing.

A solution is therefore required which is capable of consolidating our knowledge of solar radiation and the various models that make it possible to obtain relevant information about this domain (measuring instruments, repositories, estimation modeling and forecasting).

6 Opportunities of an Ontology for Solar Radiation Modeling and Forecasting

In recent years, ontologies—explicit formal specification of the terms in the domain and relations among them [5]—have emerged in Artificial Intelligence as a way to represent knowledge and integration of a particular domain. Ontologies were already present in various other fields, even in the renewable energy domain, such as in [1], in which an ontology for managing knowledge about photovoltaic systems, called PV-TONS, was developed. It represents information and knowledge about renewable energy technologies, and facilitates system decision-making in recommending appropriate choices for use in different situations.

The basic aspects that concerns all solar energy systems (photovoltaic, thermal and CSP plants) are solar radiation data and the models that make it possible to estimate and predict their values (Sect. 4 above). An ontology for solar radiation modeling and forecasting can enable a semantic description of models, and more importantly, facilitate exploration and reasoning in this domain. The possibility of operating on an adequate collection of models, formally defined and opportunely classified, can improve the model selection process as a function of the constraints imposed by

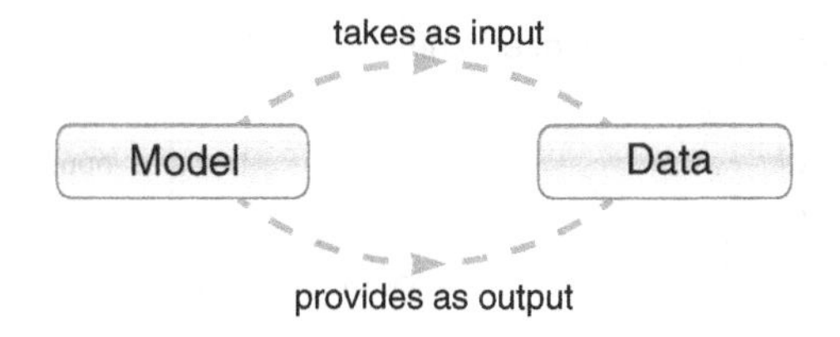

Fig. 3 Portion of ontology indicating model and data classes and their relations

particular applications and different situations. This increases the accuracy of the information produced, and it gives reasons for the uncertainty levels of the results using the reasoning mechanisms of ontologies. The explanation highlights the weaknesses of information processing, and then suggests where to intervene to obtain even more reliable information. With regard, on the other hand, to the atmospheric data and meteorological variables required in many solar radiation models, many repositories already contain the necessary information, and the use of ontologies can facilitate the retrieval of such information and multisource information extraction.

Finally, ontologies enable us to realize systems for sharing and re-use of knowledge and, since this knowledge is formally represented and thus machine interpretable, it can be communicated and shared between software agents.

7 An Ontology for Solar Radiation Modeling and Forecasting

In attempting to construct an ontology for solar radiation modeling and forecasting, we have focused on the concept of the model. A model is clearly characterized by the variable that it is capable of quantifying. Other important aspects consist of the input parameters required, (see Fig. 3).

This general definition immediately clarifies the two main classes on which the whole ontology is based, namely, the models and the data. Model and Data clearly represent general concepts. From these two classes it is then possible to create hierarchies of subclasses or additional relations. The hierarchies specify particular aspects of general classes, whilst relations connect them with other concepts [10]. One or more reference methodologies can be associated, for example, to each solar radiation model. Specifically, in this ontology, we considered numerical, stochastic, probabilistic methodology, fuzzy logic, neural network approaches, as well as satellite and ground based imaging techniques. The need to specify the sites or the types of climates in which the models were evaluated, and in general the performance of the models, also suggests the relation with the concepts of geographical applicability and uncertainty (see Fig. 4). Geographical applicability implies a variation in the level of uncertainty inherent in the model, and has repercussions on the overall performance of the model.

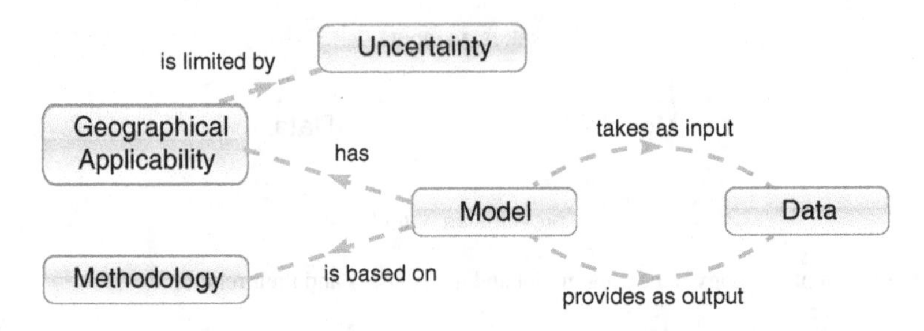

Fig. 4 Portion of ontology indicating model relations with data, geographical performance and methodology classes

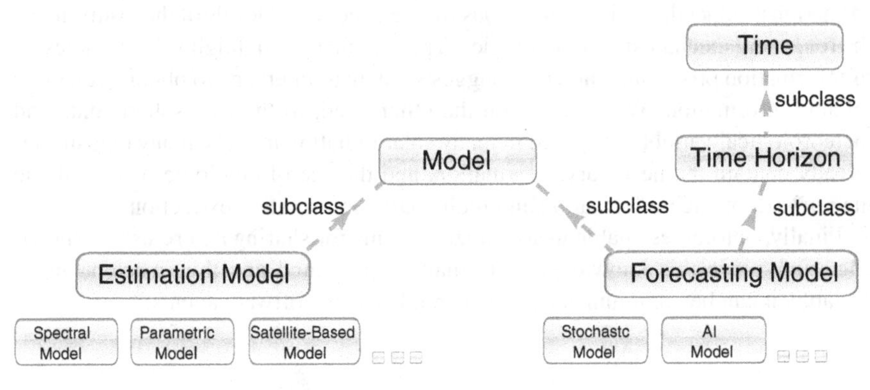

Fig. 5 Portion of ontology indicating model hierarchies

As for the specialization of the model class, the first distinction to consider is definitely that between estimation models and forecasting models. The former are able to calculate values using certain parameters, whilst the latter predict, with different time horizons, future quantities. Other types of specializations then divide the models into categories, depending on the methodology used, the modeled values, the input parameters, and so on. Next, we defined the class of physical and empirical models, the ones based on satellite images, the ones that calculate solar radiation in clear sky conditions only and in any sky conditions, the spectral models and others (see Fig. 5).

As far as Data is concerned, it is important to collect not only the type and the value, but attention should also be paid to the following key considerations, Fig. 6:

- Temporal resolution. Data can range from annually averaged to 1 s samples.
- Spatial resolution. Ground-based measurements, for example, are site-specific. On the other hand, current satellite-remote sensing estimates can be representative of 10 km × 10 km or less or larger areas, such as 1° by 1° latitude-longitude grids.
- Spatial coverage. The area represented by the data can range from a single station, a sample geographic region, or a global perspective.

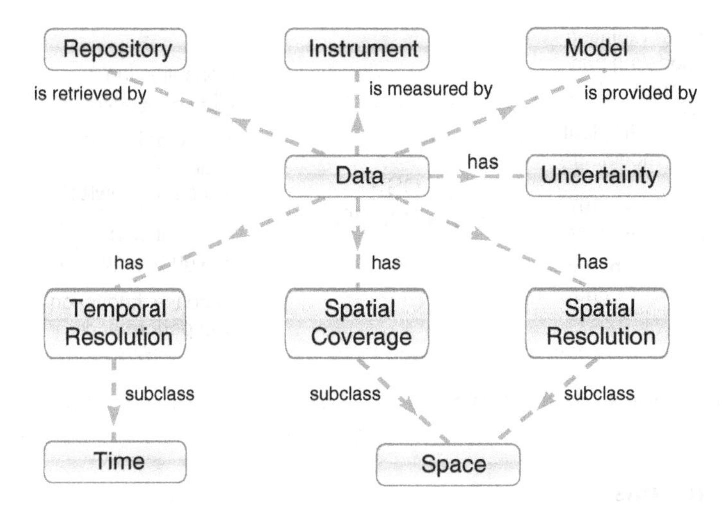

Fig. 6 Portion of ontology indicating data relations with temporal and spatial resolution, spatial coverage, uncertainty and data source classes

- Sources of the data. As well as a model, data can be acquired by measuring instruments, retrieved from a database or provided by different organizations.
- Estimated uncertainty. It is caused by limitations in the measuring instruments, as well as the processes of acquisition and modeling.

This ontological scheme was further specified by inserting the most popular radiation models used to calculate the values of solar radiation as a function of the required spatial and temporal resolution. For the representation, we used OWL language using the popular Protégé Desktop 4.3 (http://protege.stanford.edu/), but the ontology could also be expressed in other ontology languages.

In the immediate future, we will continue to test and develop our existing ontology and extend it to include further elements. This ontology needs to be combined with others to create a comprehensive ontology for the CSP domain in order to collect, integrate and structure all knowledge on CSP systems (see Fig. 7).

Such ontology should:

- consolidate existing skills, such as studies, projects and specialized activities related to industrialization and the development of production processes;
- encourage the sharing and re-use of knowledge and research results reached between research, design and manufacturing centers;
- facilitate the design and management of plants;
- increase innovation, performance and, subsequently, the production of energy from renewable sources.

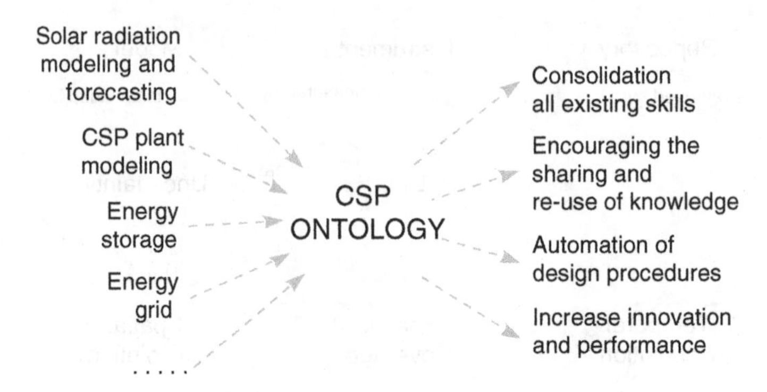

Fig. 7 Comprehensive ontology for CSP

8 Conclusions

For their development and management, concentrated solar power plants require continuous monitoring and forecasting of the solar radiation that reaches the specific local area. This issue has prompted the development of numerous models.

Our study aimed to explore the possibility of using ontologies to represent solar radiation modeling and forecasting, and the solar radiation information, in order to obtain a broad picture of the solar radiation resource in developing, planning and managing CSP plants. To this end, we have developed an ontology that, although very simple, affords sufficient expressiveness to create a complete knowledge base in this domain. The ontological and structural organization of the knowledge enables us to fully explore the solar radiation modeling and forecasting domain and can also be used to select the models best able to calculate the values of solar radiation with the required spatial and time resolution, as a function of the constraints imposed by the various situations in question. It also improves the quality of the data obtained, supports decision-making and optimizes design processes. Furthermore, it is a machine interpretable representation and facilitates sharing and re-use of the knowledge between software agents.

Acknowledgments This work has been partially supported by the PO FESR 2007/2013 grant G73F12000110004 funding the CESTA project.

References

1. Abanda, F.H., Tah, J.H., Duce, D.: Pv-tons: a photovoltaic technology ontology system for the design of pv-systems. Eng. Appl. Artif. Intell. **26**(4), 1399–1412 (2013)
2. Badescu, V.: Modeling Solar Radiation at the Earth's Surface: Recent Advances. Springer, Berlin (2008)

3. Davies, J., Fensel, D., Harmelen, F.V.: Towards the Semantic Web. Wiley Online Library, Chichester (2003)
4. De Paola, A., Lo Re, G., Milazzo, F., Ortolani, M.: Adaptable data models for scalable ambient intelligence scenarios. In: 2011 International Conference on Information Networking (ICOIN), pp. 80–85 (2011)
5. Gruber, T.R., et al.: A translation approach to portable ontology specifications. Knowl. acquisition **5**(2), 199–220 (1993)
6. Gueymard, C.: Mathermatically integrable parameterization of clear-sky beam and global irradiances and its use in daily irradiation applications. Sol. Energy **50**(5), 385–397 (1993)
7. Inman, R.H.: Solar forecasting review. Master's thesis, University of California, San Diego (2012)
8. Lalomia, A., Lo Re, G., Ortolani, M.: A hybrid framework for soft real-time wsn simulation. In: 13th IEEE/ACM International Symposium on Distributed Simulation and Real Time Applications, 2009 DS-RT '09, pp. 201–207 (2009)
9. Leckner, B.: The spectral distribution of solar radiation at the earth's surface–elements of a model. Sol. Energy **20**(2), 143–150 (1978)
10. Noy, N.F., McGuinness, D.L., et al.: Ontology development 101: A guide to creating your first ontology (2001)
11. Panwar, N., Kaushik, S., Kothari, S.: Role of renewable energy sources in environmental protection: a review. Renew. Sustain. Energy Rev. **15**(3), 1513–1524 (2011)
12. Paulescu, M., Paulescu, E., Gravila, P., Badescu, V.: Weather Modeling and Forecasting of PV Systems Operation. Springer, London (2013)
13. Paulescu, M., Schlett, Z.: A simplified but accurate spectral solar irradiance model. Theoret. Appl. Climatol. **75**(3–4), 203–212 (2003)
14. Prescott, J.: Evaporation from a water surface in relation to solar radiation. Trans. R. Soc. S. Aust. **64**(1940), 114–118 (1940)
15. Ribino, P., Oliveri, A., Lo Re, G., Gaglio, S.: A knowledge management system based on ontologies. In: International Conference on New Trends in Information and Service Science, NISS '09, pp. 1025–1033 (2009)
16. Richter, C., Teske, S., Short, R.: Concentrating Solar power: Global Outlook 2009. Greenpeace International, Netherlands (2009)
17. Riva Sanseverino, E., Di Silvestre, M., Ippolito, M., De Paola, A., Lo Re, G.: An execution, monitoring and replanning approach for optimal energy management in microgrids. Energy **36**(5), 3429–3436 (2011)
18. Sen, Z.: Solar Energy Fundamentals and Modeling Techniques: Atmosphere, Environment, Climate Change and Renewable Energy. Springer, London (2008)
19. Stoffel, T., Renné, D., Myers, D., Wilcox, S., Sengupta, M., George, R., Turchi, C.: Concentrating solar power: Best practices handbook for the collection and use of solar resource data (csp). Technical report National Renewable Energy Laboratory (NREL), Golden, CO (2010)
20. Yang, K., Huang, G., Tamai, N.: A hybrid model for estimating global solar radiation. Sol. Energy **70**(1), 13–22 (2001)
21. Yang, K., Koike, T., Ye, B.: Improving estimation of hourly, daily, and monthly solar radiation by importing global data sets. Agri. For. Meteorol. **137**(1), 43–55 (2006)
22. Zhang, H., Baeyens, J., Degrève, J., Cacères, G.: Concentrated solar power plants: Review and design methodology. Renew. Sustain. Energy Rev. **22**, 466–481 (2013)

Designing Ontology-Driven Recommender Systems for Tourism

Pierluca Ferraro and Giuseppe Lo Re

Abstract Nowadays, Internet users may experience some difficulty in finding the information they need from the huge multitude of existing web pages. A possible solution to this problem might lie in delegating some of the search tasks to machines, or in other words, in building a Semantic Web in which information could be processed automatically by intelligent software agents. Given the constantly increasing growth of the tourism industry, it might be particularly helpful to develop virtual assistants capable of planning trips on the basis of a user's interests. If so, adopting Semantic Web technologies would make it possible to provide a more customized service to each user and thus satisfy their requests better. Hypothesizing such a scenario, this chapter describes an adaptive recommender system which adopts a semantic approach to assist the user both in the travel planning phase and, on-site, during the trip. Finally, the software system provides a module that infers the user's interests and preferences using data mining techniques aimed at improving the quality of the suggestions made by the system.

1 Introduction

Given the volume of documents and data present on the web, Internet users nowadays tend to be overwhelmed by the huge amounts of information found online. This problem, known as *information overload*, compels Internet users to rely on the results provided by search engines [2]. Unfortunately, most of the data on the web is in the form of unstructured text, which may be easy for a human being to understand, but is very difficult for a computerized system to analyze. This severely limits the indexing capability of search engines, which are based on keyword retrieval, thus leaving the task of creating intelligent queries to their users [8]. Furthermore, today, most web

P. Ferraro · G. Lo Re (✉)
DICGIM, University of Palermo, Viale delle Scienze ed. 6, 90128 Palermo, Italy
e-mail: giuseppe.lore@unipa.it

S. Gaglio and G. Lo Re (eds.), *Advances onto the Internet of Things*, 339
Advances in Intelligent Systems and Computing 260, DOI: 10.1007/978-3-319-03992-3_24,

pages are dynamically generated by querying databases that are not accessible to search engines, so a considerable amount of data (the so-called *Deep Web*) is completely ignored. The web can therefore be seen as the largest of document archives, but one requiring continuous effort by human beings to be explored, since only a small portion of the data can be analyzed automatically.

In an attempt to overcome these limitations, many researchers have highlighted the need to structure information in order to simplify data comprehension by computers and to improve interoperability between different systems. For instance, intelligent search engines may adopt a semantic approach, analyzing documents by considering the concepts they contain, instead of relying exclusively on a keyword analysis [25].

The first step towards a web consisting of data connected by semantic relationships implies the creation of well-structured documents enriched with meta-data facilitating definition of the terms used in HTML pages and the relationships between related concepts, and thus allowing them to be interpreted automatically by intelligent software agents.

In the field of the tourism industry, which is rapidly evolving, the advent of the Semantic Web could provoke a serious revolution. Increasingly people prefer to act autonomously, planning their own trips personally without the help of a travel agent, using current web resources to learn about the best opportunities available to them [6].

However, most of the data in current travel portals (if we exclude those for flights and hotels) are not in a structured form. They don't therefore lend themselves to automatic processing by a software agent designed to support a user's choices using customized suggestions. The adoption of semantic technologies may enable conversion of all the data into a form appropriate for machine-processing and thus for analysis by intelligent software agents, designed to allow users to plan an entire trip in just a few clicks.

The remainder of this chapter is organized as follows. Section 2 analyzes the key technologies of the Semantic Web and its layered architecture. Section 3 highlights the benefits that may arise from the adoption of recommender systems in the field of tourism. Section 4 outlines the general architecture of the system proposed here, and finally, Sect. 5 reports some conclusions.

2 The Semantic Web

The Semantic Web has been defined by its creators, Tim Berners-Lee, James Hendler and Ora Lassila, as "an extension of the current one, in which information is given well-defined meaning, better enabling computers and people to work in cooperation" [3]. The goal is to build a communication infrastructure in which the meaning of data is not only accessible to human users, but also to computers that are able to analyze and understand them, making them available for further processing.

One of the most ambitious goals of the Semantic Web is the exchange of information between different agents acting independently or in collaboration with other software systems [11]. These agents could then exploit the semantic knowledge

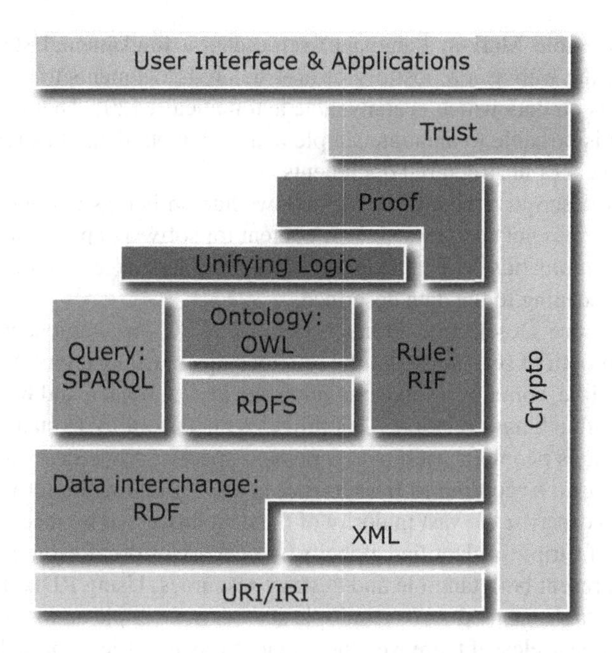

Fig. 1 Semantic Web architecture. Image taken from http://www.w3.org/2001/sw/

contained in the meta-data to provide new services to users. To do this, they should have a high enough level of autonomy to understand users' goals and then plan a sequence of actions to meet those objectives, working with other agents wherever necessary [10].

To achieve this purpose, it makes sense to exploit all the research work carried out in Artificial Intelligence, especially in the field of Knowledge Representation. However, many of the techniques developed over the years use a centralized approach, which requires everyone to share the same definitions of common concepts, which is not feasible in the World Wide Web, given its open and dynamic nature [1]. The attention of researchers has therefore shifted more and more towards decentralized systems able to handle the existence of multiple ontologies. Unfortunately, the chance of running into missing and contradictory information is a price one has to pay for the versatility needed to adapt to a changing world [3].

2.1 Semantic Web Technologies

The architecture initially proposed by the creators of the Semantic Web has been modified over the years, although the main elements, which are shown in Fig. 1, remain unaltered.

XML (eXtensible Markup Language) represents a fundamental step from the conception of the web as a repository of unstructured documents to the one where the emphasis is on data which is analyzable automatically [20]. That is, using such a language, it is possible to annotate simple text with meta-data, thus removing the ambiguity typical of unstructured documents.

The use of descriptive names for tags allows human beings to understand their function, but it does not add any semantic content for software applications trying to analyze them automatically. The adoption of a further language is thus necessary in order to give meaning to the data described.

RDF (Resource Description Framework) is a W3C Recommendation for the representation both of resources and the relationships between them. A basic statement, called triple, contains a subject, a predicate and an object, and binds together two entities with a binary relation. This provides the capability to state that certain resources, such as people or web pages, possess specific properties with their corresponding values. According to Berners-Lee et al., "this structure turns out to be a natural way to describe the vast majority of the data processed by machines" [3].

Each field of a triple is identified globally by a *URI* (Uniform Resource Identifier), which can represent both tangible and abstract resources. Using RDF, it is possible to represent data distributed across multiple servers in a simple and structured way, but this would be useless if there were no efficient way to retrieve these data and use them for further processing [21].

SPARQL (SPARQL Protocol and RDF Query Language) is the standard query language for RDF. It has been a W3C Recommendation since 2008 and is considered a key technology for the Semantic Web [1]. SPARQL has a syntax similar to SQL and, in addition to being a query language, is also a protocol that provides access to RDF knowledge bases by means of appropriate endpoints.

The adoption of RDF to encode meta-data is only the first step in the realization of the Semantic Web. It may be necessary, for example, to compare the information contained in two databases that use different identifiers to denote the same concept.

The solution to this problem is represented by ontologies, which are documents that formally define the relationships among a set of terms belonging to a specific domain. The most common ontologies on the web are vocabularies according to which all the resources should be described, and take the form of taxonomies enriched by a set of inference rules [22].

RDFS (RDF Schema) is a simple ontology language based on RDF that provides the capability to express relationships between generic terms, unlike RDF, which only describes relationships between individuals [24]. RDFS introduces the concepts of classes and subclasses, through which one can make simple inferences based on hierarchies of types and properties as well as their domains and ranges.

The inference capabilities provided by RDFS are limited, but they are still of significant utility in a Semantic Web application, if it becomes necessary to merge information from multiple data sources.

OWL (Web Ontology Language) is used to describe in more detail the entities represented and the relationships between the resources of a particular domain.

The most interesting feature of OWL is the ability to create new classes from concepts already defined in the model, using the operators of union, intersection and complement, and the option of specifying properties with special features, such as symmetry or transitivity [9]. In addition, the automated reasoners available for the language can check the satisfiability of the ontological description, ensuring that anything new added to the model is not in conflict with what has already been stated, since this would make the ontology inconsistent [19]. Moreover, OWL allows developers to express equivalence relations between classes, properties and individuals, in order to connect equivalent concepts that use different identifiers.

The greater expressiveness of OWL as compared to RDFS, however, also has negative aspects, since it involves an increase in computational complexity associated with each operation of automated reasoning [9].

Although the Semantic Web is mainly directed towards Knowledge Representation systems, focusing on Description Logics, it is common to use rule-oriented languages to overcome modeling problems that are difficult to solve with RDFS and OWL alone [7].

RIF (Rule Interchange Format) is based on the observation that there are already too many rule-based languages, and it would be useless to develop a new one. RIF, however, is designed to merge existing formalisms, from Horn clause logics up to higher-order logics and production systems [24].

The upper layers of the architecture shown in Fig. 1, namely *Unifying Logic*, *Proof* and *Trust*, have never been implemented in a comprehensive manner. At the moment, therefore, the application layer is directly connected to the lower layers (OWL, RDFS, SPARQL and RDF).

Finally, the *Encryption* layer plays a particularly important role on the web, since all semantic agents should always verify the authenticity and integrity of data used to answer user queries, in order to prevent forgery and tampering [23].

2.2 Semantic Web Application Architecture

In order to better illustrate the structure of a Semantic Web application, it might be useful to analyze its main components, which are shown in Fig. 2.

Many of these components are available both as commercial products offered by software producers specialized in semantic technologies, and also as open source software developed by user communities [1]. The main elements of a typical Semantic Web application are:

- a *parser* that reads the text in one of the standard RDF formats, such as N-Triples, Turtle or RDF/XML, and interprets it as a set of triples in the data model;
- an *RDF store*, a database optimized for storing and retrieving data in the form of triples, with the ability to merge information from multiple sources;
- an *RDF query engine* to retrieve information from an RDF store, performing the desired queries in an efficient manner;

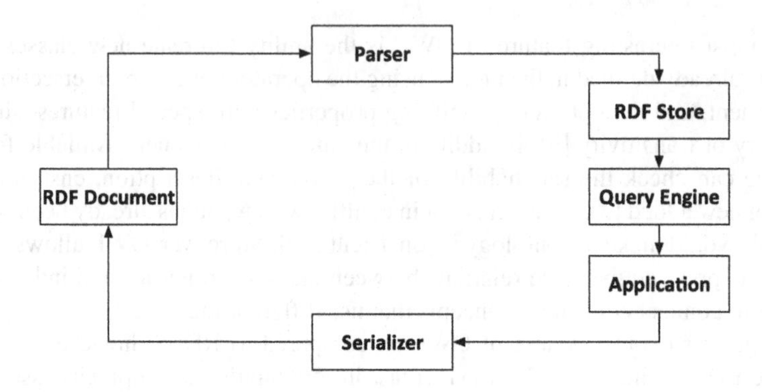

Fig. 2 Semantic Web application architecture

- an *application* written in a general-purpose programming language that processes the results of the queries and presents them to the user;
- a *serializer* that plays a dual role with respect to the parser, creating new files to represent the triples of the model in one of the standard RDF formats.

3 Travel Recommender Systems

Planning a trip down to the last detail, taking into account all the large number of variables involved, is a complex task that requires a lot of time and great attention [15]. Today, if a web user needs to plan a tour, he is highly likely to have to visit several websites to make all the necessary reservations, typing in his personal data each time and waiting for confirmation (for instance, e-mails from hotel and airline companies). Moreover, such an user will be required to make multiple payments, providing the number of his credit card to all the companies involved, thus exposing himself to multiple security threats [4]. Users are often discouraged by such obstacles and would prefer to use a single system to achieve their goals in a centralized way, preferring to acquire whole travel packages instead of collecting single parts manually [12]. Furthermore, additional software layers are necessary to ensure the quality of the services that will be provided to users [5].

For this reason, one of the most promising fields in the tourism industry is that involving recommender systems, capable of playing an important role in delivering correct and accurate information, and in addressing the choice of products, services, events and places to visit, in line with users' interests. The suggestions offered by such systems are based on user profiles, which are built by observing and analyzing their previous choices by exploiting *data mining* techniques, as discussed in Sect. 4 below. Such systems are particularly suited to the tourism sector, since they allow companies to offer solutions, services and packages tailored to the needs of users, thus increasing sales figures and overall customer satisfaction.

The quality of the suggestions made, however, is closely related to the precision with which resources and user habits are described [6]. It is thus necessary to include meta-data and semantic information inside web pages, in order to describe the relationships between the concepts belonging to the domain in question, that of tourism in this case. To this end, it is helpful to adopt ontologies which make it possible to link different models, thereby overcoming the unavoidable heterogeneity in the terminology adopted by different companies [17].

In the literature, there are several examples of tourism recommender systems that successfully exploit semantic technologies to offer customized services to users. For example, a very interesting project for planning trips is *CulTuRek* [6], which stores structured descriptions of resources as meta-data, and relies on the reviews submitted by tourists to infer the semantic relations that will be used during travel planning. *CulTuRek* exploits the semantic relationships between the resources to improve the list of suggestions returned. The process follows three main phases, namely, Retrieval, Ranking and Semantic Enrichment. The most characteristic aspect of the system is the latter step, which identifies the resources that do not fully comply with the parameters set by the user in the search query, but which are semantically related to the results of the first two phases, and which are included in the final list of suggestions for this reason. A thorough analysis of the system, with particular emphasis on the Semantic Enrichment phase, is presented by Di Bitonto et al. in [6].

Another interesting recommender system is *mITR* (Mobile Intelligent Travel Recommender), developed by Nguyen et al. to help users to create the queries that will be used to generate lists of suggestions [16]. The recommendation process used by this system evolves in cycles, and is designed to assist tourists during their trips. For this reason, the main objective of the project was to simplify the user's interactions, taking into consideration the specific conditions in which the application might be used.

4 The Proposed Architecture

An intelligent information system for tourism should satisfy the customers' requirements by providing them with the best possible recommendations for museums, exhibitions, guided tours, restaurants, nightlife, shopping, or any other field of interest. Such a system should be, above all, "context sensitive", since it should adapt its behavior to the environment and exploit all of the tools available to help tourists, without being too intrusive. For this reason, a complete solution should include at least the following three components:

- a *pre-travel virtual assistant* capable of planning trips and driving users through all the steps prior to departure;
- an *on-site recommender system* to help tourists during their journey, making the most appropriate suggestions for the particular circumstances, depending on the context, so as to manage unexpected events or sudden changes in travel plans;

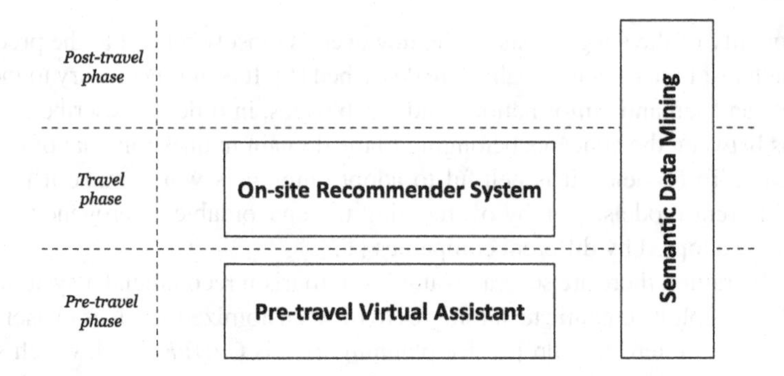

Fig. 3 The proposed architecture

- a *semantic data mining* module to keep track of users' activities and infer their interests and preferences by analyzing the way they behave when interacting with websites.

In the following section, we describe the general architecture of an integrated adaptive system that exploits Semantic Web technologies to perform such tasks effectively. The combined use of all three such components, as shown in Fig. 3, makes it possible to offer personalized services to users, assisting them in all the operations they have to perform and trying to satisfy their requests.

4.1 Pre-travel Virtual Assistant

If all the information contained in the websites of airlines, hotels and travel agencies were enriched with semantic meta-data, users could exploit them as a basis for planning journeys and tours, as well as choosing places to visit and events to attend, thereby creating custom packages that satisfy their particular requirements, using so-called *Dynamic Packaging Systems*. According to Cardoso [4], "dynamic packaging can be defined as the combining of different travel components, bundled and priced in real time, in response to the request of the consumer or booking agent". They are thus ideal to enable users to plan entire journeys in a centralized way.

The architectures of such systems, however, can be very complex and challenging, because, as mentioned in the previous sections above, most of the data on the web nowadays are not structured properly, and the information conversion process often requires the collaboration of several software agents and sub-systems. For this reason, our architecture employs wrapper modules to process information in a simple way, starting from web pages with unstructured data up to the creation of complete travel packages, as suggested by Knoblock et al. [14].

After the data have been converted and imported into the knowledge base of the virtual assistant, user interaction is managed by establishing a dialogue with the tourists: in order to plan a trip, users are required to answer a set of questions about personal interests and to provide detailed information regarding their requirements, such as time or budget constraints, and quality criteria in general.

This information will be taken into account to provide customized recommendations or to propose choices made in the past by tourists with analogous interests.

If the query does not produce any results, the user is required to relax some constraints, whereas, if the selection criteria are too general and the query produces too many results, additional questions are posed to the user in order to narrow the search, as in [16]. The semantic relationships between the resources, defined in the ontology used by the system, are then used to improve the list of suggestions.

Unlike many other similar projects, such as *CulTuRek*, which is a single system, our virtual assistant can be seen as a component of a more complex architecture. In this way, all the modules that constitute the system are able to communicate with each other in order to share user preferences, thus realizing a virtuous cycle: the more users use the system, the more accurate the suggestions will be.

4.2 On-site Recommender System

The widespread use of mobile applications for smart devices is of fundamental importance for complete recommender systems, because users need continuous assistance *on-site* during their journeys, since planned itineraries are subject to sudden changes related to unforeseen circumstances or unexpected changes in the weather conditions.

In such cases, it would make sense to adopt an application specifically designed for smartphones or tablets that takes into account the particular conditions in which it will operate, namely, small screens, a touch interface, and slow and unstable Internet connections.

Such an application should provide satisfactory answers quickly and should not require users to enter large amounts of data, given the discomfort involved in using a touch screen keyboard. Finally, the recommendations provided should be "context sensitive", using a GPS device to determine the user's location, and should take into account temporal information, in addition to the user's past choices.

For all these reasons, an application with complex user interaction is not suitable for mobile use. In order to assist tourists during their trips, it is therefore advisable to take a different approach, involving proposing an initial list of suggestions that can be modified according to the feedback from users, which will be used to improve the search query.

In our architecture, users simply specify the kind of resource they want to search for, such as, for instance, "restaurants". In response, the system constructs an initial query and proposes a list of suggestions, based on the context and on what the system knows about the user's preferences, using an approach similar to [16].

If the results satisfy the users, they may proceed with one of the proposed resources, ending the search. Otherwise, users should further specify features, which should then be modified. For instance, the user might specify the type of restaurant he is looking for. The system then updates the query, and the whole process is repeated cyclically, gradually trying to improve the list of suggestions. Unlike [16], which does not explicitly use Semantic Web technologies, our system takes advantage of the languages described in Sect. 2 above, with considerable benefits of various kinds. Many of the existing systems adopt feature vectors or other *ad hoc* data structures to represent resources.

A restaurant, for example, could be described by name, type, geographical location, average cost per person, weekly opening days and ratings assigned by other users. All such data structures can be easily converted into a set of RDF triples. In this way, for instance, a hypothetical pizza restaurant is represented in Turtle notation as:

```
tr:Vesuvio   tr:name            'Vesuvio'       ;
             rdf:type           tr:Pizzeria     ;
             tr:location        geo:Rome        ;
             tr:price           15              ;
             tr:openingDays     (1,3,5,6,7)     ;
             tr:rating          4               .
```

where the namespace `tr` is an abbreviation for travel. Such a representation offers several advantages over an *ad hoc* data structure: RDF triples represent the standard adopted by an increasing number of companies operating in the field of tourism, and their use ensures complete interoperability with other systems.

The use of RDF also enables companies to take full advantage of all the other Semantic Web technologies, which have been widely discussed in previous sections. SPARQL, in an especial way, can be used to perform complex queries on distributed datasets and makes it possible to use ontologies to define semantic relations between resources in order to infer new triples from the initial ones, and thus to offer a customized service to users.

This query language also provides the ability to specify complex conditions, which cannot be expressed in a system that uses simple *pattern matching* techniques, such as the ones described in [16].

For instance, the following SPARQL query identifies restaurants with an average cost per person of less than €15 and an evaluation of more than 4 stars, sorted by ratings:

```
SELECT    *
WHERE     { ?restaurant   tr:name      ?name    ;
                          tr:price     ?price   ;
                          tr:rating    ?rating  .
           FILTER ( ?price < 15 && ?rating > 4 ) }
ORDER BY   DESC    ( ?rating )
```

A query of this kind is unlikely to be realized by those systems that are based only on exact keywords matches and which do not use Boolean conditions with comparison operations.

The use of semantic technologies therefore provides the capability of returning improved suggestions to users and promotes interoperability between different recommender systems able to communicate with each other and work together with the aim of providing innovative services to users.

4.3 Semantic Data Mining

In order to improve the quality of the suggestions returned, to create targeted offers, and to provide custom travel packages, it is essential that a recommender system constantly keeps track of the user's behavior while browsing the web application. Data mining techniques try to analyze the behavior of users, starting from the web pages they visit and the way they react to particular offers.

Using Artificial Intelligence techniques, such as neural networks, genetic algorithms, Bayesian networks, and other *machine learning* algorithms, along with statistical analysis [18], it is possible to infer the interests and preferences of users on the basis of the patterns that characterize the way they use the web application.

That is, each request received by a web server is automatically stored in databases and log files, and all the data collected are used for further processing and analysis. Given the large amount of information stored in log files, it is essential to filter the collected data in order to avoid excessive *information overload*, and to perform accurate analyses on resource categories or groups of users. The semantic relations defined in the ontology adopted by the system contribute to focalize the data analysis process. By exploiting such relations, it is possible to consider only certain user groups, or particular categories of resources, such as itineraries or complete packages, which can be grouped, in turn, based on geographical location, price range or the time of year when the offer was published. Users, on the other hand, can be grouped on the basis of static information such as their personal data (age, sex, nationality, etc.) or according to their behavior (and thus, indirectly, their supposed interests), exploiting meta-data and semantic annotations with the aim of testing the reaction of certain categories of users to special offers.

According to Kanellopoulos et al. [13], the main feature that a data mining component should monitor is the number of visits to web pages that advertise particular resources, such as events, hotels, restaurants, shows and attractions. It is therefore necessary to define a variable $C_{ij} = \{0, 1\}$ to indicate whether the ith user visited the web page corresponding to the jth resource. Since C_{ij} assumes Boolean values, only a user's first visit to a particular web page will be considered. In this way, the sum

$$\sum_{i=1}^{n} C_{ij} \tag{1}$$

represents the popularity of the jth resource, as the number of distinct users who viewed the web page of that particular resource. Analogously, the sum

$$\sum_{j=1}^{m} C_{ij} \tag{2}$$

represents the number of pages that were viewed by the ith user, and is thus an indirect measure of the interest shown by that individual user in the entire catalog offered by the company.

Furthermore, in our system, we define a variable $R_{ij} = \{0, 1\}$ to take into account the actual bookings made by tourists. The relationship between the two variables thus allows the system to perform more complex analyses. For instance, if the value of the ratio

$$\frac{\sum_{i=1}^{n} R_{ij}}{\sum_{i=1}^{n} C_{ij}} \quad (3)$$

is too low, this means that the actual bookings are scarce in relation to the resource pages visited. This could mean that the offer appears interesting at first glance, but some details (for instance, a high price) discourage users from making a reservation.

All the data collected can be analyzed in order to be used in various ways, for instance to ascertain the quality and popularity of a particular travel package.

The offers that turn out to be unattractive can thus be eliminated or modified, for example, by providing discounts or other incentives aimed at meeting the needs of users. On the other hand, the most attractive resources could be highlighted on the home page of the website. The same data can also be used to propose travel packages tailored for individual users, based on their individual interests.

It would therefore seem obvious that data mining techniques, when combined with semantic meta-data and used to group similar users and resources, constitute the fundamental pillars on which to build modern web applications capable of offering cutting-edge services, and of facing the challenges posed by the current tourism market.

5 Conclusions

Given that the tourism industry is constantly evolving, the development of virtual assistants able to plan trips based on an individual user's interests is noteworthy, and the use of semantic technologies helps in providing customized services to each tourist to better satisfy his or her demands. In this chapter, we have analyzed the main technologies of the Semantic Web, and highlighted the benefits produced by the adoption of such an approach in the tourism industry. That is, the decision to use semantic technologies for the implementation of virtual assistants and recommender systems guarantees full interoperability with all the companies operating in the tourism field that have adopted the same approach, and enables companies to provide services and custom packages to all their potential clients and satisfy their demands.

As a proposal for an intelligent virtual assistant, we described the general architecture of an adaptive recommender system that uses a semantic approach to assist users both in the pre-travel stage and during their trip. The proposed architecture includes a capability designed to infer users' interests and preferences using data mining techniques with the aim of improving the quality of the suggestions made to users.

Acknowledgments This work has been partially supported by the PO FESR 2007/2013 grant G63F12000240004 funding the OnSicily project.

References

1. Allemang, D., Hendler, J.: Semantic web for the working ontologist: effective modeling in RDFS and OWL. Morgan Kaufmann Publishers Inc, CA (2011)
2. Benjamins, V.R., Contreras, J., Corcho, O., Gómez-Pérez, A.: Six challenges for the semantic web. In: KR2002 Semantic Web Workshop (2002)
3. Berners-Lee, T., Hendler, J., Lassila, O.: The semantic web. Sci. Am. **284**(5), 28–37 (2001)
4. Cardoso, J.: E-tourism: Creating dynamic packages using semantic web processes. In: W3C Workshop on Frameworks for Semantics in Web Services (2005)
5. Crapanzano, C., Milazzo, F., De Paola, A., Lo Re, G.: Reputation management for distributed service-oriented architectures. In: 4th IEEE International Conference on Self-Adaptive and Self-Organizing Systems Workshop (SASOW), pp. 160–165 (2010)
6. Di Bitonto, P., Laterza, M., Rossano, V., Roselli, T.: A semantic approach implemented in a system recommending resources for cultural heritage tourism. J. e-Learning Knowl. Soc. **8**(2) (2012)
7. Di Bona, D., Lo Re, G., Aiello, G., Tamburo, A., Alessi, M.: A methodology for graphical modeling of business rules. In: Proceedings of the 5th UKSim European Symposium on Computer Modeling and Simulation (EMS), pp. 102–106 (2011)
8. Ding, L., Finin, T., Joshi, A., Peng, Y., Pan, R., Reddivari, P.: Search on the semantic web. Computer **38**(10), 62–69 (2005)
9. Grau, B.C., Horrocks, I., Motik, B., Parsia, B., Patel-Schneider, P., Sattler, U.: OWL 2: the next step for OWL. Web Semant. Sci. Serv. Agents World Wide Web **6**(4), 309–322 (2008)
10. Greco, L., Lo Presti, L., Augello, A., Lo Re, G., La Cascia, M., Gaglio, S.: A decisional multi-agent framework for automatic supply chain arrangement. In: New Challenges in Distributed Information Filtering and Retrieval, pp. 215–232. Springer (2013)
11. Hendler, J.: Agents and the semantic web. IEEE Intell. Syst. **16**(2), 30–37 (2001)
12. Kanellopoulos, D., Panagopoulos, A., Karahanidis, J.: How the semantic web revolutionizes destination management systems. In: Proceedings of the International Conference on Tourism Development and Planning, pp. 21–35 (2005)
13. Kanellopoulos, D., Panagopoulos, A., Psillakis, Z.: Multimedia applications in tourism: The case of travel plans. Tourism Today, pp. 146–156 (2004)
14. Knoblock, C.A., Minton, S., Ambite, J.L., Muslea, M., Oh, J., Frank, M.: Mixed-initiative, multi-source information assistants. In: Proceedings of the 10th International Conference on World Wide Web, pp. 697–707 (2001)
15. Maedche, A., Staab, S.: Applying semantic web technologies for tourism information systems. In: Proceedings of the 9th International Conference for Information and Communication Technologies in Tourism (2002)
16. Nguyen, Q.N., Cavada, D., Ricci, F.: On-tour interactive travel recommendations. In: Information and Communication Technologies in Tourism, pp. 26–28 (2004)

17. Oliveri, A., Ribino, P., Gaglio, S., Lo Re, G., Portuesi, T., La Corte, A., Trapani, F.: Kromos: Ontology based information management for ICT societies. In: Proceedings of the 4th International Conference on Software and Data Technologies, pp. 318–325 (2009)
18. Olmeda, I., Sheldon, P.J.: Data mining techniques and applications for tourism internet marketing. J. Travel Tourism Mark. **11**(2), 1–20 (2002)
19. Parsia, B., Sirin, E., Kalyanpur, A.: Debugging OWL ontologies. In: ACM Proceedings of the 14th International Conference on World Wide Web, pp. 633–640 (2005)
20. Pattal, M., Li, Y., Zeng, J.: Web 3.0: A real personal web! more opportunities and more threats. In: Proceedings of the 3rd International Conference on Next Generation Mobile Applications, Services and Technologies (NGMAST), pp. 125–128 (2009)
21. Quilitz, B., Leser, U.: Querying distributed RDF data sources with SPARQL. In: The Semantic Web: Research and Applications, pp. 524–538. Springer (2008)
22. Ribino, P., Oliveri, A., Lo Re, G., Gaglio, S.: A knowledge management system based on ontologies. In: International Conference on New Trends in Information and Service Science, pp. 1025–1033 (2009)
23. Richardson, M., Agrawal, R., Domingos, P.: Trust management for the semantic web. In: The Semantic Web (ISWC), pp. 351–368. Springer (2003)
24. Shadbolt, N., Berners-Lee, T., Hall, W.: The semantic web revisited. IEEE Intell. Syst. **21**(3), 96–101 (2006)
25. Shah, U., Finin, T., Joshi, A., Cost, R.S., Matfield, J.: Information retrieval on the semantic web. In: Proceedings of the 11th International Conference on Information and Knowledge Management, pp. 461–468 (2002)